Biology and Ecology of Atlantic Cod

Editors

Nataliia Kulatska

Swedish University of Agricultural Sciences
Department of Aquatic Resources
Uppsala, Sweden

Daniel Howell

Institute of Marine Research
Bergen, Norway

Peter J. Wright

Marine Ecology and Conservation Consultancy
Ellon, Scotland, UK

Ingibjörg G. Jónsdóttir

Marine and Freshwater Research Institute
Hafnarfjörður, Iceland

CRC Press
Taylor & Francis Group
Boca Raton London New York

CRC Press is an imprint of the
Taylor & Francis Group, an **informa** business
A SCIENCE PUBLISHERS BOOK

Cover credit: Heringe von Kabeljau und Schellfisch verfolgt (from ger. Herring, followed by cod and haddock). Illustration: Heinrich Harder. In Leitfaden der Tierkunde. O. Schmeil, 1929.

First edition published 2025
by CRC Press
2385 NW Executive Center Drive, Suite 320, Boca Raton FL 33431

and by CRC Press
4 Park Square, Milton Park, Abingdon, Oxon, OX14 4RN

CRC Press is an imprint of Taylor & Francis Group, LLC

Library of Congress Cataloging-in-Publication Data (applied for)

ISBN: 978-0-367-63828-3 (hbk)
ISBN: 978-0-367-63829-0 (pbk)
ISBN: 978-1-003-12087-2 (ebk)

DOI: 10.1201/9781003120872

Typeset in Times New Roman
by Prime Publishing Services

Preface

Balancing societal needs with ecosystem health is essential for the sustainable use of fish resources. This is exemplified by the history of Atlantic cod stocks. This fish has shaped many societies around the North Atlantic and stock collapses have had a major impact on many coastal communities. As a predator as well as a prey in North Atlantic shelf sea food webs, such stock collapses and recoveries have significantly impacted marine communities. Additionally, climate change is increasingly impacting the distribution, productivity and dynamics of this species. Collapses in cod stocks have been linked to both overfishing and environmental conditions, highlighting the need for a thorough understanding of the biology and ecology of this species.

Our understanding of cod has greatly improved in recent decades, thanks to improvements in molecular tools as well as electronic tags, otolith microchemistry and biophysical models that help to track movements from eggs to adults. Developments in gear design and collaboration with fishers has also helped develop technical approaches to reduce cod catches in depleted stocks. Consequently, Atlantic cod is one of the most highly studied species in the world.

Despite the considerable knowledge base about Atlantic cod, there are only a few summaries around stock dynamics and management. This book provides a comprehensive overview of the biology and ecology of many of the Atlantic cod stocks and the challenges they face. For topics where it is important to compare across stocks, such as climate change, trophic effects, and regional variability in productivity, we sought comparative reviews from those involved in recent debates. However, to appreciate the range of different pressures and local adaptations relevant to cod management, regional experts were sought to provide reviews on their local cod stocks that detail the stock units, their dynamics and management. Many of these authors participated in working groups of the International Council for the Exploration of the Sea (ICES) which provides annual assessments and advice on the management of Northeast Atlantic fish stocks and these chapters greatly benefit from the work of that organisation.

The chapters include both reviews and some novel evidence. The book should appeal to both those new to fisheries management as well as scientists and managers tasked with managing this and related fish species.

N. Kulatska
D. Howell
P.J. Wright
I.G. Jónsdóttir

Contents

Introduction

Atlantic cod (*Gadus morhua*) is one of the most important commercial fish species in Northern Europe and the east coast of North America. Long-standing traditional fisheries have roots extending many centuries (Anon 1240), serving as an important foundation for numerous coastal communities on both sides of the Atlantic.

Cod has historically been highly abundant, is considered a keystone species in the ecosystem it inhabits, and plays a major role as a predator influencing both its prey and other species, whether via trophic cascades or competition.

Cod was once among the fish species with the highest catches. Even today, despite global cod catches being merely around a third of their maximum for the last 70 years (FAO 2021), it is still one of the ten most fished species in the world (FAO 2022). Due to overexploitation, often in combination with climate change, many cod stocks have declined with some notable collapses. This has further increased appreciation of the high value of cod both as a resource and also as a key element of marine and coastal ecosystems.

The importance of cod for humans and ecosystems has resulted in many publications. Google Scholar returns about 687000 results on 'Atlantic cod'. A previous synthesis of spawning and life-history information on cod stocks (Brander 1994) provided a summary of available information, but there have been major developments since that time. Two books give a good overview of the species: *Atlantic Cod: A Bio-Ecology* (Rose 2019), which describes cod biology, ecology, life history, behaviour, fisheries exploitation, and conservation but not at a stock level; and *Cod: The Ecological History of the North Atlantic Fisheries* (Rose 2007), which concentrates on cod fisheries, but mainly for North-Western Atlantic cod stocks. However, due to a greater understanding of population structure and dynamics, there have been major changes in cod stock management in recent years.

The goal of this book is to describe how major cod stocks differ across their geographic range. Stock-specific chapters first describe the stock in terms of population structure, distribution and life history. Then, differences in productivity, reproductive success, and trophodynamics are considered before discussing the challenges facing management. In addition, the book begins with chapters addressing two key issues relevant to all cod stocks.

Chapter 1 (Plasticity and Evolution in Atlantic Cod Populations During Climate Change) describes the current and future impact of climate warming on population dynamics, life history, and behaviour of cod. It integrates ecological and evolutionary perspectives for understanding local adaptations and broader responses to climate change.

Chapter 2 (Trophic Interactions) highlights the role of cod in ecosystems both as predator and prey. It describes variations in cod diet across geographic regions of cod distribution and the impact on local food web dynamics.

In Chapters 3–9, details of the local variations of cod stocks, including life-history parameters, population dynamics, and fishery management, are presented, moving across the North Atlantic, starting at the USA East coast and finishing in the Barents Sea (Figure 1).

The book concludes with Chapter 10 (Comparison of the Atlantic Cod Stocks: Biology, Fisheries, and Management), which examines variation among cod stocks both in terms of biology (growth, maturity, and stock size) and effects of fisheries (landings, fishing mortality, and stock status) to explain the current differences in stock status.

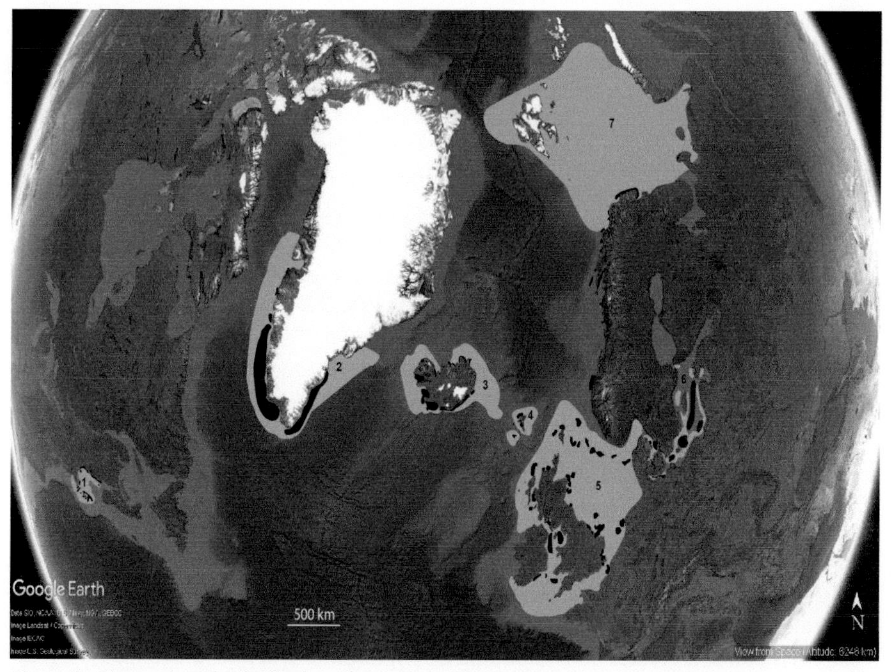

Figure 1. The geographic distribution of cod stocks is covered in this book. Feeding areas are coloured grey, while spawning areas are black. Numbers represent geographic regions: 1 – New England (Georges Bank and Gulf of Maine cod stocks; Chapter 3); 2 – Greenland (East Greenland Iceland offshore spawning, West Greenland inshore spawning cod and West Greenland offshore spawning cod stocks; Chapter 4); 3 – Iceland (Chapter 5); 4 – Faroe islands (Faroe Plateau and Faroe Bank cod stocks; Chapter 6); 5 – Northwest European shelf (Celtic Sea (including also Western Channel), Irish Sea, West of Scotland and the North Sea (including Skagerrak and eastern English Channel) cod stocks; Chapter 7); 6 – Kattegat and the Baltic Sea (Kattegat, western and eastern Baltic Sea cod stocks; Chapter 8); 7 – Northeast Arctic (Northeast Arctic cod stock; Chapter 9).

References

Anon. c. 1240. Egils saga Skallagrímssonar. Stofnun Árna Magnússonar, University of Reykjavik, Iceland.

Brander, K. (ed.). 1994. Spawning and life history information for North Atlantic cod stocks. ICES Cooperative Research Report, Vol. 205. 154 pp. https://doi.org/10.17895/ices.pub.5500.

FAO, 2021. Fishery statistics: global capture production quantity (1950–2021) [WWW Document]. URL https://www.fao.org/fishery/statistics-query/en/capture/capture_quantity.

FAO. 2022. The state of world fisheries and aquaculture 2022. Towards blue transformation. Rome, FAO. https://doi.org/10.4060/cc0461en.

Rose, G.A. 2007. Cod: the ecological history of the North Atlantic fisheries. Breakwater, St. John's, N.L.

Rose, G.A. (ed.). 2019. Atlantic cod: a bio-ecology. John Wiley & Sons.

Chapter 1

Plasticity and Evolution in Atlantic Cod Populations during Climate Change

Esben Moland Olsen,[1,2,*] *Jeffrey A. Hutchings,*[3,†] *Lauren A. Rogers,*[4] *Nils Chr. Stenseth*[5] and *Leif Asbjørn Vøllestad*[5]

In memory of Jeffrey A. Hutchings

Introduction

Climate variability has major influences on the Earth's ecosystems (Stenseth et al. 2002; Parmesan and Yohe 2003; Poloczanska et al. 2013). In marine systems, climate variability directly affects sea-surface temperature and wind stress. For instance, the large-scale North Atlantic Oscillation (NAO) drives the strength and extent of westerly winds across the North Atlantic, with significant influences on winter temperature and precipitation in both the northwest and Northeast Atlantic (Hurrell 1995). Furthermore, there is mounting evidence that human activities responsible for burning fossil fuels are influencing the world's climate via carbon emissions to the atmosphere, leading to global warming (Huber and Knutti 2011; Frölicher et al. 2018; Osman et al. 2021).

This chapter highlights research on the current and future impact of climate warming on population dynamics, life history, and behaviour of Atlantic cod (*Gadus morhua*). Our goal is not to provide a complete review of existing literature

[1] Institute of Marine Research Flødevigen, 4817 His, Norway.
[2] Centre for Coastal Research (CCR), Department of Natural Sciences, University of Agder, P.O. Box 422, 4604 Kristiansand, Norway.
[3,†] Department of Biology, Dalhousie University, 1355 Oxford Street, Halifax, Nova Scotia, Canada Deceased.
[4] Alaska Fisheries Science Center, National Marine Fisheries Service, National Oceanic and Atmospheric Administration, 7600 Sand Point Way NE, Seattle WA 98115, United States.
[5] Centre for Ecological and Evolutionary Synthesis (CEES), Department of Biosciences, University of Oslo, P.O. Box 1066, Blindern, 0316 Oslo, Norway.
* Corresponding author: esben.moland.olsen@hi.no

on the topic. Rather, we specifically make the case for integrating ecological and evolutionary perspectives for understanding local adaptations and broader responses to climate change. It is increasingly recognised that ecology and evolution may interact on contemporary time scales, resulting in ongoing eco-evolutionary dynamics at population, community, and ecosystem levels (Hendry 2017). Fish populations, including cod, can evolve within decades in response to natural and anthropogenic agents of selection (Reznick et al. 1990; Olsen et al. 2004a; Conover et al. 2009; Uusi-Heikkilä et al. 2015). Hence, we argue that an evolutionary perspective is needed to understand better how cod will respond to ongoing and future climate change. We focus particularly on the concept of phenotypic plasticity. With some exceptions, such as the fixed age-at-maturity in pink salmon (*Oncorhynchus gorbuscha*), fish tend to be highly plastic, meaning that the phenotype being expressed by an individual will depend on the environment that it experiences during its lifetime, and potentially that experienced by its parents. Understanding the mechanisms and constraints related to such plasticity is therefore relevant for studies on the effects of climate change (Salinas and Munch 2012; Ciannelli et al. 2015). Plasticity may be adaptive or non-adaptive and costly, depending on the context (DeWitt et al. 1998; Auld et al. 2010). Further, plasticity can be considered a heritable phenotypic trait, capable of evolving in response to selection (Hutchings et al. 2007; Hutchings 2011; Uusi-Heikkilä et al. 2017). Dobzhansky (1937) opined that "...what is inherited in a living being is not this or that morphological character, but a definite norm of reaction to environmental stimuli."

Demography and Population Dynamics

Experimental studies suggest Atlantic cod thrives best when water temperatures are below 15°C (Björnsson and Steinarsson 2002). However, cod in the southern range of the species' distribution, such as the North Sea and Skagerrak, can experience water temperatures up to 20°C in summer (Neat and Righton 2007; Freitas et al. 2021). These observations point towards potentially strong effects of climate on cod demography and population dynamics. For instance, the 1960s and 1970s were characterised by a negative NAO and cold, dry winters in northern Europe, and this period witnessed record high recruitment of cod and other gadoids in the North Sea, known as the "gadoid outburst" (Cushing 1984; Hislop 1996). Since then, the North Sea has warmed by approximately 1°C and cod recruitment has been poor (Olsen et al. 2011; Rogers et al. 2017). Under the influence of human-induced global warming, the North Sea is heading towards a further temperature increase of at least 2°C and accelerated negative development of the southern cod populations towards the end of this century (Clark et al. 2003; Butzin and Pörtner 2016).

Climate change is likely to influence cod population dynamics directly through physiological and energy allocation mechanisms (Otterlei et al. 1999; Pörtner et al. 2001; Pörtner et al. 2008) and indirectly through effects on predators or prey. In the North Sea, there is evidence of a climate-driven plankton effect on cod recruitment (Beaugrand et al. 2003). Throughout the eastern North Atlantic, there is additional evidence of a biogeographical shift in planktonic copepod assemblages associated with ocean warming, where warm-water species extend northwards, and cold-water species decrease (Beaugrand et al. 2002). In particular, nauplii stages of *Calanus finmarchicus,* an energy-rich and often dominant food source for fish larvae (Munk 1997), are associated with a high probability of Atlantic cod occurrence; smaller copepods are less important (Beaugrand and Kirby 2010). In the North Sea and Skagerrak region, *C. finmarchicus* has decreased in abundance while the more southern species *C. helgolandicus* has increased (Beaugrand et al. 2003; Falkenhaug et al. 2022). The latter is apparently less nutritious, peaks in abundance later in the season, and is not a common prey for cod larvae (Heath and Lough 2007; Falkenhaug et al. 2022). Therefore, the poor recruitment of cod is probably related to both the quality and timing of food abundance for cod larvae (Beaugrand et al. 2003).

Beaugrand and Kirby (2010) suggested that for the southernmost populations of Atlantic cod, climate change alone would be sufficient to decrease stocks to the point of triggering a collapse (see also Núnez-Riboni et al. 2019). Within the species' geographic range, however, different populations of Atlantic cod experience a broad range of climatic conditions. At the northern limit of its distribution, reaching as far as 80°N by the Svalbard archipelago, Atlantic cod can be found in partly enclosed Arctic lakes where water temperature during summer is no more than 2–7°C (Hardie et al. 2008). Most of these lake populations are actually outside the species' contemporary coastal and offshore distribution. The relict cod populations probably persist because permanent water stratification keeps the lake temperatures somewhat above what is found in open waters at the same latitudes (Hardie et al. 2008). One might expect, therefore, that the effects of climate variation will depend very much on where each cod population is located within the species' geographic distribution. Indeed, analyses including a range of cod populations have documented a positive effect of warm periods on recruitment in northern populations and a negative effect of warm periods in southern populations (Brander and Mohn 2004; Stige et al. 2006; Kjesbu et al. 2014; Free et al. 2019). It should be noted that such relationships between climate and recruitment could also be associated with changes in spawning stock biomass and appear to be changing over time in some areas. For instance, in the southern part of the Northeast Atlantic, a positive phase of the North Atlantic Oscillation (NAO) has had an increasingly negative effect on recruitment, which might be related to demographic changes in the cod populations caused by fishing (Stige et al. 2006). Accounting for changes in spawning stock biomass, a study on Northeast Arctic cod concluded that warm temperatures during the larval period were necessary but insufficient for good recruitment (Ohlberger et al. 2014a).

The Atlantic cod is an important food fish, and many populations, such as those found in the North Sea and off Newfoundland and Labrador, have suffered from a much too intense harvest pressure (Hutchings and Myers 1994; Cook et al. 1997). These population declines and collapses beg the question of whether overfishing and climate change can be treated as independent additive stressors on cod demography and population dynamics. Brander (2007) argues that this is not likely the case; indeed, Hutchings and Myers (1994) made the same argument for Newfoundland cod in the early 1990s. There are probably strong interactions between the effects of fishing and the environment, where depleted population size resulting from intense fishing increases the susceptibility of populations to natural and anthropogenic environmental variability (Brander 2005; Ottersen et al. 2006; Winter et al. 2020).

This is linked to the way fishing is practised. Fishing is often selective and typically removes more of the larger and faster-growing individuals in harvested populations of Atlantic cod (Swain et al. 2007; Olsen and Moland 2011; Denechaud et al. 2020). Larger individuals are targeted because they provide better profit to the fisher and because smaller individuals are often protected by management regulations with the goal of letting the fish spawn at least once before they are harvested. For instance, the minimum legal size of coastal cod in Southern Norway is currently 40 cm, a size at which many cod from this region will already be sexually mature (Olsen et al. 2004b).

As a result of selective and intensive fishing practices, cod populations are now often severely age- and size-truncated, meaning relatively few young age classes dominate them compared to a more natural situation (Jackson et al. 2001; Limburg et al. 2008; Moland et al. 2013). In Atlantic cod, small and young spawners have relatively low fecundity, producing smaller offspring and spawning during a shorter season than older and larger cod (Hutchings and Myers 1993; Trippel 1998; Barneche et al. 2018). Such juvenescence can destabilise the dynamics of populations and constrain their natural capacity to buffer environmental changes, including climate change (Anderson et al. 2008; Ohlberger et al. 2014b; Morrongiello et al. 2019). The variance-dampening effect of maintaining life-history diversity and population diversity in exploited fish has been termed the portfolio effect, and fisheries may benefit from this effect in the form of more temporally stable harvestable yields (Schindler et al. 2010).

Brander (2007) concludes that an important step towards reducing the impact of climate change would be to reduce fishing mortality in populations considered fully exploited or overexploited. Also, several studies have argued that management needs to address the protection of phenotypic diversity (old-growth age structure), genetic diversity and population diversity in exploited populations to ensure long-term sustainable population levels (Berkeley et al. 2004; Hutchings et al. 2007; Barneche et al. 2018; Olsen et al. 2023). In practical terms, this could mean a shift away from the traditional management philosophy of encouraging selective fisheries and instead a move towards more balanced forms of fishing across species, populations and phenotypes (Zhou et al. 2010; Garcia et al. 2012). On the other hand, Jacobsen et al. (2014) pointed out that even if a balanced harvesting strategy would likely produce the highest total maximum sustainable yield, it would also be a fishery mainly bringing small fish to market. Lastly, the Barents Sea cod found off Northern Norway and Russia is currently at a historically high level, and Kjesbu et al. (2014) argue that this can be ascribed to synergies between a warming climate extending the potential feeding areas to the north and a sharpening of traditional fishery management regulations involving a strictly enforced harvest control rule. In a recent analysis of a 75 y long time series on the same Northeast Arctic cod population, Ohlberger et al. (2022) confirm that increased temperature has a positive effect on population productivity but also present compelling evidence that fisheries-induced truncations of the natural age structure have been responsible for substantially reduced productivity. Indeed, their study indicates that the productivity of Barents Sea cod was three times higher when the age structure was at its maximum compared to when it was most truncated by fishing. Another recent analysis concluded that a truncated age structure makes the population more vulnerable to environmental fluctuations (Ottersen and Holt 2023).

Mobile animals such as cod may, in principle, respond to unfavourable temperatures in three ways: (1) moving horizontally or vertically to habitats offering more favourable temperatures; (2) phenotypically plastic responses within the less favourable habitat; and (3) evolutionary adaptation. Next, we explore phenotypic trait responses in cod populations facing climate change and the importance of considering adaptive phenotypic plasticity as a mechanism for tackling environmental variability.

Distribution Shifts and Behavioural Responses

As recent warming coincides with a northward shift of the distribution of cod in the Northeast Atlantic (Perry et al. 2005; Engelhard et al. 2014; Kjesbu et al. 2014), this could be explained by behavioural responses as well as spatially structured demographic responses where, for instance, survival declines in southern parts of the range or increases in northern parts. However, few studies have actually quantified individual-based cod behaviour in relation to temperature in the wild. One large-scale study used data-storage tags to investigate cod behaviour throughout both the northern and southern parts of the North Sea (Neat and Righton 2007; see also Righton et al. 2010; Neat et al. 2014). Interestingly, the main findings from this study did not support the notion that a shift in the geographical distribution of cod is caused by adult fish moving northward to avoid the warmer temperatures in the southern North Sea. Rather, the data revealed that cod utilised surprisingly warm waters. Many tagged cod remained in the southern North Sea during late summer and experienced temperatures up to 19°C, far beyond what is thought to be an optimal temperature for growth of larger cod (Björnsson and Steinarsson 2002). As pointed out by the authors, this finding begs the question of whether the thermal optima documented in laboratory studies adequately reflect what occurs in the wild and whether temperature is actually a strong driver of Atlantic cod habitat choice. One weakness of many laboratory experiments is that fish are not acclimated appropriately, and the fish is thus investigated in the wrong context (Niemelä and Dingemanse 2014; Wootton et al. 2022). Both thermal optima and upper and lower temperature limits may vary with size, season and context.

Studies from the Norwegian Skagerrak coast (bordering the North Sea region) shed additional light on these questions by analysing fine-scale horizontal and vertical movements of cod in relation to thermal stratification of the water column. Tagged Atlantic cod typically performed diel vertical migrations such that shallow nearshore habitats, holding dense populations of potential prey species (Johannessen 2014), were exploited during dark hours while the cod retreated to deeper waters during daytime (see also, Espeland et al. 2010). However, during the summer, when the surface waters heated to 16–20°C, the fish ceased to utilise the shallow habitats. This habitat-displacement effect was stronger for larger fish when compared to smaller fish, in conjunction with predictions from laboratory studies on thermal preferences (Björnsson and Steinarsson 2002). Intriguingly, the cod temporarily resumed their shallow-water migrations during summer when strong offshore winds sometimes caused upwelling events that replaced the warm surface water with colder water from below. The authors concluded that future elevations in sea-surface temperature would deprive juvenile and adult cod in coastal areas of key feeding habitats for extended periods (Freitas et al. 2016). In support of this, Dulvy et al. (2008) reported that North Sea cod shifted their distribution towards significantly deeper habitats (approximately 10 m per decade) during the warming period from 1980 to 2000.

Finally, climate change is also expected to have a clear influence on Atlantic cod phenology, which is the seasonal timing of important life-history events such as spawning. Higher temperatures will increase the oocyte development rate, likely resulting in a shift towards spawning earlier in the season (Kjesbu et al. 2010). In support of this, McQueen and Marshall (2017) detected a trend towards earlier spawning in North Sea cod during three decades, potentially explained by rising sea temperatures during oocyte development.

Life-History Responses: Growth and Maturation

Across the geographical range of cod, life-history traits such as age-at-maturity and growth vary in a fairly predictable way with ambient water temperatures (Brander 1995, 2005). At their northern limits, cod grow slowly and mature late (e.g., age 6 to 10 years in Northeast Arctic cod). In contrast, in relatively warm regions such as the North Sea, cod grow much more quickly and mature as early as age 2 (Brander 2005). This suggests that, within a region, changes in life history are likely to accompany shifts in the underlying environment.

Growth is a fundamental process that is linked to age and size at maturation, fecundity, and possibly survival in Atlantic cod and other ectotherms (Campana 1996; Olsen et al. 2008). In general, the biochemical reactions that determine growth in ectothermic organisms are also strongly dependent on temperature (Pörtner et al. 2008; Björnsson and Steinarsson 2002; Arroyo et al. 2022). Growth is, therefore, one of the main pathways through which climate change might affect cod. Indeed, studies have already linked interannual variation in size-at-age to changes in temperature (Armstrong et al. 2004; Brander 2005). In northern cod populations, such as those found around Flemish Cap, the Faroe Islands, West Greenland, and the Barents Sea, body size at a given age tends to co-vary with temperature through time, becoming larger during warm phases (Brander 1995; Denechaud et al. 2020; Ruiz-Díaz et al. 2022). The direct effects of temperature on growth may result in changes in optimal life-history strategies. For instance, a dynamic energy allocation model for Northeast Arctic cod predicts larger size-at-age, increased investment in reproduction, and riskier foraging behaviour under a 2°C warming scenario (Holt and Jørgensen 2014). Because faster growth is typically associated with earlier age at maturation (Neuheimer and Grønkjaer 2012), changes in maturation schedules may also be expected with warming temperatures.

While most studies have found that growth rates of cod increase with warming temperatures, there can be considerable complexity in responses at the individual or population level, depending on location, season, size, and the extent of warming. The expected relationship between temperature and growth tends to be dome-shaped, with maximum growth occurring at intermediate temperatures (Björnsson and Steinarsson 2002). Further, the dome is asymmetrical, so growth performance drops quickly as optimal temperatures are exceeded. Thus, the effect of warming will depend on the current temperature conditions experienced and whether warming shifts the organisms toward or past their optimum (Ohlberger 2013). As suggested by Hutchings et al. (2007), this leads to different expectations of the effect of warming on cod growth throughout the species' geographical range. In fact, there are signs that the recent increase in ocean temperature could already be slowing down the growth of cod and other marine fish within warmer regions of their distribution (Neuheimer et al. 2011; Rogers et al. 2011).

Experimental studies have shown that the optimal temperature for growth of cod varies by size, with smaller cod achieving maximum growth at a higher temperature than larger cod, provided that food is unlimited (Björnsson and Steinarsson 2002). This means that the effects of warming on different age- or size classes may differ and that smaller individuals, in general, may be better able to cope with warmer environments (Atkinson 1994; Lindmark et al. 2022).

A detailed study of Atlantic cod on the Norwegian Skagerrak coast found that the effects of temperature on the body size of juvenile cod depended on the season, such that warmer springs resulted in larger juveniles, whereas warmer summers resulted in smaller juveniles (Rogers et al. 2011). Numerous explanations are possible for this pattern. Spring temperatures are cooler, so an increase will bring cod closer to their growth optimum, whereas summer temperatures in the Skagerrak often exceed the optimum for juvenile growth. However, it is likely that the mechanisms are more complicated than a simple thermal growth response. Spawn timing is linked to temperature (Hutchings and Myers 1994; Neuheimer and MacKenzie 2014; but see Morgan et al. 2013). Thus, warmer springs could lead to earlier spawning and a longer opportunity for growth before cod are measured as juveniles. The temperature could also affect cod growth indirectly through effects on prey such as zooplankton (Beaugrand et al. 2003). If food resources are limited, the optimal temperature becomes lower as individuals struggle to meet metabolic demands. The effects of climate change on cod body size depend not only on the biology of the population in question but also on the seasonality of warming and its effects on their prey. A full understanding of body-size responses to warming requires the integration of effects at the individual, population, community, and ecosystem levels (Ohlberger 2013).

Phenotypic Plasticity: The Concept of Reaction Norms

Local adaptation is integral to conservation, reflecting the consequences of natural and human-induced selection. From a conservation and management perspective, understanding the factors that affect population persistence in the face of anthropogenic challenges, such as climate change, is of fundamental interest. Key questions concern the ability of marine fish populations to respond to environmental change and the spatial scale at which adaptive responses to environmental change are realised. The former question depends on the level of phenotypic plasticity expressed by individuals within a population, whereas the latter is reflected by genetic differences in plasticity at the population level, such as different population responses to the unidirectional temperature shifts (i.e., warming) that are associated with climate change.

The phenotypic variability expressed by any trait is a function of the environment, an individual's genotype, and how the genotype interacts with the environment (Box 1). The interaction between genotypes and the environment (often termed the GxE interaction) is a reflection of phenotypic plasticity—the ability of a genotype to produce different phenotypes across an environmental gradient (Schlichting and Pigliucci 1998; Sultan and Stearns 2005). Plasticity can be heuristically and graphically described as a norm of reaction—a linear or nonlinear function that expresses how the phenotypic value of a trait changes with the environment (Figure 1). By providing information about the magnitude of trait plasticity and the presence of G×E interactions (de Jong 2005), norms of reaction have great potential to increase our understanding of how genotypes, and ultimately populations and species, respond adaptively to natural and human-induced environmental variability.

Ideally, reaction norms would be constructed at the genotypic level, such that each linear or nonlinear function would describe the phenotypic change of a single genotype along an environmental gradient. However, this is not easy to do in fish. More practically, reaction norms are constructed at the level of a group, such as a family or population (Hutchings 2011). At the population level, reaction norms can be used to predict how individuals will respond, on average, to specific changes in an environmental variable. Genetic differences in reaction norms reflect differences in the ability of populations to respond to environmental change and how they do so. Thus, if temperature was the environmental variable under study, one could construct thermal reaction norms for traits such as survival, growth and fertilisation success to address whether different cod populations are likely to respond differently to the warming temperatures forecasted to be associated with climate change.

The construction of reaction norms is ideally done under experimental conditions in which one controls all environmental variability except the environmental variable of interest. For example, by rearing populations from birth under conditions that are the same in every respect but temperature, one can determine whether populations differ genetically in their phenotypically plastic responses to temperature. These types of experiments are called common-garden experiments. They represent one of the most powerful means of assessing the genetic basis of phenotypic variation (Imsland and Jónsdóttir 2003; Conover et al. 2006; Hutchings et al. 2007). Although common-garden experiments have been used in studies of plasticity and evolution for more than a century (Schlichting and Pigliucci 1998), their application to studies of plasticity in marine fishes has been uncommon (some examples include those by Schultz et al. 1996; Conover et al. 2006; Baumann and Conover 2011; Hurst et al. 2012).

In Atlantic cod, common-garden experiments have been used to explore evidence of genetic differences in traits among different populations. Purchase and Brown (2000, 2001) observed differences in the growth rate of larvae and the food conversion efficiency of juveniles between cod from the Grand Banks and the Gulf of Maine. Rearing Northeast Arctic cod and Norwegian coastal cod under the same environmental conditions, Svåsand et al. (1996) documented evidence of genetic differences in growth, hepatosomatic index, and the proportional allocation of body tissue to gonads. Salvanes et al. (2004) used a common-garden protocol to quantify genetic differences in growth and condition factors between different

Box 1: Quantitative genetics—an underused tool in marine ecology?

Common garden and reciprocal transplant experiments have long been used to untangle the importance of genetic and environmental effects on phenotype expression. Such quantitative genetic experiments and associated analysis have a long history, and a number of good textbooks (Roff 1992; Stearns 1992; Falconer and Mackay 1996; Roff 1997; Hutchings 2021) and reviews are available (Shaw and Shaw 2014). Especially interesting is how the quantitative genetic approaches and present-day genomics are getting merged (Visscher et al. 2008; Savolainen et al. 2013).

In general, the phenotypic variance of a trait (Vp) can be decomposed and expressed as follows:

$$Vp = Vg + Ve + Vg*e,$$

where Vg is the variance due to genotypic effects, Ve is the variance due to environmental variation, and Vg*e is the variance due to the genotype-environment interaction. Given that the genotype-environment interaction (usually only denoted G*E) is 0 the broad sense heritability can be estimated as:

$$H^2 = Vg/Vp.$$

G*E is, however, of utmost importance for the study of genetic variation among genotypes, families, and populations and will be discussed in more detail later.
The genotypic variance captured by Vg also contains genetic factors irrelevant to adaptive evolution. Therefore, Vg is often partitioned into several components:

$$Vg = Va + Vi + Vd$$

where Va is the variance of additive genetic effects (the breeding values). Vi is the epistatic interactions between alleles at different loci, and Vd is the interactions between alleles at the same locus (dominance effects). Given Vd = 0 and Vi = 0, we can conveniently estimate heritability in the narrow sense as:

$$h^2 = Va/Vp.$$

In all these models, we assume that there is no genotype-environment covariance. This can safely be done given appropriate experimental designs. However, such co-variation may strongly impact results when estimating parameters based on studies of natural (wild) populations.

The G*E interaction captures the effects of plasticity, the fact that a genotype can express different genotypes across an environmental gradient. The plastic response is commonly represented using the norm of reaction approach.

Quantitative genetic experiments were classically statistically challenging, requiring large and balanced experimental designs (Falconer and Mackay 1996; Roff 1997). However, recent developments in general linear mixed modelling, including Bayesian approaches, have facilitated greater flexibility and have made it possible to analyse complex pedigrees with deep familial links using various Animal Model approaches (Kruuk 2004; Kruuk et al. 2008; Wilson et al. 2009; Hadfield 2010; see Charmantier et al. 2014 for recent applications). The rapid development of high-throughput sequencing and genotyping tools facilitates the analysis of quantitative genetic traits in natural populations.

However, these kinds of quantitative experiments seem to be underutilized in the marine field.

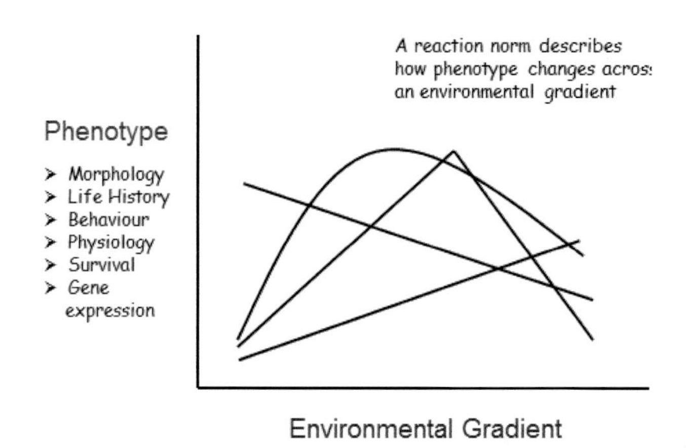

Figure 1. Four hypothetical, population-level reaction norms. Each line illustrates the change in a phenotypic variable along an environmental gradient, such as temperature. In common-garden experiments, shape differences between reaction norms can be used to detect genetic differences in how spatially distinct populations respond to environmental change.

groups of Norwegian coastal cod, providing evidence of population substructure within this single ICES management unit. Otterå et al. (2006) found evidence for a north-south gradient in spawning time among populations of Norwegian costal cod reared under identical conditions, with later dates of spawning in the north (see also, Otterå et al. 2012). Examining a physiologically related trait under common-garden conditions, Goddard et al. (1999) concluded that there are genetic differences in antifreeze protein production among cod populations, hypothesising that these represent adaptive responses to differing winter environments.

Studies of genetic differentiation in Atlantic cod were comprehensively reviewed by Imsland and Jónsdóttir (2003). Although they discussed evidence of genetic differences among populations (such as the examples provided above), they provided no examples of studies that had documented genetic variation in plasticity (as reflected by crossing norms of reaction), suggesting that such work had yet to be undertaken. Since Imsland and Jónsdóttir's (2003) paper, however, several new studies have found empirical support for genetic variability in phenotypic plasticity in Atlantic cod, some of which are described below (see also Grabowski et al. 2009; Larsen et al. 2012).

The first and most detailed experimental evidence of genetic variation in plasticity in Atlantic cod, reflected by differences in the slopes of reaction norms, appears to be work undertaken in the Northwest Atlantic. Based on an analysis of homologous morphological landmarks, Marcil et al. (2006) studied the influence of temperature and food supply on the body shape of larval cod from four spatially divergent populations. They documented evidence of genetic differences in plasticity at both large (> 1,000 km) and small (< 100 km) distances. Population differences in morphology and allometry were manifest by differences in body depth, head size, and length of caudal peduncle. Of the two variables examined, they reported that the influence of temperature on morphology was greater than that of food supply.

Genetic differences in plasticity have also been documented for the effects of temperature and food supply on larval cod growth rate and survival. Using a common-garden experimental protocol similar to that employed by Marcil et al. (2006), Hutchings et al. (2007) found that how growth rate, particularly survival, varies as a function of food and temperature differs genetically among cod populations. Oomen and Hutchings (2016) extended Hutchings et al.'s (2007) work on thermal plasticity, documenting additional genetic differences, and at a smaller geographical scale, in the survival responses of cod to changing temperatures.

Lastly, Purchase et al. (2010) quantified reaction norms in sperm performance of cod from the Northwest Atlantic and found that the sperm generally swam fastest at intermediate temperatures. The study by Purchase et al. (2010) also documented a significant interaction between genotype and the environment, meaning that this trait has genetic differences in phenotypic plasticity.

It is somewhat surprising that experimental studies of plasticity have been comparatively rare, given the considerable interest in studying environmental correlates of cod productivity, recruitment, and population growth. It can be anticipated, however, that with the growing availability of genomic tools and an increasing awareness of the need to forecast cod population responses to climate change, there could be a surge in research activity on plasticity in Atlantic cod and on the genetic and genomic basis for how individuals respond to environmental change (Oomen and Hutchings 2022).

Challenges and Opportunities in the Years Ahead

Plasticity plays a major role in explaining contemporary phenotypic responses to environmental change in aquatic as well as terrestrial systems (Schlichting and Pigliucci 1998; Lande 2009; Chevin et al. 2010). Also, phenotypic variation in life-history traits and behaviour typically contains a heritable component, meaning that these traits can evolve in response to natural and anthropogenic agents of selection (Mousseau and Roff 1987; Roff 2002). In Atlantic cod, common-garden studies have found additive genetic variation (i.e., heritability) for traits such as seasonal timing of spawning, somatic growth and juvenile survival (Gjerde et al. 2004; Otterå et al. 2012). Further, there is genomic evidence that cod populations on both sides of the North Atlantic are evolving in response to ocean temperature (Bradbury et al. 2010; Therkildsen et al. 2013). Populations of cod will, therefore, likely respond to future climate change through phenotypic plasticity as well as contemporary evolution. We argue that reaction norms hold particular promise as a tool for understanding the relative contribution of temporal changes in cod phenotypes, given that a reaction norm specifically describes plasticity, and a change in the slope or position of the reaction norm itself can represent evidence for a genetic response. However, although there is every reason to predict that cod are capable of some level of adaptation to climate change, there is limited evidence of such adaptation in fish to date (Crozier and Hutchings 2014; Merilä and Hendry 2014). Indeed, a recent common-garden study found that fish may struggle to adapt to the speed of current and projected climate change, linked to clear limits in adaptive plasticity at upper thermal tolerance (Morgan 2020).

The likelihood that one can reliably attribute phenotypic or genetic change to climate change depends initially on whether the environmental variable of interest, such as temperature, can truly be characterised as being part of a change in climate. In other words, there needs to be evidence that there has been a shift in the probability distribution of the occurrence of the values of the variable of interest. Although this could mean a change in temporal variability of the environmental variable (e.g., the mean temperature remains constant while the maxima and minima increase or decrease over time), this is comparatively difficult to detect. In practice, from a climate-change perspective, one is interested primarily in a directional shift in the variable of interest that is not part of naturally occurring aperiodic or cyclic change. Otherwise, one is documenting responses by organisms, such as cod, to directional environmental change rather than climate change. That said, the documentation of responses to a directional temperature change would nonetheless be informative about the capacity to respond to climate change.

Thus, the first challenge in any study of the effects of climate change on fish must be the demonstration that the environmental variable of interest has been shifting relative to its past statistical distribution. A related constraint pertains to the spatial scale across which data are available. One requires data across a spatial scale that is meaningful relative to the spatial distribution of the population of interest to detect adaptive genetic or plastic responses to climate change. Unfortunately, in the marine realm, oceanographic data are often available at the global scale and tend to be sparse at small biologically meaningful scales. Another constraint in detecting climate-related shifts lies in the length of time over which environmental data are available.

Despite these caveats, unidirectional shifts that are not part of natural fluctuations are evident in the marine environment. Data available for the Gulf of St. Lawrence, a large 236,000 km^2 ecosystem in eastern Canada, reveal a steady increase in sea-surface temperature since the 1870s (Hutchings et al. 2013). Since 1985, the rate of change in sea-surface temperature has been notably rapid; temperatures have increased by almost 2° C; the two warmest temperatures in the past 140 years are estimated to have occurred in the past decade (Galbraith et al. 2010; for a related example, see Otero et al. 2014). This warming is forecast to continue through the 21st century (Arora et al. 2011). In addition to temperature, directional changes are evident for other oceanographic features at relatively small spatial scales. Stratification of the vertical layers of the ocean, a primary determinant of nutrient availability in upper waters, appears to be strengthening in the Northwest Atlantic (the greater the stratification, the slower the flow of nutrients from deeper to shallower waters). This is evident, for example, on the eastern Canadian Scotian Shelf (Petrie et al. 2009).

Two additional challenges in demonstrating that climate change is directly responsible for plastic or genetic change in fishes pertain to the strength of selection generated by climate change and the time required for a measurable genetic or plastic response (irrespective of whether those responses are adaptive). The greater the intensity of selection, the smaller the percentage of actual breeders from the pool of potential breeders in a population, and the greater the likelihood that a plastic or genetic response will be realised (and, thus, detectable) over relatively few generations (Endler 1986; Siepielski et al. 2009; Kingsolver et al. 2012). It is clear that sufficiently high selection intensities can yield measurable selection responses in a few generations of fish (Reznick et al. 1990). Laboratory experiments on Atlantic silverside (*Menidia menidia*), for example, have demonstrated how intense selection (breeders representing 10% of the potential spawning population) can generate measurable selection responses in life-history traits with moderate heritability (e.g., $h^2 = 0.2$) over relatively few generations (four generations for the silverside experiment; Conover and Munch 2002). Similarly, intense directional selection for traits considered desirable in fish farms can also yield demonstrable genetic change in 3–5 generations (Debes and Hutchings 2014).

A key question is whether climate change, in general, is likely to be associated with strong selection intensities in fishes. If the gradual change evident for many climate-related environmental variables in the oceans is typical, selection intensities are unlikely to be sufficiently strong to generate measurable plastic or genetic responses that can be reliably attributed to climate change. A further requirement would be that selection differentials be quite large (the greater the selection differential, the greater the selection response and the greater the probability of detecting a response, all else being equal). Nonetheless, even under strong selection intensities and high selection differentials, which might not be common correlates of climate change in many aquatic environments, genetic change will likely only be detectable in species with comparatively brief generation times. Reliable detection of a truly plastic response is also likely to require a greater magnitude of environmental change than that typical of climate change. Common-garden experiments indicate that the probability of detecting statistically significant phenotypical plastic change can be low (e.g., Wood and Fraser 2015) unless the species experiences temperatures close to its maximum or minimum limits.

It is important to note that a gradual change in one climate-change-related variable might well be associated with a comparatively rapid change in another climate-change-related variable. So, rather than resulting in a gradual change in fitness, a gradual change in a physical environmental variable can be associated with a rapid, possibly even a threshold or stepwise change in fitness. One example of this might be associations between seasonal changes in precipitation and water flow in freshwater or anadromous fishes migrating from lakes or the ocean, respectively, to spawn in rivers and streams. What could be described as a gradual reduction in precipitation might be associated with rapid changes in the fitness consequences associated with the timing of a spawning migration into a river/stream and the timing of reproduction itself (Hendry et al. 2003; Carlson and Quinn 2007). In other words, gradual change in one climate-change-related variable can be associated with rapid changes in others, leading to relatively high selection intensities and detectable responses to selection.

Another point of consideration is the fact that the effects of climate change on organisms will rarely act in isolation from other selection pressures. For example, climate change might negatively influence the ability of exploited fish populations to respond adaptively to the predation pressures exerted by fisheries either by genetic or phenotypically plastic changes in life-history traits (Hutchings and Fraser 2008). The challenge when climate change compounds or acts synergistically with other variables is to separate the genetic or plastic responses to climate change from those resulting from other selection pressures.

Bearing in mind the potential caveats listed above, when attempting to link phenotypic change in cod populations directly to climate change, emerging knowledge from marine ecology, life-history evolution, and modelling strongly suggest that cod populations are responding to current temperature change and that these responses give hints about what scenarios we might see under future climate change. We particularly encourage future studies to tease out the strength, direction and relative contributions of phenotypic plasticity versus evolutionary adaptations to a change in temperature and associated environmental variables. This will inform about the speed of potential responses (i.e., within-generation plastic changes or among-generation evolutionary responses). However, it is clearly not a straightforward task.

As discussed above, common-garden studies represent a powerful tool to estimate reaction norms and should be further employed in studies on climate change. One option is the space-for-time approach, where populations from a range of climatic regions are reared under identical conditions to mimic the influence of temperature change on the evolution of phenotypic plasticity (reaction norms). As pointed out by Crozier and Hutchings (2014), within-population shifts in reaction norms over time estimated from replicated common-garden studies represent another largely unexplored approach. By sacrificing some precision (but gaining increased realism), reaction norms may, in certain cases, also be estimated from field data. For instance, the probabilistic maturation reaction norm approach, estimating the probabilities of maturing at age and size, has been developed for studies on fisheries-induced evolution (e.g., Olsen et al. 2004a). When applied to multiple cod populations along a north-south gradient, this approach has hinted at countergradient variation in reaction norms, where the capacity for early maturation appears to be greatest in cold, northern regions in which the environmental opportunities for growth are poorest (Olsen et al. 2005). Recently, a new analytical framework has been developed to ease the investigation of countergradient and also co-gradient variation (Albecker et al. 2022).

Probabilistic maturation reaction norms can also be used to quantify and visualise the frequency of skipped spawning as a function of fish size and condition (Skjæraasen et al. 2012), where temperature might influence population productivity via both these explanatory variables. Crozier and Hutchings (2014) also highlight the value of quantifying selection differentials over time series for detecting evolutionary responses to climate change. Such selection differentials can be estimated directly from individual mark-recapture data (Olsen et al. 2012) and also from traditional fishery- and survey-type data (Swain et al. 2007).

Recent advances in tracking technology, such as acoustic telemetry, allow for repeated observations of individual fish behaviour across different environments (e.g., temperatures) and represent a promising tool for quantifying behavioural reaction norms in the wild (Dingemanse et al. 2010). In a recent application of this approach, Villegas-Ríos et al. (2018) found that timid Atlantic cod were more plastic in their behaviour compared to bolder individuals and reduced their fjord home range area in response to increased sea temperature.

Lastly, genomic tools offer rapidly expanding opportunities for the years to come. For instance, gene expressions can be considered phenotypes (Figure 1), allowing the construction of genomic reaction norms to infer adaptive responses and plastic responses to climate change (Oomen et al. 2022). Furthermore, whole-genome sequencing can identify "supergenes" and how these can be linked to eco-evolutionary dynamics, including the dynamics of sympatric ecotypes of cod (Matschiner et al. 2022; Sodeland et al. 2022). The occurrence of such genetic diversity has probably been underestimated and may contribute to, for instance, genetic clines where coastal populations of Atlantic cod receive genes from offshore cod (Jorde et al. 2021).

Although evolutionary changes will generally be slower than plastic changes, they should not be overlooked. Intraspecific genetic diversity and the capacity for adaptive change, in essence, represent insurance for population viability and recovery during difficult times, thus providing the basis for "evolutionary rescue" in a changing world (Carlson et al. 2014).

References

Albecker, M.A., Trussell G.C., and Lotterhos, K.E. 2022. A novel analytical framework to quantify co-gradient and countergradient variation. Ecol. Lett. 25: 1521–1533.

Anderson, C.N.K., Hsieh, C., Sandin, S.A., et al. 2008. Why fishing magnifies fluctuations in fish abundance. Nature 452: 835–839.

Arora, V.K., Scinocca, J.F., Boer, G.J., et al. 2011. Carbon emission limits required to satisfy future representative concentration pathways of greenhouse gases. Geophys. Res. Lett. 38: L05805.

Armstrong, M.J., Gerritsen, H.D., Allen, M., et al. 2004. Variability in maturity and growth in a heavily exploited stock: cod (*Gadus morhua* L.) in the Irish Sea. ICES J. Mar. Sci. 61: 98–112.

Arroyo, J.I., Diez, B., Kempes, C.P., et al. 2022. A general theory for temperature dependence in biology. Proc. Natl. Acad. Sci. 119: e2119872119.

Atkinson, D. 1994. Temperature and organism size: a biological law for ectotherms? Adv. Ecol. Res. 25: 1–58.

Auld, J.R., Agrawal, A.A., and Relyea, R.A. 2010. Re-evaluating the costs and limits of adaptive phenotypic plasticity. Proc. R. Soc. B 277: 503–511.

Barneche, D.R., Robertson, D.R., White, C.R., et al. 2018. Fish reproductive-energy output increases disproportionately with body size. Science 360: 642–645.

Baumann, H., and Conover, D.O. 2011. Adaptation to climate change: contrasting patterns of reaction-norm-evolution in Pacific versus Atlantic silversides. Proc. R. Soc. B 278: 2265–2273.

Beaugrand, G., Brander, K.M., Lindley, J.A., et al. 2003. Plankton effect on cod recruitment in the North Sea. Nature 426: 661–664.

Beaugrand, G., and Kirby, R.R. 2010. Climate, plankton and cod. Glob. Change Biol. 16: 1268–1280.

Beaugrand, G., Reid, P.C., Ibañez, F., et al. 2002. Reorganization of North Atlantic marine copepod biodiversity and climate. Science 296: 1692–1694.

Berkeley S.A., Hixon, M.A., Larson, R.J., et al. 2004. Fisheries sustainability via protection of age structure and spatial distribution of fish populations. Fisheries 29: 23–32.

Björnsson, B., and Steinarsson, A. 2002. The food-unlimited growth rate of Atlantic cod (*Gadus morhua*). Can. J. Fish. Aquat. Sci. 59: 494–502.

Bradbury, I.R., Hubert, S., Higgins, B., et al. 2010. Parallel adaptive evolution of Atlantic cod on both sides of the Atlantic Ocean in response to temperature. Proc. R. Soc. B 277: 3725–3734.

Brander, K.M. 1995. The effect of temperature on growth of Atlantic cod (*Gadus morhua* L.). ICES J. Mar. Sci. 52: 1–10.

Brander, K.M. 2005. Cod recruitment is strongly affected by climate when stock biomass is low. ICES J. Mar. Sci. 62: 339–343.

Brander, K.M. 2007. Global fish production and climate change. Proc. Natl. Acad. Sci. USA 104: 19709–19714.
Brander, K., and Mohn, R. 2004. Effect of the North Atlantic Oscillation on recruitment of Atlantic cod (*Gadus morhua*). Can. J. Fish. Aquat. Sci. 61: 1558–1564.
Butzin, M., and Pörtner, H.-O. 2016. Thermal growth potential of Atlantic cod by the end of the 21st century. Glob. Change Biol. 22: 4162–4168.
Campana, S.E. 1996. Year-class strength and growth rate in young Atlantic cod *Gadus morhua*. Mar. Ecol. Prog. Ser. 135: 21–26.
Carlson, S.M., Cunningham C.J., and Westley, P.A.H. 2014. Evolutionary rescue in a changing world. Trends Ecol. Evol. 29: 521–530.
Carlson, S.M., and Quinn, T.P. 2007. Ten years of varying lake level and selection on size-at-maturity in sockeye salmon. Ecology 88: 2620–2629.
Charmantier, A., Garant, D., and Kruuk, L.E.B. 2014. Quantitative Genetics in the Wild. Oxford University Press, Oxford.
Chevin, L.-M., Lande, R., and Mace, G.M. 2010. Adaptation, plasticity and extinction in a changing environment: towards a predictive theory. PLoS Biol. 8: e1000357.
Ciannelli, L., Bailey, K., and Olsen, E.M. 2015. Evolutionary and ecological constraints of fish spawning habitats. ICES J. Mar. Sci. 72: 285–296.
Clark, R.A., Fox, C.J., Viner, D., et al. 2003. North Sea cod and climate change—modelling the effects of temperature on population dynamics. Glob. Change Biol. 9: 1669–1680.
Conover, D.O., Clarke, L.M., Munch, S.B., et al. 2006. Spatial and temporal scales of adaptive divergence in marine fishes and the implications for conservation. J. Fish Biol. 69: 21–47.
Conover, D.O., and Munch, S.B. 2002. Sustaining fisheries yields over evolutionary time scales. Science 297: 94–96.
Conover, D.O., Munch, S.B., and Arnott, S.A. 2009. Reversal of evolutionary downsizing caused by selective harvest of large fish. Proc. R. Soc. B 276: 2015–2020.
Cook, R.M., Sinclair, A., and Stefánsson, G. 1997. Potential collapse of North Sea cod stocks. Nature 385: 521–522.
Crozier, L.G., and Hutchings, J.A. 2014. Plastic and evolutionary responses to climate change in fish. Evol. Appl. 7: 68–87.
Cushing, D.H. 1984. The gadoid outburst in the North Sea. J. Cons. Int. Explor. Mer. 41: 159–166.
Cushing, D.H. 1990. Plankton production and year-class strength in fish populations: an update of the match/mismatch hypothesis. Adv. Mar. Biol. 26: 249–292.
Daufresne, M., Lengfellner, K., and Sommer, U. 2009. Global warming benefits the small in aquatic ecosystems. Proc. Natl. Acad. Sci. USA 106: 12788–12793.
Debes, P.V., and Hutchings, J.A. 2014. Effects of domestication on parr maturity, growth and vulnerability to predation in Atlantic salmon. Can. J. Fish. Aquat. Sci. 71: 1371–1384.
de Jong, G. 1995. Phenotypic plasticity as a product of selection in a variable environment. Am. Nat. 145: 493–512.
Denechaud, C., Smolinski, S., Geffen, A.J., et al. 2020. A century of fish growth in relation to climate change, population dynamics and exploitation. Glob. Change Biol. 26: 5661–5678.
DeWitt, T.J., Sih, A., and Wilson, D.S. 1998. Costs and limits of phenotypic plasticity. Trends Ecol. Evol. 13: 77–81.
Dingemanse, N.J., Kazem, A.J.N., Réale, D., et al. 2010. Behavioural reaction norms: animal personality meets individual plasticity. Trends Ecol. Evol. 25: 81–89.
Dobzhansky, T. 1937. Genetics and the origin of species. Columbia, New York.
Dulvy, N.K., Rogers, S.I., Jennings, S.V., et al. 2008. Climate change and deepening of the North Sea fish assemblage: a biotic indicator of warming seas. J. Appl. Ecol. 45: 1029–1039.
Endler, J.A. 1986. Natural Selection in the Wild. Princeton University Press, Princeton.
Engelhard, G.H., Righton, D.A., and Pinnegar, J.K. 2014. Climate change and fishing: a century of shifting distribution in North Sea cod. Glob. Change Biol. 20: 2473–2483.
Espeland, S.H., Olsen, E.M., Knutsen, H.J., et al. 2008. New perspectives on fish movement: kernel and GAM smoothers applied to a century of tagging data on coastal cod. Mar. Ecol. Prog. Ser. 372: 231–241.

Espeland, S.H., Thoresen, A.G., Olsen, E.M., et al. 2010. Diel vertical migration patterns in juvenile cod from the Skagerrak coast. Mar. Ecol. Prog. Ser. 405: 29–37.
Falconer, D.S., and Mackay, T.F.C. 1996. Introduction to Quantitative Genetics. Longman, New York.
Falkenhaug, T., Broms, C., Bagøien, E., et al. 2022. Temporal variability of co-occurring *Calanus finmarchicus* and *C. helgolandicus* in Skagerrak. Front. Mar. Sci. 9: 779335.
Free, C.M., Thorson, J.T., Pinsky, M.L., et al. 2019. Impacts of historical warming on marine fisheries production. Science 363: 979–983.
Freitas, C., Olsen, E.M., Moland, E., et al. 2015. Behavioural responses of Atlantic cod to sea temperature changes. Ecol. Evol. 5: 2070–2083.
Freitas, C., Olsen, E.M., Knutsen, H., et al. 2016. Temperature-associated habitat selection in a cold-water marine fish. J. Anim. Ecol. 85: 628–637.
Freitas, C., Villegas-Ríos, D., Moland, E., et al. 2021. Sea temperature effects on depth use and habitat selection in a marine fish community. J. Anim. Ecol. 90: 1787–1800.
Frölicher, T.L., Fisher, E.M., and Gruber, N., et al. 2018. Marine heatwaves under global warming. Nature 560: 360–364.
Galbraith, P.S., Larouche, P., Gilbert, D., et al. 2010. Trends in sea-surface and CIL temperatures in the Gulf of St. Lawrence in relation to air temperature. DFO Atl. Zone Mon. Prog. Bull. 9: 20–23.
Garcia, S.M., Kolding, J., Rice, J., et al. 2012. Reconsidering the consequences of selective fisheries. Science 335: 1045–1047.
Gjerde, B., Terjesen, B.F., Barr, Y., et al. 2004. Genetic variation for juvenile growth and survival in Atlantic cod (*Gadus morhua*). Aquacult. 236: 167–177.
Goddard, S.V., Kao, M., and Fletcher, G.L. 1999. Population differences in antifreeze production cycles of juvenile Atlantic cod (*Gadus morhua*) reflect adaptations to overwintering environment. Can. J. Fish. Aquat. Sci. 56: 1991–1999.
Grabowski, T.B., Young, S.P., Libungan, L.A., et al. 2009. Evidence of phenotypic plasticity and local adaptation in metabolic rates between components of the Icelandic cod (*Gadus morhua* L.) stock. Environ. Biol. Fish. 86: 361–370.
Hadfield, J.D. 2010. MCMC methods for multi-response generalized linear mixed models: the MCMCglmm R package. J. Stat. Softw. 33: 1–22.
Hardie, D.C., Renaud, C.B., Ponomarenko, V.P., et al. 2008. The isolation of Atlantic cod, *Gadus morhua* (Gadiformes), populations in northern meromictic lakes—a recurrent Arctic phenomenon. J. Ichthyol. 48: 230–240.
Heath, M.R., and Lough, R.G. 2007. A synthesis of large-scale patterns in the planktonic prey of larval and juvenile cod (*Gadus morhua*). Fish. Oceanogr. 16: 169–185.
Hendry, A.P. 2017. Eco-evolutionary dynamics. Princeton University Press, Princeton.
Hendry, A.P., Bohlin, T., Jonsson, B., et al. 2003. To sea or not to sea? Anadromy versus non-anadromy in salmonids. pp. 92–125. *In*: Hendry, A.P., and Stearns, S.C. (eds.). Evolution Illuminated. Salmon and their Relatives. Oxford University Press, Oxford.
Hislop, J.R.G. 1996. Changes in North Sea gadoid stocks. ICES J. Mar. Sci. 53: 1146–1156.
Holt, R.E., and Jørgensen, C. 2014. Climate warming causes life-history evolution in a model for Atlantic cod (*Gadus morhua*). Conserv. Physiol. 2: cou050.
Huber, M., and Knutti, R. 2011. Anthropogenic and natural warming inferred from changes in Earth's energy balance. Nature Geosci. 5: 31–36.
Hurrell, J.W. 1995. Decadal trends in the North Atlantic oscillation: regional temperatures and precipitation. Science 269: 676–679.
Hurst, T.P., Munch, S.B., and Lavelle, K.A. 2012. Thermal reaction norms for growth vary among cohorts of Pacific cod (*Gadus macrocephalus*). Mar. Biol. 159: 2173–2183.
Hutchings, J.A. 2011. Old wine in new bottles: reaction norms in salmonid fishes. Heredity 106: 421–437.
Hutchings, J.A. 2021. A primer of life histories. Ecology, evolution, and application. Oxford University Press, Oxford.
Hutchings, J.A., Côte, I.M., Dodson, J.J., et al. 2012. Climate change, fisheries, and aquaculture: trends and consequences for Canadian marine biodiversity. Environ. Rev. 20: 220–311.
Hutchings, J.A., and Fraser, D.J. 2008. The nature of fisheries- and farming-induced evolution. Mol. Ecol. 17: 294–313.

Hutchings, J.A., and Myers, R.A. 1993. Effect of age on the seasonality of maturation and spawning of Atlantic cod, *Gadus morhua*, in the Northwest Atlantic. Can. J. Fish. Aquat. Sci. 50: 2468–2474.

Hutchings, J.A., and Myers, R.A. 1994. What can be learned from the collapse of a renewable resource? Atlantic cod, *Gadus morhua*, of Newfoundland and Labrador. Can. J. Fish. Aquat. Sci. 51: 2126–2146.

Hutchings, J.A., Swain, D.P., Rowe, S., et al. 2007. Genetic variation in life-history reaction norms in a marine fish. Proc. R. Soc. B 274: 1693–1699.

Imsland, A.K., and Jónsdóttir, Ó.D.B. 2003. Linking population genetics and growth properties of Atlantic cod. Rev. Fish Biol. Fisher. 13: 1–26.

Jackson, J.B.C., Kirby, M.X., Berger, W.H., et al. 2001. Historical overfishing and the recent collapse of coastal ecosystems. Science 293: 629–638.

Jacobsen, N.S., Gislason, H., and Andersen, K.H. 2014. The consequences of balanced harvesting of fish communities. Proc. R. Soc. B 281: 20132701.

Johannessen, T. 2014. From an Antagonistic to a Synergistic Predator Prey Perspective. Bifurcations in Marine Ecosystems. Elsevier, New York.

Kingsolver, J.G., Diamond, S.E., Siepielski, A.M., et al. 2012. Synthetic analyses of phenotypic selection in natural populations: lessons, limitations and future directions. Evol. Ecol. 26: 1101–1118.

Kjesbu, O.S., Righton, D., Krüger-Johnsen, M.A., et al. 2010. Thermal dynamics of ovarian maturation in Atlantic cod (*Gadus morhua*). Can. J. Fish. Aquat. Sci. 67: 605–625.

Kjesbu, O.S., Bogstad, B., Devine, J.A., et al. 2014. Synergies between climate and management for Atlantic cod fisheries at high latitudes. Proc. Natl. Acad. Sci. USA 111: 3478–3483.

Knutsen, H., Olsen, E.M., Jorde, P.E., et al. 2011. Are low but statistically significant levels of genetic differentiation in marine fishes "biologically meaningful"? A case study of coastal Atlantic cod. Mol. Ecol. 20: 768–783.

Kruuk, L.E.B. 2004. Estimating genetic parameters in natural populations using the "animal model". Phil. Trans. R. Soc. B 359: 873–890.

Kruuk, L.E.B., Slate, J., and Wilson, A.J. 2008. New answers for old questions: the evolutionary quantitative genetics of wild animal populations. Ann. Rev. Ecol. Syst. 39: 525–548.

Lande, R. 2009. Adaptation to an extraordinary environment by evolution of phenotypic plasticity and genetic assimilation. J. Evol. Biol. 22: 1435–1446.

Larsen, P.F., Nielsen, E.E., Meier, K., et al. 2012. Differences in salinity tolerance and gene expression between two populations of Atlantic cod (*Gadus morhua*) in response to salinity stress. Biochem. Genet. 50: 454–466.

Limburg, K.E., Walther, Y., Hong, B., et al. 2008. Prehistoric versus modern Baltic Sea cod fisheries: selectivity across the millennia. Proc. R. Soc. B 275: 2659–2665.

Lindmark, M., Ohlberger, J., and Gårdmark, A. 2022. Optimum growth temperature declines with body size within fish species. Glob. Change Biol. 28: 2259–2271.

Marcil, J., Swain, D.P., and Hutchings, J.A. 2006. Genetic and environmental components of phenotypic variation in body shape among populations of Atlantic cod (*Gadus morhua* L.). Biol. J. Linn. Soc. 88: 351–365.

Matschiner, M., Barth, J.M.I., Tørresen, O.K., et al. 2022. Supergene origin and maintenance in Atlantic cod. Nature Ecology and Evolution 6: 469–481.

McQueen, K., and Marshall, C.T. 2017. Shifts in spawning phenology of cod linked to rising sea temperatures. ICES J. Mar. Sci. 74: 1561–1573.

Meager, J.J., Skjæraasen, J.E., Karlsen, Ø., et al. 2012. Environmental regulation of individual depth on a cod spawning ground. Aquat. Biol. 17: 211–221.

Merilä, J., and Hendry, A.P. 2014. Climate change, adaptation, and phenotypic plasticity: the problem and the evidence. Evol. Appl. 7: 1–14.

Moland E., Olsen, E.M., Knutsen, H., et al. 2013. Lobster and cod benefit from small-scale northern marine protected areas: inference from an empirical before-after control-impact study. Proc. R. Soc. B 280: 20122679.

Morgan, M.J., Wright, P.J., and Rideout, R.M. 2013. Effect of age and temperature on spawning time in two gadoid species. Fish. Res. 138: 42–51.

Morgan, R., Finnøen, M.H., Jensen, H., et al. 2020. Low potential for evolutionary rescue from climate change in a tropical fish. Proc. Natl. Acad. Sci. USA 117: 33365–33372.

Morrongiello, J.R., Sweetman, P.C., and Thresher, R.E. 2019. Fishing constrains phenotypic responses of marine fish to climate variability. J. Anim. Ecol. 88: 1645–1656.
Mousseau, T.A. and Roff, D.A. 1987. Natural selection and the heritability of fitness components. Heredity 59: 181–197.
Munk, P. 1997. Prey size spectra and prey availability of larval and small juvenile cod. J. Fish. Biol. 51(Suppl. A): 340–351.
Neat, F. and Righton, D. 2007. Warm water occupancy by North Sea cod. Proc. R. Soc. B 274: 789–798.
Neat, F.C., Bendall, V., Berx, B., et al. 2014. Movement of Atlantic cod around the British Isles: implications for finer scale stock management. J. Appl. Ecol. 51: 1564–1574.
Neuheimer, A.B., and Grønkjær, P. 2012. Climate effects on size-at-age: growth in warming waters compensates for earlier maturity in an exploited marine fish. Glob. Change Biol. 18: 1812–1822.
Neuheimer, A.B., and MacKenzie, B.R. 2014. Explaining life history variation in a changing climate across a species' range. Ecology 95: 3364–3375.
Neuheimer, A.B., Thresher, R.E., Lyle, J.M., et al. 2011. Tolerance limit for fish growth exceeded by warming waters. Nature Clim. Change 1: 110–113.
Niemelä, P.T. and Dingemanse, N.J. 2014. Artificial environments and the study of 'adaptive' personalities. Trends Ecol. Evol. 29: 245–247.
Núnez-Riboni, I., Taylor, M.H., Kempf, A., et al. 2019. Spatially resolved past and projected changes of the suitable thermal habitat of North Sea cod (*Gadus morhua*) under climate change. ICES J. Mar. Sci. 76: 2389–2403.
Ohlberger, J. 2013. Climate warming and ectotherm body size—from individual physiology to community ecology. Funct. Ecol. 27: 991–1001.
Ohlberger, J., Rogers, L.A., and Stenseth, N.C. 2014a. Stochasticity and determinism: how density-independent and density-dependent processes affect population variability. PLoS ONE 9: e98940.
Ohlberger, J., Thackeray, S.J., Winfield, I.J., et al. 2014b. When phenology matters: age—size truncation alters population response to trophic mismatch. Proc. R. Soc. B 281: 20140938.
Ohlberger, J., Langangen, Ø., and Stige, L.C. 2022. Age structure affects population productivity in an exploited fish species. Ecol. Appl. 32: e2614.
Olsen, E.M., Heino, M., Lilly G.R., et al. 2004a. Maturation trends indicative of rapid evolution preceded the collapse of northern cod. Nature 428: 932–935.
Olsen, E.M., Knutsen, H., Gjøsæter, J., et al. 2004b. Life-history variation among local populations of Atlantic cod from the Norwegian Skagerrak coast. J. Fish. Biol. 64: 1725–1730.
Olsen, E.M., Heupel, M.R., Simpfendorfer, C.A., et al. 2012. Harvest selection on Atlantic cod behavioral traits: implications for spatial management. Ecol. Evol. 2: 1549–1562.
Olsen, E.M., Knutsen, H., Gjøsæter, J., et al. 2008. Small-scale biocomplexity in coastal Atlantic cod supporting a Darwinian perspective on fisheries management. Evol. Appl. 1: 524–533.
Olsen, E.M., Lilly, G.R., Heino, M., et al. 2005. Assessing changes in age and size at maturation in collapsing populations of Atlantic cod (*Gadus morhua*). Can. J. Fish. Aquat. Sci. 62: 811–823.
Olsen, E.M., and Moland, E. 2011. Fitness landscape of Atlantic cod shaped by harvest selection and natural selection. Evol. Ecol. 25: 695–710.
Olsen, E.M., Ottersen, G., Llope, M., et al. 2011. Spawning stock and recruitment of North Sea cod shaped by food and climate. Proc. R. Soc. B 278: 504–510.
Olsen, E.M., Karlsen, Ø., and Skjæraasen, J.E. 2023. Large females connect Atlantic cod spawning sites. Science 382: 1181–1184.
Oomen, R.A., and Hutchings, J.A. 2016. Genetic variation in plasticity of life-history traits between Atlantic cod (*Gadus morhua*) populations exposed to contrasting thermal regimes. Can. J. Zool. 94: 257–264.
Oomen, R.A., and Hutchings, J.A. 2022. Genomic reaction norms inform predictions of plastic and adaptive responses to climate change. J. Anim. Ecol. 91: 1073–1087.
Oomen, R.A., Knutsen, H., Olsen, E.M., et al. 2022. Warming accelerates the onset of the molecular stress response and increases mortality of larval Atlantic cod. Integr. Comp. Biol. 62: 1784–1801.
Osman, M.B., Tierney, J.E., Zhu, J., et al. 2021. Globally resolved surface temperatures since the Last Glacial Maximum. Nature 599: 239–244.

Otero, J., L'Abeé-Lund, J.H., Castro-Santos, T., et al. 2014. Basin-scale phenology and effects of climate variability on global timing of initial seaward migration of Atlantic salmon (*Salmo salar*). Glob. Change Biol. 20: 61–75.

Otterlei, E., Nyhammer, G., Folkvord, A., et al. 1999. Temperature- and size-dependent growth of larval and early juvenile Atlantic cod (*Gadus morhua*): a comparative study of Norwegian coastal cod and northeast Arctic cod. Can. J. Fish. Aquat. Sci. 56: 2099–2111.

Ottersen, G., Hjermann, D.Ø., and Stenseth, N.C. 2006. Changes in spawning stock structure strengthen the link between climate and recruitment in a heavily fished cod (*Gadus morhua*) stock. Fish. Oceanogr. 15: 230–243.

Ottersen, G., and Holt, R.E. 2023. Long-term variability in spawning stock age structure influences climate–recruitment link for Barents Sea cod. Fish. Oceanogr. 32: 91–105.

Otterå, H., Agnalt, A.-L., and Jørstad, K.E. 2006. Differences in spawning time of captive Atlantic cod from four regions of Norway, kept under identical conditions. ICES J. Mar. Sci. 63: 216–223.

Otterå, H., Agnalt, A.-L., Thorsen, A., et al. 2012. Is spawning time of marine fish imprinted in the genes? A two-generation experiment on local Atlantic cod (*Gadus morhua* L.) populations from different geographical regions. ICES J. Mar. Sci. 69: 1722–1728.

Parmesan, C., and Yohe, G. 2003. A globally coherent fingerprint of climate change impacts across natural systems. Nature 421: 37–42.

Petrie, B., Pettipas, R.G., Petrie, W.M., et al. 2009. Physical oceanographic conditions on the Scotian Shelf and in the Gulf of Maine during 2009. Can. Sci. Advis. Secret. Res. Doc. 2009/039.

Poloczanska, E.S., Brown, C.J., Sydeman, W.J., et al. 2013. Global imprint of climate change on marine life. Nature Clim. Change 3: 919–925.

Pörtner, H.O., Berdal, B., Blust, R., et al. 2001. Climate induced temperature effects on growth performance, fecundity and recruitment in marine fish: developing a hypothesis for cause and effect relationships in Atlantic cod (*Gadus morhua*) and common eelpout (*Zoarces viviparus*). Cont. Shelf Res. 21: 1975–1997.

Pörtner, H.O., Bock, C., Knust, R., et al. 2008. Cod and climate in a latitudinal cline: physiological analyses of climate effects in marine fishes. Climate Res. 37: 253–270.

Purchase, C.F. and Brown, J.A. 2000. Interpopulation differences in growth rates and food conversion efficiencies of young Grand Banks and Gulf of Maine Atlantic cod (*Gadus morhua*). Can. J. Fish. Aquat. Sci. 57: 2223–2229.

Purchase, C.F. and Brown, J.A. 2001. Stock-specific changes in growth rates, food conversion efficiencies, and energy allocation in response to temperature change in juvenile Atlantic cod. J. Fish. Biol. 58: 36–52.

Purchase, C.F., Butts, I.A.E., Alonso-Fernández, A., et al. 2010. Thermal reaction norms in sperm performance of Atlantic cod (*Gadus morhua*). Can. J. Fish. Aquat. Sci. 67: 498–510.

Reznick, D.A., Bryga, H., and Endler, J.A. 1990. Experimentally induced life-history evolution in a natural population. Nature 346: 357–359.

Righton, D.A., Andersen, K.H., Neat, F., et al. 2010. Thermal niche of Atlantic cod *Gadus morhua*: limits, tolerance and optima. Mar. Ecol. Prog. Ser. 420: 1–13.

Roff, D.A. 1992. The Evolution of Life Histories. Theory and Analysis. Chapman & Hall, New York.

Roff, D.A. 1997. Evolutionary Quantitative Genetics. Chapman & Hall, New York.

Roff, D.A. 2002. Life History Evolution. Sinauer, Sunderland.

Rogers, L.A., Stige, L.C., Olsen, E.M., et al. 2011. Climate and population density drive changes in cod body size throughout a century on the Norwegian coast. Proc. Natl. Acad. Sci. USA 108: 1961–1966.

Rogers, L.A., Olsen, E.M., Knutsen, H., et al. 2014. Habitat effects on population connectivity in a coastal seascape. Mar. Ecol. Prog. Ser. 511: 153–163.

Rogers, L., Storvik, G., Knutsen, H., et al. 2017. Fine-scale population dynamics in a marine fish species inferred from dynamic state-space models. J. Anim. Ecol. 86: 888–898.

Ruiz-Díaz, R., Dominguez-Petit, R., and Saborido-Rey, F. 2022. Atlantic cod growth history in Flemish Cap between 1981 and 2016: the impact of fishing and climate on growth performance. Front. Mar. Sci. 9: 876488.

Salinas, S., and Munch, S.B. 2012. Thermal legacies: transgenerational effects of temperature on growth in a vertebrate. Ecol. Lett. 15: 159–163.

Salvanes, A.G.V., Skjæraasen, J.E., and Nilsen, T. 2004. Subpopulations of coastal cod with different behaviour and life history strategies. Mar. Ecol. Prog. Ser. 267: 241–251.

Savolainen, O., Lascoux, M., and Merilä, J. 2013. Ecological genomics of local adaptation. Nature Rev. Gen. 14: 807–820.

Schindler D.E., Hilborn, R., Chasco, B.C.P., et al. 2010. Population diversity and the portfolio effect in an exploited species. Nature 465: 609–613.

Schlichting, C.D., and Pigliucci, M. 1998. Phenotypic Evolution: A Reaction Norm Perspective. Sinauer, Sunderland.

Shaw, R.G., and Shaw, F.H. 2014. Quantitative genetic study of the adaptive process. Heredity 112: 13–20.

Siepielski, A.M., J.D. DiBattista and S.M. Carlson. 2009. It's about time: the temporal dynamics of phenotypic selection in the wild. Ecol. Lett. 12: 1261–1276.

Skjæraasen, J.E., Nash, R.D.M., Korsbrekke, K., et al. 2012. Frequent skipped spawning in the world's largest cod population. Proc. Natl. Acad. Sci. USA 109: 8995–8999.

Stearns, S.C. 1992. The Evolution of Life Histories. Oxford University Press, Oxford.

Sultan, S.E., and Stearns, S.C. 2005. Environmentally contingent variation: phenotypic plasticity and norms of reaction. pp. 303–332. *In*: Hallgrimsson, B., and Hall, B. (eds.). Variation: A Central Concept in Biology. Elsevier, Amsterdam.

Stenseth, N.C., Mysterud, A., Ottersen, G., et al. 2002. Ecological effects of climate fluctuations. Science 297: 1292–1296.

Stige, L.C., Ottersen,G., Brander, K., et al. 2006. Cod and climate: effect of the North Atlantic Oscillation on recruitment in the North Atlantic. Mar. Ecol. Prog. Ser. 325: 227–241.

Svåsand, T., Jørstad, K.E., Otterå, H., et al. 1996. Differences in growth performance between Arcto-Norwegian and Norwegian coastal cod reared under identical conditions. J. Fish Biol. 49: 108–119.

Swain, D.P., Sinclair, A.F., and Hanson, J.M. 2007. Evolutionary response to size-selective mortality in an exploited fish population. Proc. R. Soc. B 274: 1015–1022.

Therkildsen, N.O., Hemmer-Hansen, J., Als, T.D., et al. 2013. Microevolution in time and space: SNP analysis of historical DNA reveals dynamic signatures of selection in Atlantic cod. Mol. Ecol. 22: 2424–2440.

Trippel, E.A. 1998. Egg size and viability and seasonal offspring production of young Atlantic cod. Trans. Am. Fish. Soc. 127: 339–359.

Uusi-Heikkilä, S., Whiteley, A.R., Kuparinen, A., et al. 2015. The evolutionary legacy of size-selective harvesting extends from genes to populations. Evol. Appl. 8: 597–620.

Uusi-Heikkilä, S., Sävilammi, T., Leder, E., et al. 2017. Rapid, broad-scale gene expression evolution in experimentally harvested fish populations. Mol. Ecol. 26: 3954–3967.

Villegas-Ríos, D., Réale, D., Freitas, C., et al. 2018. Personalities influence spatial responses to environmental fluctuations in wild fish. J. Anim. Ecol. 87: 1309–1319.

Visscher, P.M., Hill, W.G., and Wray, N.R. 2008. Heritability in the genomics era—concepts and misconceptions. Nature Rev. Gen. 9: 255–266.

Wilson, A.J., Réale, D., Clements, M.N., et al. 2009. An ecologist's guide to the animal model. J. Anim. Ecol. 79: 13–26.

Winter, A.-M., Richter, A., and Eikeset, A.M. 2020. Implications of Allee effects for fisheries management in a changing climate: evidence from Atlantic cod. Ecol. Appl. 30: e01994.

Wood, J.L.A. and Fraser, D.J. 2015. Similar plastic responses to elevated temperature among different-sized brook trout populations. Ecology 96: 1010–1019.

Wootton, H.F., Morrongiello, J.R., Schmitt, T., et al. 2022. Smaller adult fish size in warmer water is not explained by elevated metabolism. Ecol. Lett. 25: 1177–1188.

Zhou, S., Smith, A.D.M., Punt, A.E., et al. 2010. Ecosystem-based fisheries management requires a change to the selective fishing philosophy. Proc. Natl. Acad. Sci. USA 107: 9485–9489.

Chapter 2
Trophic Interactions

Bjarte Bogstad

Introduction

The role of cod in the ecosystems in the North Atlantic has been reviewed by Pálsson (1994), Link et al. (2009) and Link and Sherwood (2019), and this chapter relies heavily on their reviews. The chapter will also summarise some of their conclusions and add results from more recent research as well as knowledge from ecosystems that were not (or barely) included in those reviews, such as the areas around the Faroes and Greenland.

The chapter concentrates on the interactions between cod and its prey, predators and competitors and discusses how these have contributed to the large changes in cod population size seen in many ecosystems and what effect these changes have had on these interactions. Throughout its life, cod is both a predator and a prey, as well as a cannibal, and thus it is difficult to place it in a trophic level within the food web. Also, cod prey on both demersal/benthic and pelagic prey species. The chapter starts with a review of information about the cod diet and its role as a predator before providing knowledge about predation on cod. Then, it moves on to a discussion of the role of cod in the various ecosystems and reviews how interactions between cod and other species are currently taken into account in management.

The present and past role of cod in the ecosystem, as well as the magnitude of the fluctuations in cod stock size and geographical distributions, vary strongly between ecosystems, which provides a lot of useful contrast.

Predation by Cod—Diet Composition and Total Consumption

In order to quantify what cod eats, diet data and information about cod stomach evacuation rate and/or bioenergetics are needed. For most cod stocks, extensive stomach sampling programmes have gathered much information about the cod diet, which will be summarised here.

Institute of Marine Research, P.O box 1870 Nordnes, NO-5817 Bergen, Norway.
Email: bjarte.bogstad@hi.no

Cod is known to be an omnivorous feeder with a broad diet consisting mainly of fish and crustaceans. However, other organisms such as benthic fauna, polychaetes, jellyfish and cephalopods are also found in the diet. For example, Mehl (1991) reported more than 200 prey species in cod diet in the Barents Sea. The diet shifts ontogenetically, as described in the following. Heath and Lough (2007) reviewed published studies of the diet composition of larval and pelagic juvenile cod around the northern North Atlantic. They found that larvae at the northern edge of the latitudinal range of cod depend primarily on the development stages of the copepod Calanus finmarchicus, while those at the southern edge depend on *Paracalanus* and *Pseudocalanus* species. Juvenile cod preyed on a wider range of taxa than larvae, but euphausiids were their main target prey.

After settling to the bottom, cod mainly feed on small crustaceans such as krill and amphipods and various benthic organisms until cod reach a size of at least 20 cm. As cod grow, the diet shifts to being dominated by small fish and large crustaceans; this shift does, in some areas, occur at a length of 20–30 cm (Link et al. 2009). In other areas, the shift occurs at larger lengths, e.g., in the North Sea, where the increase in the proportion of fish is more gradual (50% fish prey at about 50 cm) and in Greenlandic waters, where stronger piscivorous feeding behaviour occurs when cod reach 80 cm length (Werner et al. 2019). In some areas, there is a further shift from small fish as main prey to larger fish prey (including increasing cannibalism), around 60–70 cm cod length, e.g., in the Barents Sea (Holt et al. 2019).

In some areas, one or two fish species dominate the diet. Capelin (*Mallotus villosus*) is dominant in the Barents Sea (Bogstad and Mehl 1997; Holt et al. 2019) and around Iceland (Pálsson and Björnsson 2011) and was also prevalent as prey in the Newfoundland/Labrador area before the cod collapse (Lilly 1991). Capelin was also an important prey for some cod stocks in Canadian waters farther south. In all areas where capelin is an important prey, the predation by cod on capelin is particularly intense from January to March when capelin approaches the coast for spawning. Herring (*Clupea harengus*) and sprat (*Sprattus sprattus*) are the dominant fish prey for cod in the Baltic (Kulatska et al. 2019; Neuenfeldt et al. 2020).

In other, more species-rich areas such as the North Sea, the Celtic Sea and off the northeastern US, one or two individual fish species are not as dominant in the diet. Among the fish prey found frequently in cod stomachs in many areas are herring, sprat, sand eel (*Ammodytes* spp.), Norway pout (*Trisopterus esmarkii*), redfish (*Sebastes* spp.), and flatfishes, e.g., long rough dab (*Hippoglossoids platessoides*). Cannibalism is an important part of the diet for larger cod in some areas, e.g., the Barents Sea and the Baltic (Link et al. 2009).

Some invertebrates are also particularly important in the diet of larger cod (> 20–30 cm) in some areas. Northern shrimp (*Pandalus borealis*) are important prey both in the Barents Sea and in Icelandic and Newfoundland/Labrador waters, as well as in Greenlandic waters (Werner et al. 2019) and on the Flemish cap (Pérez-Rodríguez et al. 2011). In the Baltic, *Saduria* and *Mysis* spp. are important prey, while euphausiids and hyperiids are important for larger cod in Greenlandic waters (Werner et al. 2019).

Cod can eat fish prey up to at least half its length, but even large cod may feed partly on small organisms (e.g., Holt et al. 2019; Barents Sea; Scharf et al. 2000, northeast US).

Predators of Cod

In its first year of life, cod is vulnerable to many predators, which is unsurprising given the extremely high mortality in the first year of life (see, e.g., Bogstad et al. 2016 for calculations of cod mortality at different life stages). Cod eggs and larvae are preyed on by several pelagic fish species, e.g., herring and sprat, and this has been suggested to be an important trophic interaction, e.g., in the Baltic (Köster and Möllmann 2000). Invertebrates, such as crustaceans, cephalopods, medusae and ctenophores, have also been identified as predators on cod during its first months of life when it is distributed pelagically (Pálsson 1994).

After the first months of life, when cod as 0-group gradually moves into deeper waters and settles to the bottom, various fish species are the main predators on young cod, but marine mammals and, in some areas also, seabirds are important. Among the fish species identified to have notable proportions of small cod in their diet are skates, gurnards, Greenland halibut (*Reinhardtius hippoglossoides*), whiting (*Merlangius merlangus*) and Atlantic halibut (*Hippoglossus hippoglossus*) (Pálsson 1994; Temming et al. 2007; Link et al. 2002).

Several marine mammals have also been identified as predators of cod. Among the seals, the most important ones are the harp seal (*Pagophilus groenlandicus*), grey seal (*Halichoerus gryphus*) and harbour seal (*Phoca vitulina*). Among the whales, the most important predators are minke whale (*Balaenoptera acutorostrata*) as well as several species of toothed whales, such as dolphin species (*Lagenorhynchus* spp.) and harbour porpoise (*Phocoena phocoena*) (Skern-Mauritzen et al. 2022). Note that Skern-Mauritzen et al. (2022) refer to the consumption of gadoids and not of cod, but given the general dominance of cod relative to other gadoids, it is likely that a considerable part of the consumed gadoids consists of cod. They also found that marine mammal consumption of gadoids was somewhat less than removal of gadoids by fisheries, although of the same order of magnitude. Birds prey notably on young cod in several areas, e.g., the North Sea, and great cormorants (*Phalacrocorax carbo*) prey on cod along the Norwegian coast (Lorentsen et al. 2021).

Role of Cod in the Ecosystem

General Features Across Ecosystems

In general, cod is, or has been, the dominant piscivorous fish in most ecosystems. Cod grow to a larger size (maximum length observed in the Barents Sea is 169 cm; Boitsov et al. 1996) than other piscivorous gadoids such as hake (*Merluccius merluccius*) and saithe (*Pollachius virens*), and they also have a large gape size relative to body size. Overlap with hake is limited even in the southern parts of the distributional range of cod, but saithe overlaps to some extent with cod in many areas, e.g., in the northern North Sea, although their spawning and nursery areas differ in most areas.

The proportion of fish in the haddock (*Melanogrammus aeglefinus*) diet is relatively low in most areas (Tam et al. 2016), thus the competition between haddock and cod for food is fairly limited.

Cod cannibalism, discussed above, maybe the most important fish predation on cod in many areas. As cod grows to a length above 20–30 cm, there are few fish predators capable of preying on them, and those that are, e.g., Atlantic halibut and Greenland shark, are not abundant. Although there are many predators on cod, there are only a few, if any, examples of predators being heavily dependent on cod as prey. The closest to this may be the special case of cod cannibalism in enclosed areas such as meromictic lakes (Hardie et al. 2008), where cod may be the only fish species present.

The role of cod within an ecosystem may become more apparent when cod biomass declines to low levels, causing an increase in its prey abundance, as has happened with most stocks in the Northwest Atlantic as well as the North West European shelf stocks (Celtic Sea, Irish Sea, Scottish West coast and North Sea) and the Baltic, but also when cod stock size increases and cod move into new feeding areas as observed in the Barents Sea (Kjesbu et al. 2014).

The main interactions between cod and its main prey vary between areas, but they are generally reasonably well understood and can be related to diet data and knowledge about the biology of cod and its prey. The competitive interactions between cod and other top predators like marine mammals (seals and whales) are, on the other hand, less well known and understood and are also very different between areas as the ratio of food consumption by cod to marine mammals varies considerably between areas. Skern-Mauritzen et al. (2022) have recently quantified predation by marine mammals in the Northeast Atlantic. They give average figures for total consumption by marine mammals in the Barents and Iceland Seas as 7.1 and 13.4 million tonnes annually, although with large uncertainty. For the Barents Sea, this figure is comparable to estimates of annual consumption by cod (Howell et al. 2023), while for the Iceland Sea, where the cod stock is smaller than in the Barents Sea, and the consumption by marine mammals is larger, the prey consumption by marine mammals would likely be several times larger than the prey consumption by cod. The role of marine mammals (mainly seals) in Canadian waters has been a much–studied subject in recent decades, as discussed below. Birds as predators are generally considerably less important for cod than marine mammals.

As key prey species of cod such as capelin and herring show strong population fluctuations, cascading effects of such fluctuations in the ecosystem may occur, and cod as a key species in the ecosystem could play an important role here through food web dynamics affecting both cod recruitment and capelin, the main prey of cod in some areas (Hjermann et al. 2007, 2010), as well as the effects of cod no longer dominating the ecosystem (Frank et al. 2005). Effects of environmental changes are closely linked to species interactions involving cod, and it may be difficult to sort out their relative importance. A study addressing this issue was made by Koen-Alonso et al. (2021), who did comparative modelling of cod-capelin dynamics in the Newfoundland/Labrador and Barents Sea areas. They found that the Newfoundland cod stock would be expected to rebuild if enough capelin were available.

Barents Sea

In the Barents Sea, capelin is the main prey for cod, and predation by herring on capelin larvae and resulting capelin collapses has been identified as an important interaction in the ecosystem (Gjøsæter et al. 2009, 2016). The main interactions between cod and other species are illustrated in Figure 1. Cod is clearly the dominant fish predator in this area.

Prey switching when preferred prey disappeared was observed in the Barents Sea in the late 1980s when cod switched to krill and amphipods following a capelin collapse. In that period, there was little alternative fish prey to switch to, individual growth of cod decreased considerably, and cannibalism increased (Mehl and Sunnanå 1991). During later capelin collapses, alternative fish prey was available, so effects on cod growth and other ecosystem effects of capelin collapses were relatively minor compared to the capelin collapse in the mid-1980s (Gjøsæter et al. 2009). Cannibalism is also found to be an important factor in regulating cod population dynamics in the area (Yaragina et al. 2009).

Young Norwegian spring-spawning herring was an important prey for cod in the Barents Sea prior to the herring collapse around 1970 (Townhill et al. 2021) but has not reached the same levels of importance in the diet after the recovery of the stock in the 1980s and 1990s. In contrast to Iceland and Newfoundland/Labrador, there is no clear inverse relationship between cod and northern shrimp abundance in the

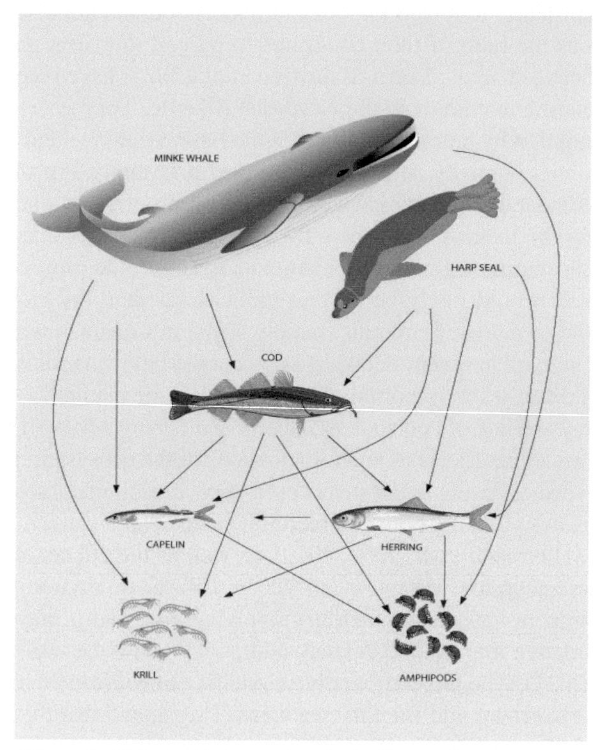

Figure 1. Main trophic interactions in the Barents Sea ecosystem.

Barents Sea. Northern shrimp abundance in the Barents Sea was high when the cod stock was low in the first part of the 1980s, but after that, the northern shrimp stock in the area has been relatively stable despite considerable variation in cod abundance. The recent borealisation of the area has caused cod to move into areas previously dominated by small Arctic fish species (Fossheim et al. 2015), and this has caused a decline in the abundance of such species (Johannesen et al. 2020).

Although cod is prey for some marine mammals, it may also be a trophic competitor. In the Barents Sea, cod, harp seal and minke whale are the most important top predators. Bogstad et al. (2015) found that in the decade 2004–2013, when the abundance of cod tripled and reached a record high level, the growth and condition of cod remained rather stable, although some decrease was seen in size at the age of large, mature cod. During the same period, the abundance of harp seals declined, whereas the minke whale stock was stable. However, the body condition (blubber thickness) of these two mammal stocks decreased, with the strongest decrease observed for harp seals. It has not been investigated whether there was a reversal in the trend in the body condition of the mammals in the years after 2013, when the cod abundance decreased again.

Icelandic Waters

The dominant interaction involving cod in Icelandic waters has been between cod, capelin and northern shrimp, and changes in the abundance and distribution of capelin and northern shrimp greatly impact food availability for cod. In periods of low cod stock size, an increase in the northern shrimp stock has been observed, likely due to a release in predation pressure (Stefánsson et al. 1998). The stocks of northern shrimp collapsed around the year 2000 and have stayed low since, and one of the

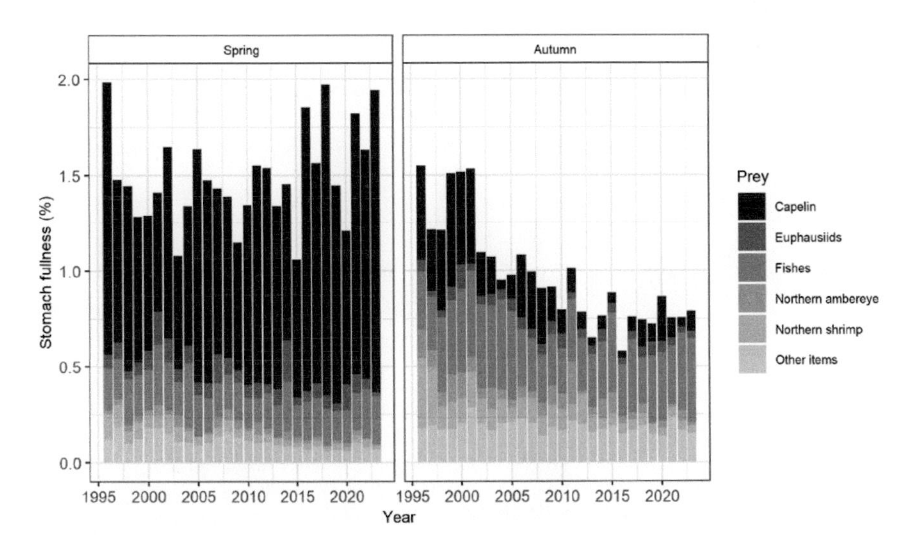

Figure 2. Stomach fullness (%) (stomach content weight as a percentage of predator body weight, calculated according to Lilly and Rice 1987) of cod (20–100 cm) in Icelandic waters in March and October from 1996–2023.

driving factors is thought to be increased predation by gadoids such as cod (ICES 2022a; MFRI 2023). Northern shrimp has been of less importance in the cod diet after 2000 (Figure 2; see also Jónsdóttir et al. 2024).

Capelin is the main prey for cod (Figure 2; see also Jónsdóttir et al. 2024), with most predation taking place during the capelin spawning migration onto the continental shelf. A relationship between capelin abundance and cod individual growth was found until 2002, but after that, this relationship broke down (Pálsson and Björnsson 2011). In autumn, the feeding intensity on capelin is not as apparent as late in the winter, and the diet composition is more diverse, with a higher reliance on other fish species. The ecosystem is relatively similar to the Barents Sea, but there have been fewer fluctuations in the cod stock abundance and not as strong fluctuations in capelin abundance as in the Barents Sea.

North Sea

In the North Sea, cod abundance has also declined considerably from very high levels in the 1960-1970s, with the spawning stock biomass (SSB) in the 2010s being less than half the level in the 1960s–1970s and no strong year classes after 1997 (ICES 2022b). Since there are many abundant piscivorous fish species in the North Sea, it is more difficult to identify the effects of the release of predation by cod. However, Figure 3 (ICES 2021) shows how prey consumption by various predators has changed over time. Predation by cod on most of these prey species declined to low levels after 2000 compared to before 2000, with herring as an exception.

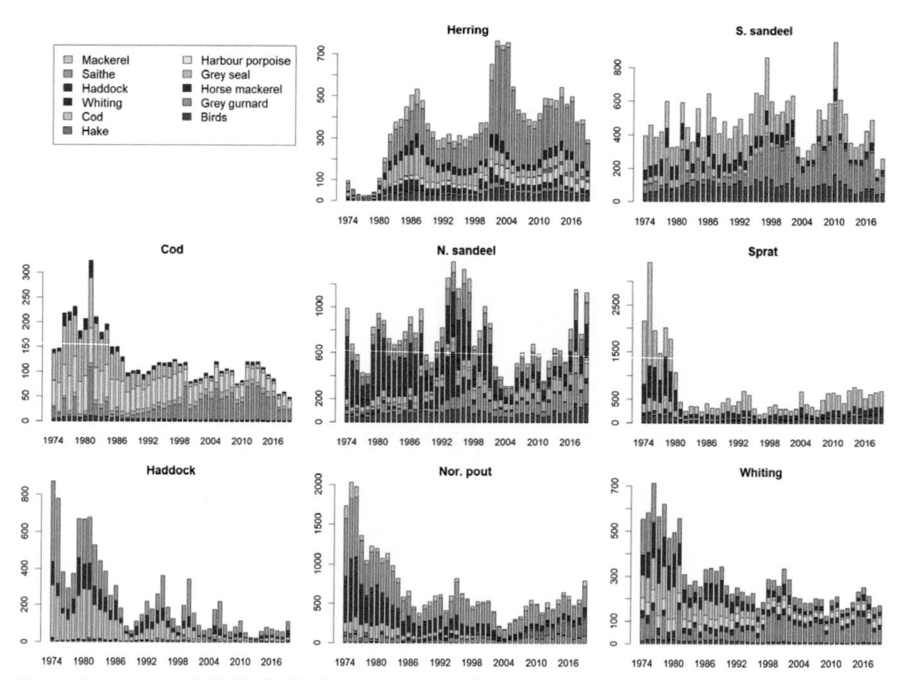

Figure 3. (Figure 5.1.21 ICES 2021) Eaten biomass (1,000 tonnes) of the individual prey species by predator (groups) in the North Sea.

Figure 3 also shows that more cod and other prey (haddock and whiting) have been eaten by grey gurnard (*Eutrigla gurnardus*) recently, but grey gurnard prey mainly at age 0 while cod eat bigger prey. It should be noted that the overlap between cod and other stocks is likely to have changed, as the decline in cod abundance has been strongest in the southern North Sea. Since most of the stomach data used in the SMS model (Lewy and Vinther 2004) are from the 'Years of the Stomach,' in 1981 and 1991, the model results have hardly been compared to recent observations of diet. There are now signs of recovery of the cod stock in the northern and western parts of the North Sea, and it will be very interesting to follow the effect this has on the main prey species for cod. Hake could also become a significant competitor to cod in the North Sea.

Baltic

Cod is, or at least was, the main fish predator in the Baltic, with herring, sprat and *Saduria* as its main prey. Cod cannibalism may, in certain periods, be important as a component of the diet and in population dynamics. The main interactions are shown in Figure 4.

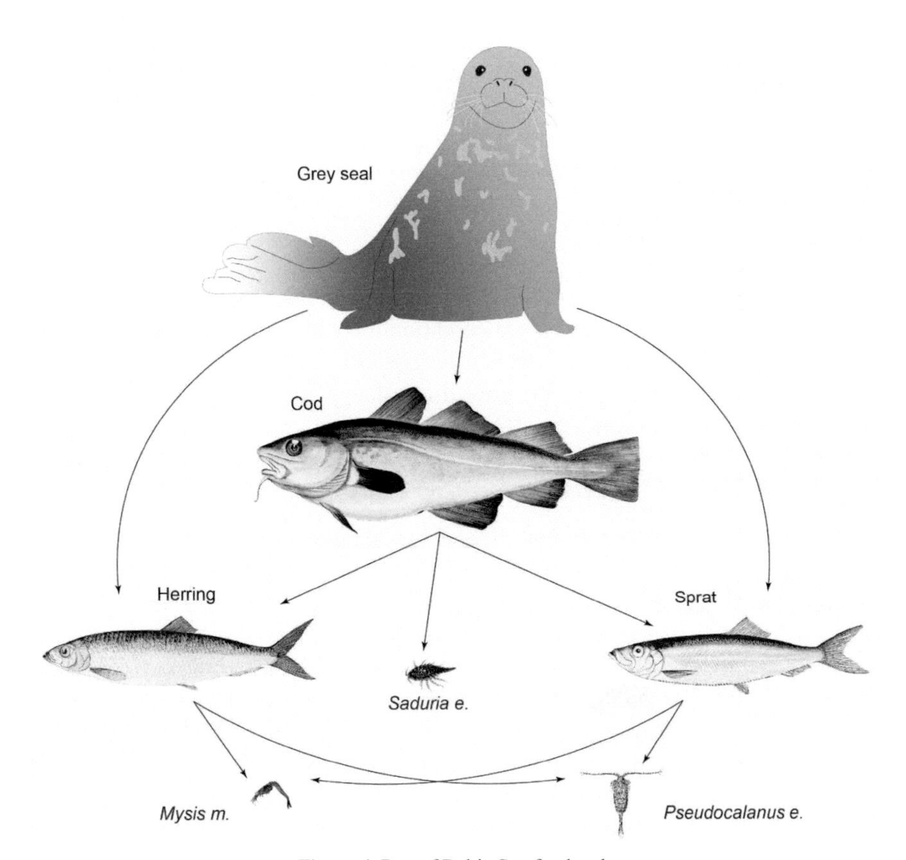

Figure 4. Part of Baltic Sea food web.

For a number of years, the Eastern Baltic cod has been in distress with decreasing abundance, declining nutritional conditions, disappearance of larger fish, high natural mortality, and no signs of recovery of the population. The decrease in feeding conditions is mostly due to a decrease in prey availability for cod due to less spatial overlap. This is influenced, e.g., by increased areas of hypoxia in the bottom waters, which decreases the suitable habitat for cod. A notable factor not described in other areas is that feeding levels of cod 21–30 cm decreased after 2005, creating a bottleneck (Neuenfeldt et al. 2020): cod has to reach a size of 30–35 cm in order to prey on herring and sprat, but is unable to grow due to decreased feeding. This is likely mainly due to decreased benthic prey availability and increased hypoxic areas. The situation is made even worse by an increased grey seal population and, as a result, increased infection loads of the parasitic nematode *Contracaecum osculatum* (Rudolphi).

Several recent papers further elaborated on the situation for Baltic cod (e.g., Kulatska et al. 2019, 2021; Neuenfeldt et al. 2020). The total prey availability in the area has changed less as herring and sprat abundance have been relatively stable in the 2000s (ICES 2023a). The recent increase in flatfish stocks in the Baltic (ICES 2023a) may also be related to the low cod abundance.

Historically, the abundances of the three key species in the Baltic have gone through dramatic changes, with the Baltic being cod-dominated prior to the late 1980s, after which there was an outburst of sprat, which likely was related to the release of predation pressure by cod (Alheit et al. 2005; Casini 2013; Österblom et al. 2007).

The western Baltic cod stock has also been declining lately, although the reasons for this are less clear. Processes other than those captured by the available data on fisheries catches and assumed natural mortality are, for the time being, influencing the SSB of the western Baltic cod stock. The sources for the presumed additional mortality are presently unclear but could involve increased natural mortality (e.g., due to increased predation, decreased condition linked to heat stress and hypoxia in summer, a shift in distribution towards the Eastern Baltic, and unreported catches or local/regional fish kills due to upwelling of hypoxic waters) as described by Receveur et al. (2022) and ICES (2023a).

Newfoundland and Labrador (Northern Cod)

All cod stocks within the Newfoundland and Labrador area collapsed (or suffered severe declines) in the early 1990s. These collapses were part of a broader ecosystem collapse, which involved declines in many groundfish species and capelin and increases in shellfish (northern shrimp and snow crab *Chionoecetes opilio*). Northern cod had capelin as a key prey before the collapse and switched to northern shrimp afterwards. While cod numbers subsequently increased, total biomass never rebuilt to pre-collapse levels. These ecosystem changes have been linked to the compounding effects of overfishing and a regime shift in the early 1990s (in extremely cold conditions). Groundfish started to show signals of rebuilding in the mid-late 2000s, coinciding with modest improvements in capelin and declines in shellfish. Predation by harp seals has not been identified as a significant driver of the dynamics of northern cod (Buren et al. 2014).

Scotian Shelf

Similar to the other cod stocks in the Northwest Atlantic, the populations of cod along the Scotian Shelf of Eastern Canada underwent a major crash in the 1990s and have shown little recovery since then, even in cases where a cod fishing moratorium has been in place since the crash. The available stock assessments indicate a persistently high level of natural mortality of older fish, though the main driver of this mortality across the Scotian Shelf stocks has been a subject of some debate (e.g., Bernier et al. 2018). The low population levels and high natural mortality are accompanied by the disappearance of small-scale fall spawning events, although the condition of fish remains stable and has shown signs of improvement in some areas since the crash. Although the cod stocks in the area remain in a perpetual state of low productivity, populations of redfish and Atlantic halibut are reaching record levels, as are benthic invertebrates and seals. Both increasing and decreasing population trends for several species in this complex ecosystem have been tied to warming waters, including the establishment of new species expanding their northern range (Bernier et al. 2018).

Gulf of St. Lawrence (GSL)

There are two distinct cod stocks in the GSL, one in the north (nGSL, NAFO 3Pn4RS) and one in the south (sGSL, NAFO 4TVn), separated by a channel that limits mixing. Both stocks collapsed in the early 1990s and have failed to recover. Herring is the most abundant forage species in the sGSL and the main fish prey for cod, especially larger cods. However, when the relative abundance of herring and capelin was likely reversed in the 1960s, capelin was a more important prey item. In the nGSL, the prominence of capelin in cod diets was greater, but capelin was not necessarily a dominant prey item, particularly for larger cod. In the sGSL, predation by grey seals is the most likely driver for the lack of cod recovery (Swain and Benoit 2015; Neuenhoff et al. 2019), and predation by harp seals can become a factor in the nGSL under certain environmental conditions (Bousquet et al. 2014).

The collapse of cod in the sGSL and nGSL was associated with predation release for many small-bodied fish species and northern shrimp, evidence of the formerly dominant role of cod (Benoit and Swain 2008; Morrisette et al. 2009).

Gulf of Maine/Georges Bank

The prominence of cod in this ecosystem has diminished. Link and Garrison (2002a, b) show that the amount of energy flowing through the cod population has declined notably, with cod consuming about one-third of what the population did on Georges Bank. Cod was the dominant piscivore in the ecosystem before ca. 1985, but that is no longer true. After 1985, spiny dogfish (*Squalus acanthias*) became the dominant piscivorous species. Essentially, in the late 1980s and 1990s, there was less cod, and the cod that was present was smaller and thus less piscivorous than in the 1970s and early 1980s. This resulted in the cod population in the late 1980s and 1990s consuming about one-third of fish compared to the period before 1985. However, in 2002, cod was still one of the top ten fish predators in the ecosystem. Cod stocks in the area have stayed at a low level since then (TRAC 2020).

Greenland

In Greenland waters, the main interaction between cod and other stocks, which have been identified, is the effect of cod as a predator on northern shrimp (Wieland 2005; Wieland et al. 2007). The change in the distribution of cod and northern shrimp resulted in a drastic decrease in the spatial overlap between the two species since 2000 (Wieland et al. 2007).

Faroes

Cod, along with saithe and monkfish (*Lophius piscatorius*), are the dominant piscivorous fish in the area (Rae 1967; Du Buit 1982; Ofstad 2013). Norway pout, haddock, sandeels, blue whiting (*Micromesistius poutassou*) and crustaceans are the main prey items for cod around the Faroe Islands (Rae 1967; Sørensen 2021). Condition factors of cod and haddock are directly affected by the amount of sand eels and Norway pout (Sørensen 2021). The Norway pout stock shows strong fluctuations, inversely related to sand eels that are partly controlled by the primary production (Eliasen et al. 2011). Cannibalism seems to be a major factor regulating cod recruitment, especially when cod are in poor condition (Steingrund et al. 2010). It could be expected that cannibalism would increase with increasing size of the cod stock and hence decrease recruitment, but for some reason, the opposite seems to be the case, i.e., a positive correlation between recruitment and total stock size (Steingrund et al. 2010). This could partly be explained by recent unpublished stomach content data that show that individual cod prey more efficiently on Norway pout when the cod stock is large. Density-dependent processes may eventually reduce cod recruitment at large stock sizes of cod. Hence, complex trophic relationships may govern the optimal size of the cod stock and, consequently, the maximum sustainable yield.

Consumption Estimates—Comparing Those From Different Predators With Impact From Fisheries

For a review of methodologies for calculating cod consumption, we refer to Link and Sherwood (2019). For comparisons of stomach fullness and food consumption across ecosystems, it is worth noting that the stomach evacuation rate increases by 10–15% for each 1°C increase in temperature (e.g., Dos Santos and Jobling 1995). Cod is a competitor to the fishery for several pelagic stocks in many areas, e.g., capelin, herring and sprat. When comparing the amount of prey removed by cod (and other predators) with the fishery, it is important to take into account that the fishery normally targets considerably larger prey than the cod. Consequently, comparing prey biomass fished and consumed at face value gives an incorrect view of the relative impact on prey stocks. Even multispecies models with age or length structure of the prey and fishery and thus comparison of fishing vs predation mortality by age/length group does not explicitly estimate the relative importance of fishing vs predation on the prey stock development. Kulatska et al. (2021) outlined an approach to include both immediate and delayed effects of competition between cod and fisheries.

Cod diet and stomach fullness vary considerably in time (seasonally and interannually) and space. Due to the large variability in the diet in time and space, as well as seasonal migrations by cod, consumption estimates are associated with considerable uncertainty, as even with good spatial coverage of the stock at the time of a survey, it may be difficult to estimate seasonal variability in a good way. Representative sampling to account for possible differences in diet in time and space is demanding. Seasonal feeding migrations are important for most cod stocks, but they also feed during spawning time, as documented for North Sea cod (Daan 1973), Northeast Arctic cod (Michalsen et al. 2008) and cod in the Northwest Atlantic (Krumsick and Rose 2012).

One issue related to studies of seasonal variation which easily may be overlooked is predation processes on a smaller scale as the occurrence of 'hot spots' (e.g., Temming et al. 2007) where in that case, predation by whiting on 0-group cod over a short period had potentially a strong impact on total predation pressure. The effect of variable depth through the distribution area may also have more of an impact than previously thought (e.g., Fall et al. 2021), and diet likely varies throughout the water column (e.g., Skaret et al. 2020).

Multispecies Models and Use of Species Interactions in Fisheries Management

As a key species in the ecosystem, cod is included in multispecies models for many areas. Some of those are also used in a management context by incorporating predation mortality from cod and other predators in existing single-species assessment models, making them extended single-species models according to the model classification by Plaganyi (2007). This is the case, e.g., for capelin in the Barents Sea and Iceland (ICES 2023b), where predation by cod (in Iceland also by haddock and saithe) on spawning capelin is included in the short-term (half-year or less) prediction of spawning stock biomass from the time of survey to spawning. A target escapement strategy manages these capelin stocks, and the modelled predation by cod in this period has a considerable impact on the catch advice and is of comparable magnitude to the actual catches. In the Barents Sea, predation by cod on cod and haddock is also included in the assessment as additional natural mortality (Howell et al. 2023).

Key runs of multispecies models are regularly updated by the ICES WGSAM, e.g., for the North Sea and Baltic (ICES 2021, 2024 using the SMS model). The predation mortalities from multispecies models are used in single-species assessments of the prey stocks in these areas. Species interactions should be taken into account when exploring how best to manage an ecosystem with strong interactions, i.e., in determining reference points and management strategies. Among the many studies in this field, two recent ones are highlighted here, both introducing new features. Using a case study of the Irish Sea, Bentley et al. (2021) illustrated how stock-specific ecosystem indicators can be used to set an ecosystem-based fishing mortality reference point (F_{ECO}) within the "Pretty Good Yield" ranges for fishing mortality which form the present precautionary approach adopted by ICES. This is done for four species, including cod. They propose that this new target, F_{ECO}, can be used to scale fishing mortality down when the ecosystem conditions for the stock are poor and up when conditions are good. Voss et al. (2022) developed an ecological-economic model for the Baltic Sea, including cod, herring and sprat, to advance the understanding of optimal fisheries management and related trade-offs between user groups. They challenged the current MSY management objective in a multispecies setting (MMSY) and suggested that an economic multispecies management objective (MMEY) might be more useful for setting future management targets. They suggest an easy-to-implement new management approach, called robust management, which can better deal with variability and time trends in recruitment, as observed for cod, in order to safeguard the Central Baltic fishery resources.

Conclusions

Cod is still the King of several North Atlantic ecosystems. It has been doing badly in several of them recently, but there does not seem to be any other fish species that has occupied the throne; it could rather be said that the throne is vacant for the time being. Whether cod will return to its former glory and power in several ecosystems remains to be seen.

Acknowledgements

I would like to thank (in alphabetical order) Irene Andrushchenko, Hugues Benoit, Manon Cassista-Da Ros, Ingibjörg G. Jónsdóttir, Mariano Koen-Alonso, Nataliia Kulatska, Stefan Neuenfeldt, Jon Solmundsson, Petur Steingrund, Karl-Michael Werner and Peter Wright for useful input to this chapter. The input was provided through initial discussions before the writing started and/or by comments on a draft version.

References

Alheit, J., Möllmann, C., Dutz, J., et al. 2005. Synchronous ecological regime shifts in the central Baltic and the North Sea in the late 1980s. ICES J. Mar. Sci., 62: 1205–1215, 10.1016/j.icesjms.2005.04.024.

Anon. 2023. Report of the Joint Russian-Norwegian Working Group on Arctic Fisheries (JRN-AFWG) 2023. IMR-PINRO Report Series 7: 189.

Benoit, H.P., and Swain, D.P. 2008. Impacts of environmental change and direct and indirect harvesting on the dynamics of a marine fish community. Can J. Fish. Aq. Sci. 65(10): 2088–2104. Doi:10.1139/F08-112.

Bentley, J.W., Lundy, M.G., Howell, D., et al. 2021. Refining fisheries advice with stock-specific ecosystem information. Front. Mar. Sci., 8: 602072. Doi: 10.3389/fmars.2021.602072.

Bernier, R.Y., Jamieson, R.E., and Moore, A.M. (eds.). 2018. State of the atlantic ocean synthesis report. Can. Tech. Rep. Fish. Aquat. Sci. 3167: iii + 149 p.

Boitsov, Yu.A., Lebed, N.I., Ponomarenko, V.P., et al. 1996. The Barents Sea cod (biological and fisheries outline). Murmansk, PINRO press. 285 pp. (In Russian)

Bogstad, B., and Mehl, S. 1997. Interactions between Atlantic cod (*Gadus morhua*) and its prey species in the Barents Sea. pp. 591–615. In proceedings of the international symposium on the role of forage fishes in marine ecosystems. Alaska Sea Grant College Program Report No. 97–01. University of Alaska Fairbanks.

Bogstad, B., Gjøsæter, H., Haug, T., et al. 2015. A review of the battle for food in the Barents Sea: Cod vs. marine mammals. Frontiers in Ecology and Evolution, section Interdisciplinary Climate Studies. Front. Ecol. Evol. 3: 29. Doi: 10.3389/fevo.2015.00029.

Bogstad, B., Yaragina, N.A., and Nash, R.D.M. 2016. The early life-history dynamics of Northeast Arctic cod: levels of natural mortality and abundance during the first three years of life. Canadian Journal of Fisheries and Aquatic Science 73(2): 246–256. Doi: 10.1139/cjfas-2015-0093.

Bousquet, N., Chassot, E., Duplisea, D.E., et al. 2014. Forecasting the major influences of predation and environment on cod recovery in the northern Gulf of St. Lawrence. PloS One 9: 1–16. https://doi.org/10.1371/journal.pone.0082836.

Buren, A., Koen-Alonso, M., and Stenson, G. 2014. The role of harp seals, fisheries and food availability in driving the dynamics of northern cod. Marine Ecology Progress Series 511: 265–284.

Casini, M. 2013. Spatio-temporal ecosystem shifts in the Baltic Sea: top-down control and reversibility potential. *In*: Daniels, J.A. (ed.). Advances in Environmental Research, Vol. 28, Nova Science Publishers Inc, Hauppauge NY

Daan, N. 1973. A quantitative analysis of the food intake of North Sea cod, *Gadus morhua*. Netherlands Journal of Sea Research, 6: 479–517.

Du Buit, M.-H. 1982. Essai sur la prédation de la morue (*Gadus morhua*, L.), l'eglefin (*Melanogrammus aeglefinus* (L.)) et du lieu noir (*Pollachius virens* (L.)) aux Faeroes. Cybium, 8: 13–19.

Dos Santos, J., and Jobling, M. 1995. Test of a food consumption model for the Atlantic cod. ICES J. Mar. Sci. 52: 209–219.

Eliasen, K., Reinert, J., Gaard, E., et al. 2011. Sandeel as a link between primary production and higher trophic levels on the Faroe shelf. Marine Ecology Progress Series, 438: 185–194.

Fall, J., Johannesen, E., Englund, G., et al. 2021. Predator-prey overlap in three dimensions: cod benefit from capelin coming near the seafloor. Ecography 44: 802–815. https://doi.org/10.1111/ecog.05473.

Fossheim, M., Primicerio, R., Johannesen, E., et al. 2015. Recent warming leads to a rapid borealization of fish communities in the Arctic. Nature Clim Change 5: 673–677 https://doi.org/10.1038/nclimate2647.

Frank, K.T., Petrie, B., Choi, J.S. et al. 2005. Trophic cascades in a formerly cod-dominated ecosystem. Science 308: 1621–1623.

Gjøsæter, H., Bogstad, B., and Tjelmeland, S. 2009. Ecosystem effects of three capelin stock collapses in the Barents Sea. *In*: Haug, T., Røttingen, I., Gjøsæter, H. et al. (eds.). Fifty Years of Norwegian-Russian Collaboration in Marine Research. Thematic issue No. 2, Marine Biology Research 5(1): 40–53. Doi: 10.1080/17451000802454866.

Gjøsæter, H., Hallfredsson, E.H., Mikkelsen, N., et al. 2016. Predation on early life stages is decisive for year class strength in the Barents Sea capelin (*Mallotus villosus*) stock. ICES Journal of Marine Science 73(2): 182–195. Doi: 10.1093/icesjms/fsv177.

Hardie, D.C., Renaud, C.B., Ponomarenko, V.P. et al. 2008. The isolation of Atlantic cod, *Gadus morhua* (Gadiformes), populations in Northern Meromictic lakes—A recurrent arctic phenomenon. J. Ichthyol. 48: 230–240. https://doi.org/10.1134/S0032945208030053.

Heath, M.R. and Lough, R.G. 2007. A synthesis of large-scale patterns in the planktonic prey of larval and juvenile cod (*Gadus morhua*). Fisheries Oceanography, 16: 169–185. https://doi.org/10.1111/j.1365-2419.2006.00423.x

Hjermann, D.Ø., Bogstad, B., Eikeset, A.M., et al. 2007. Food web dynamics affect Northeast Arctic cod recruitment. Proceedings of the Royal Society, Series B 274: 661–669.

Hjermann, D.Ø., Bogstad, B., Dingsør, G.E., et al. 2010. Trophic interactions affecting a key ecosystem component: a multi-stage analysis of the recruitment of the Barents Sea capelin. Canadian Journal of Fisheries and Aquatic Science 67: 1363–1375.

Holt, R.E., Bogstad, B., Durant, J.M., Dolgov, A.V., and Ottersen, G. 2019. Barents Sea cod (*Gadus morhua*) diet composition: long-term interannual, seasonal, and ontogenetic patterns. ICES Journal of Marine Science 76(6): 1641–1652, doi:10.1093/icesjms/fsz082.

ICES 2021. Working Group on Multispecies Assessment Methods (WGSAM; outputs from 2020 meeting). ICES Scientific Reports. Report. https://doi.org/10.17895/ices.pub.7695.

ICES 2022a. Icelandic Waters ecoregion –Ecosystem overview. In Report of the ICES Advisory Committee, 2022. ICES Advice 2022, Section 11.1, https://doi.org/10.17895/ices.advice.21731663.

ICES 2022b. Working Group on the Assessment of Demersal Stocks in the North Sea and Skagerrak (WGNSSK). ICES Scientific Reports. https://doi.org/10.17895/ices.pub.19786285.v3.

ICES 2023a. Baltic Fisheries Assessment Working Group (WGBFAS). ICES Scientific Reports. Report. 5: 58, 606 pp. https://doi.org/10.17895/ices.pub.23123768.v3.

ICES 2023b. Benchmark workshop on capelin (WKCAPELIN). ICES Scientific Reports. 5: 62. 282 pp. https://doi.org/10.17895/ices.pub.23260388.

ICES 2024. Working Group on Multispecies Assessment Methods (WGSAM; outputs from 2022 meeting). ICES Scientific Reports. 6: 13, 218 pp https://doi.org/10.17895/ices.pub.22087292.v1.

Johannesen, E., Yoccoz, N.G., Tveraa, T., et al. 2020. Resource-driven colonization by cod in a high Arctic food web. Ecology and Evolution. 10: 14272–14281.

Jónsdóttir, I.G., Pampoulie, C., Hjörleifsson, E., et al. 2024. Icelandic cod stock. This book.

Karlson, A.M.L., Gorokhova, E., Gårdmark, A. et al. 2020. Linking consumer physiological status to food-web structure and prey food value in the Baltic Sea. Ambio 49: 391–406. https://doi.org/10.1007/s13280-019-01201-1.

Kjesbu, O.S., Bogstad, B., Devine, J.A., et al. 2014. Synergies between climate and management for Atlantic cod fisheries at high latitudes. Proceedings National Academy of Science 111(9): 3478–3483. https://doi.org/10.1073/pnas.1316342111.

Koen-Alonso, M., Lindstrøm, U., and Cuff, A. 2021. Comparative modeling of cod-capelin dynamics in the Newfoundland-Labrador Shelves and Barents Sea Ecosystems. Frontiers in Marine Science 8: 139.

Köster, F.W., and Möllmann, C. 2000. Trophodynamic control by clupeid predators on recruitment success in Baltic cod? ICES Journal of Marine Science 57: 310–323.

Krumsick, K.J., and Rose, G.A. 2012. Atlantic cod (*Gadus morhua*) feed during spawning off Newfoundland and Labrador. ICES Journal of Marine Science, 69: 1701–1709.

Kulatska, N., Neuenfeldt, S., Beier, U., et al. 2019. Understanding ontogenetic and temporal variability of Eastern Baltic cod diet using a multispecies model and stomach data, Fisheries Research 211: 338–349, ISSN 0165-7836, https://doi.org/10.1016/j.fishres.2018.11.023.

Kulatska, N., Woods, P.J., Elvarsson, B.Þ., et al. 2021. Size-selective competition between cod and pelagic fisheries for prey. ICES Journal of Marine Science 78(5): 1872–1886, https://doi.org/10.1093/icesjms/fsab094.

Lewy, P., and Vinther, M. 2004. A stochastic age-length-structured multispecies model applied to North Sea stocks. ICES CM 2004/ FF: 20, 33 pp.

Lilly, G.R. 1991. Interannual variability in predation by cod (*Gadus morhua*) on capelin (*Mallotus villosus*) and other prey off southern Labrador and northern Newfoundland. ICES mar. Sci. Symp. 193: 133–146.

Lilly, G.R., and Rice, J.C. 1987. Food of Atlantic cod (*Gadus morhua*) on the northern Grand Bank in spring. NAFO Scientific Council Research Document 83/IX/87.

Link, J.S., Bogstad, B., Sparholt, H., et al. 2009. Role of Cod in the Ecosystem. Fish and Fisheries 10(1): 58–87.

Link, J.S., and Garrison, L.P. 2002a. Tropic ecology of Atlantic cod *Gadus morhua* on the Northeast US Continental shelf. Marine Ecology Progress Series 227: 109–123.

Link, J.S., and Garrison, L.P. 2002b. Changes in piscivory associated with fishing induced changes to the finfish community on Georges Bank. Fisheries Research 55: 71–86.

Link, J.S., Bolles, K., and Milliken, C.G. 2002. The feeding of flatfish in the northeast United States continental shelf ecosystem. Journal of Northwest Atlantic Fisheries Science, 30: 1–17.

Link, J.S., and Sherwood, G.D. 2019. Chapter 6. Feeding, growth, and trophic ecology. pp. 219–286. *In:* Rose, G.A. (ed.) Atlantic Cod: A Bio-ecology. Wiley. DOI:10.1002/9781119460701.

Lorentsen, S.-H., Anker-Nilssen, T., Barrett, R.T., et al. 2021. Population status, breeding biology and diet of Norwegian Great Cormorants. ARDEA 109(3): 299–312.

Mehl, S. 1991. The Northeast Arctic cod stock's place in the Barents Sea ecosystem in the 1980s: an overview. Polar Research 10(2): 525–534.

Mehl, S., and Sunnanå, K. 1991. Changes in growth of northeast Arctic cod in relation to food consumption in 1984–1988. ICES Marine Science Symposia 193: 109–112.

Michalsen, K., Johannesen, E., and Bogstad, B. 2008. Feeding of mature cod (*Gadus morhua*) on the spawning grounds in Lofoten. ICES Journal of Marine Science 65: 571–580.

Morrisette, L., Castonguay, M., Savenkoff, C., et al. 2009. Contrasting changes between the northern and southern Gulf of St. Lawrence ecosystems associated with the collapse of groundfish stocks. Deep-Sea Research II 56: 2117.2131.

MRI 2023. Fisheries advice for northern shrimp in Icelandic waters. https://www.hafogvatn.is/static/extras/images/29-raekja_uthaf1388228.pdf.

Neuenfeldt, S., Bartolino, V., Orio, A., et al. 2020. Feeding and growth of Atlantic cod (*Gadus morhua* L.) in the eastern Baltic Sea under environmental change. ICES Journal of Marine Science 77(2): 624–632. https://doi.org/10.1093/icesjms/fsz224.

Neuenhoff, R.D., Swain, D.P., Cox, S.P., et al. 2019. Continued decline of a collapsed population of Atlantic cod (*Gadus morhua*) due to predation-driven Allee effects. Can. J. Fish. Aq. Sci. 76: 168–184. Dx.doi.org/10.1138/cjfas-2017–0190.

Ofstad, L.H. 2013. *Anglerfish Lophius piscatorius* L. in Faroese waters - Life history, ecological importance and stock status. PhD-thesis 2013. University of Tromsø, Faculty of Biosciences, fisheries and economics department of arctic and marine biology.

Österblom, H., Hansson, S., Larsson, U., et al. 2007. Human-induced trophic cascades and ecological regime shifts in the Baltic Sea. Ecosystems, 10: 877–889, 10.1007/s10021-007-9069-0.

Pálsson, Ó.K. 1994. A review of the trophic interactions of cod stocks in the North Atlantic. ICES mar. Sci. Symp. 198: 553–575.

Pálsson, Ó.K., and Björnsson, H. 2011. Long-term changes in trophic patterns of Iceland cod and linkages to main prey stock sizes. ICES Journal of Marine Science, 68: 1488–1499.

Pérez-Rodríguez, A., Koen-Alonso, M., González-Iglesias, C., et al. 2011. Analysis of Common Trends in the Feeding Habits of Main Demersal Fish Species on the Flemish Cap (Nafo, NAFO SCR D), 1–19. Available at: https://www.nafo.int/Portals/0/PDFs/sc/2011/scr11-077.pdf.

Plaganyi, E.E. 2007. Models for an ecosystem approach to fisheries. FAO Fisheries Technical paper 477, Rome, Italy.

Rae, B.B. 1967. The food of cod on Faroese grounds. Marine Research, 6. 23 pp.

Receveur, A., Bleil, M., Funk, S., et al. 2022. Western Baltic cod in distress: decline in energy reserves since 1977. ICES Journal of Marine Science, 79(4): 1187–1201, https://doi.org/10.1093/icesjms/fsac042.

Scharf, F.S., Juanes, F., and Rountree, R.A. 2000. Predator size - prey size relationships of marine fish predators: interspecific variation and effects of ontogeny and body size on trophic-niche breadth. Marine Ecology progress Series, 208: 229–248.

Skaret, G., Johansen, G.O., Johnsen, E., et al. 2020. Diel vertical movements determine spatial interactions between cod, pelagic fish and krill on an Arctic shelf bank. Mar. Ecol. Prog. Ser., 638: 13–23. https://doi.org/10.3354/meps13254.

Skern-Mauritzen, M., Lindstrøm, U., Biuw, M. et al. 2022. Marine mammal consumption and fisheries removals in the Nordic and Barents Seas. ICES Journal of Marine Science, 79(5): 1583–1603, https://doi.org/10.1093/icesjms/fsac096.

Stefánsson, G., Skúladottir, U., and Steinarsson, B.Æ. 1998. Aspects of the ecology of a boreal system. ICES Journal of marine Science, 55: 859–862.

Steingrund, P., Mouritsen, R., Reinert, J., et al. 2010. Total stock size and cannibalism regulate recruitment in cod (*Gadus morhua*) on the Faroe Plateau. ICES Journal of Marine Science, 67: 111–124.

Swain, D.P., and Benoît, H.P. 2015. Extreme increases in natural mortality prevent recovery of collapsed fish populations in a Northwest Atlantic ecosystem. Marine Ecology Progress Series, 519: 165–182.

Sørensen, B. 2021. Growth and spatial distribution of cod and haddock on the Faroe Plateau and relationship with the amount of forage fish. Biology thesis, University of Copenhagen, Faculty of Science, Department of Biology, November 2021. 73 pp.

Tam, J.C., Link, J.S., Large, S.I. et al. 2016. A Trans-Atlantic Examination of Haddock (*Melanogrammus aeglefinus*) Food Habits. Journal of Fish Biology 88(6): 2203–2218. doi:10.1111/jfb.12983.

Temming, A., Floeter, J., and Ehrich, S. 2007 Predation hot spots: large scale impact of local aggregations. Ecosystems 10: 865–876. https://doi.org/10.1007/s10021-007-9066-3.

Townhill, B.L., Holt, R.E., Bogstad, B., et al. 2021. Diets of the Barents Sea cod (*Gadus morhua*) from the 1930s to 2018, Earth Syst. Sci. Data, 13: 1361–1370, https://doi.org/10.5194/essd-13-1361-2021.

TRAC. 2020. Eastern Georges Bank Cod. TRAC Status Report 2020/01.

Voss, R., Quaas, M., and Neuenfeldt, S. 2022. Robust, ecological–economic multispecies management of Central Baltic fishery resources. ICES Journal of Marine Science, 79(1): 169–181. https://doi.org/10.1093/icesjms/fsab251.

Werner, K-M., Taylor, M., Diekmann, R., et al. 2019. Evidence for limited adaptive responsiveness to large-scale spatial variation of habitat quality. Mar Ecol Prog Ser 629: 179–191. https://doi.org/10.3354/meps13120.

Wieland, K. 2005. Changes in recruitment, growth and stock size of Northern shrimp (*Pandalus borealis*) at West Greenland: temperature and density-dependent effects at released predation pressure. ICES J. Mar. Sci., 62: 1454–1462.

Wieland, K., Storr-Paulsen, M., and Sünksen, K. 2007. Response in stock size and recruitment of Northern shrimp (*Pandalus borealis*) to changes in predator biomass and distribution in West Greenland waters. J. Northw. Atl. Fish.Sci., 39: 21–33.

Yaragina, N.A., Bogstad, B., and Kovalev, Yu.A. 2009. Variability in cannibalism in Northeast Arctic cod (*Gadus morhua*) during the period. pp. 1947–2006. *In:* Haug, T., Røttingen, I., Gjøsæter, H., et al. (Guest eds.). Fifty Years of Norwegian-Russian Collaboration in Marine Research. Thematic issue No. 2, Marine Biology Research, 5(1): 75–85. Doi: 10.1080/17451000802512739.

Chapter 3

New England Cod Stocks

Lisa A. Kerr[1,*] and *Steven X. Cadrin*[2]

Overview

In the Northwest Atlantic, Atlantic cod (*Gadus morhua*) are found from Greenland to Cape Hatteras, North Carolina; within U.S. waters, they are most commonly found on Georges Bank and in the western Gulf of Maine (Fahay et al. 1999; Collette and Klein-Macphee 2002). Atlantic cod is an iconic species in New England and historically was one of the principal stocks in the New England groundfish fishery and a mainstay of the regional economy. The most recent stock assessments of cod in the Gulf of Maine and on Georges Bank indicate that these stocks are overfished despite low catch limits in recent years (NEFSC 2021). The groundfish fishery targets multiple demersal species. Therefore, the depletion of the cod resource is a problem not only for targeting cod but also for the complex of species targeted by the groundfish fishery. In addition, ocean warming has impacted cod productivity and is projected to continue to impact this important marine resource (Pershing et al. 2015; Hare et al. 2016). The complex spatial structure and population diversity of Atlantic cod are known to be key factors in the resilience and persistence of this fishery resource (Kerr et al. 2014; Zemeckis et al. 2014). Currently, these features are compromised in U.S. waters due to historical exploitation that resulted in the extirpation of some spawning components (Ames 2004). This loss of population richness may be an underlying factor limiting the capacity of cod populations to rebuild and motivated an initiative to revise spatial assessment and management units.

[1] University of Maine, Gulf of Maine Research Institute, 350 Commercial Street, Portland, ME 04101, USA.

[2] University of Massachusetts Dartmouth, School for Marine Science and Technology, Department of Fisheries Oceanography, 836 South Rodney French Boulevard, New Bedford, MA 02744, USA.

* Corresponding author: lisa.kerr1@maine.edu

Stock Status and Current Management Structure

Atlantic cod in U.S. waters have been assessed and managed as two distinct spatial management units since 1972: (i) Gulf of Maine and (ii) Georges Bank (Serchuk and Wigley 1992; Figure 1). The United States and Canada jointly manage cod on the eastern portion of Georges Bank through the Transboundary Resources Assessment Committee (Wang et al. 2011). These stock boundaries were defined in part by the fishing grounds for cod at the time and by an international boundary decision but did not represent current perceptions of spatial population structure (McBride and Smedbol 2022). Over the last decade, Atlantic cod stocks have been assessed every one to two years. Gulf of Maine cod was most recently assessed using an analytical stock assessment model (i.e., statistical catch-at-age model, ASAP; Legault and Restrepo 1999) and an empirical approach (i.e., survey index smoother; Legault et al. 2023) was applied to the Georges Bank stock. Gulf of Maine cod is considered to be overfished, and overfishing status differs among models that assume different natural mortality rates, denoted as M (i.e., overfishing is occurring based on the M = 0.2 model, while it is not indicated by the M-ramp model) (NEFSC 2021). Georges Bank Atlantic cod stock status cannot be quantitatively determined due to a lack of biological reference points associated with the empirical approach, but NOAA Fisheries determined the stock is overfished, and overfishing is occurring based on the previous determination from the 55th Stock Assessment Workshop conducted in 2012 (NEFSC 2013). The Northeast Regional Coordinating Council instituted an enhanced stock assessment process wherein research track stock assessments provide a vehicle for comprehensive evaluation of new data streams and model changes. A research track assessment is currently underway for Atlantic cod using new spatial units as the basis for assessment.

The New England Fishery Management Council manages Atlantic cod in U.S. Waters (NEFMC 1985) through the Northeast Multispecies Fishery Management Plan, which determines management strategies and actions for fisheries that catch 13 groundfish species, including Atlantic cod (NEFMC 2016). Several of these species co-occur to varying degrees based on the overlap of their environmental preferences

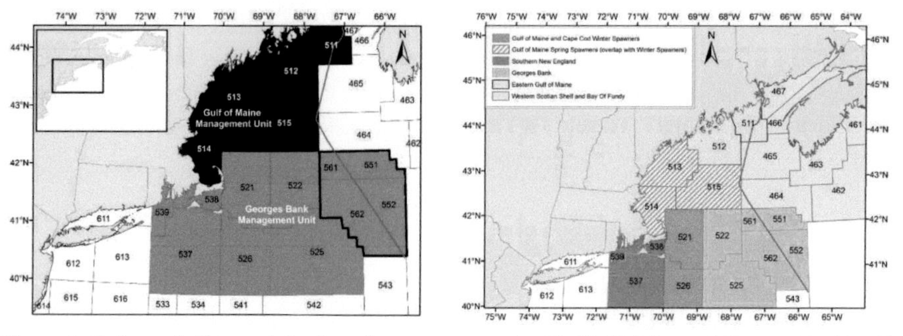

Figure 1. Left panel: Current Atlantic cod management units (Gulf of Maine and Georges Bank) with transboundary management unit outlined in black and the Hague line shown in light gray. Right panel: Biological population structure of cod based on the work on the Atlantic Cod Stock Structure Working Group (McBride and Smedbol 2022).

in space and time, and this is reflected in the mixed-species composition of landings from the groundfish fishery (Murawski and Fin 1988). Amendments to the groundfish fishery management plan have been implemented over time, making significant changes to cod management. For example, in the mid-1990s, amendments to the management plan were focused on reducing fishing pressure across stocks using measures to reduce effort (i.e., reduced days-at-sea fishing) and spatial management (i.e., areas closed to the fishery). In 2010, the management council introduced sector management whereby the total allowable catch (TAC) for each groundfish stock is divided among groups of fishermen ("sectors") according to the recent catch history of each sector's membership. As estimated stock abundances and TACs change, the allocation of the stock to fishermen changes. Under a multispecies quota system, stocks with low allocations can constrain the ability of a vessel to harvest stocks for which they have ample allocation. These are known as "choke" stocks. In 2021, there was a shift to a target of 100% at-sea monitoring of the fishery under Amendment 23 to increase catch accountability and mitigate illegal discarding of choke stocks like Atlantic cod (NEFMC 2022).

Historical Perspective on the Resource and Fishery

The history of the New England cod fishery has several periods defined by fishing technology, jurisdictions, stock conditions, and management strategies. The New England fishery for Atlantic cod was a primary industry that played an important role in the cultural and economic development of the region (Kurlansky 1997). Bartholomew Gosnold explored New England in 1602, caught a "great store of codfish," and named the peninsula between Georges Bank and western Gulf of Maine 'Cape Cod' (Merriman 1982). Cod was initially targeted off coastal New England by handline fisheries in the early 1600s and set gillnets were introduced in the late 1800s (Bigelow and Schroeder 1953; Jensen 1967; Murawski et al. 1997). The fishery expanded to offshore fishing grounds using schooners, dories and longlines in the 1700s, and landings of cod from Georges Bank peaked at 62,000 tons in 1895 (Serchuk and Wigley 1992). In recognition of the socioeconomic value of cod fisheries, a "sacred cod" was hung in the Massachusetts State House as "a memorial of the importance of the cod fishery to the welfare of this Commonwealth" (Massachusetts House of Representatives 1895).

The development of otter trawling in the 1920s further expanded the fishery to target multiple demersal species (Jensen 1967). Annual landings of cod from Georges Bank generally decreased from 41,000 tons in 1930 to 8,100 tons in 1953 as fishermen shifted targeting to haddock, then increased to 53,00 tons in 1966 primarily from foreign distant-water fleets (Serchuk and Wigley 1992). Meanwhile, annual commercial landings of cod from the Gulf of Maine varied from 6,000 to 14,500 tons from 1932 to 1950 (NEFSC 2013). A smaller-scale recreational fishery for cod also developed in the early 1900s, primarily in the Gulf of Maine and off southern New England.

The International Commission for the Northwest Atlantic Fisheries (ICNAF) was established in 1950 and regulated minimum mesh sizes for trawls and gillnets as well as minimum fish sizes, spawning closures and annual quotas for cod and several other New England groundfish (Kulka 2012). European distant-water trawl fisheries targeted cod on Georges Bank in the 1960s (Warner 1984), and their catch of cod increased to be greater than U.S. domestic catch (NEFSC 2013). In 1976, the U.S. Fishery Conservation and Management Act claimed jurisdiction of an exclusive economic zone, excluded foreign fisheries and formed regional fishery management councils. A fishery management plan was developed for cod, haddock and yellowtail flounder by the New England Fishery Management Council in 1977 (Wang and Rosenberg 1997), but TACs were not well enforced in that period (Murawski et al. 1999). After excluding distant-water foreign fisheries, landings from U.S. and Canadian fleets on Georges Bank increased from 27,000 tons in 1977 to 57,000 tons in 1982 (Serchuk and Wigley 1992). During the same period (1977–1982), commercial landings of cod from the Gulf of Maine were 11,000 to 13,000 tons per year (NEFSC 2013).

An input control system was implemented for the U.S. groundfish fishery in 1982 (including limited entry, limited days at sea, fishing gear, minimum fish sizes and closed areas), and an international boundary between U.S. and Canadian waters was defined in 1984 (Figure 1). Annual landings of cod from Georges Bank initially decreased to 26,000 in 1986, then increased to 42,500 tons in 1990. From 1982 to 1993, commercial landings in the Gulf of Maine varied from 8,000 to 18,000 tons per year, and recreational landings ranged from 600 to 3,500 tons per year (NEFSC 2013; Figure 2).

In the 1990s, New England cod stocks were determined to be overfished (NEFSC 1994, 2013). The decline of cod and other groundfish off New England as well as cod in other areas of the Northwest Atlantic, resulted from overfishing and environmental changes (Sinclair and Murawski 1997). In response, rebuilding plans for Georges Bank and the Gulf of Maine cod imposed restrictions on fishing efforts,

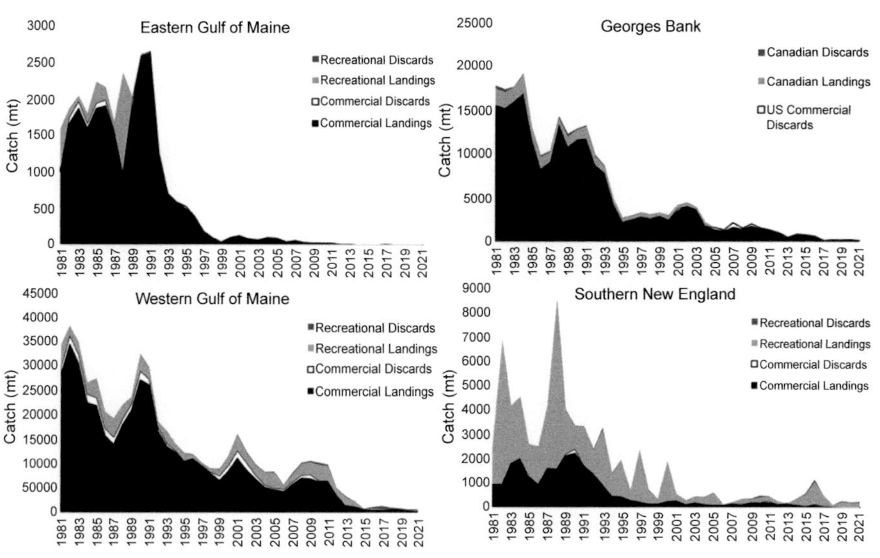

Figure 2. Total catch of New England cod in recently revised spatial assessment units, 1981–2021.

including limited days at sea and trip limits and seasonal and year-round closures to limited access to spawning cod fish (NEFSC 2013). In the late 1990s, commercial landings of cod from Georges Bank were 7,000–10,000 tons per year; commercial landings from the Gulf of Maine decreased from 8,000 tons in 1994 to 1,400 tons in 1999, but recreational landings from the Gulf of Maine increased from 700 tons in 1994 to 2,000 tons in 2001 (NEFSC 2013). The revised stocks have the same general trends in catch (Figure 2).

In 2007, the U.S. Fishery Conservation and Management Act was revised to require annual catch limits to avoid overfishing and rebuild depleted stocks. In response, management of the New England groundfish fishery transitioned to an output control and catch share system in 2010 (Murphy et al. 2018). The fishery currently includes commercial and recreational fleets in the U.S., and a commercial fishery in Canada. The U.S. commercial fishery is managed as two components (sectors and a common pool), with the recreational fishery (comprised of private anglers and for-hire charter and party) generally managed through seasons, bag limits, minimum fish sizes and recently maximum fish sizes. In 2012, after two years of the annual catch limit and catch shares system, the Secretary of Commerce declared a fishery failure and an economic disaster for the New England groundfish fishery. The total cod catch has been less than 2,000 tons since 2017 on Georges Bank, and a record low of 500 tons in 2019 in the Gulf of Maine (NEFSC 2022).

The fishing gears that target New England cod (e.g., otter trawls and set gillnets; He et al. 2021) retain multiple demersal species, but catch limits for each species are derived from single-species assessments and harvest control rules. As a result, annual catch limits for some species create "choke" stocks with low allocation relative to their availability and the availability of other species in the fishery. Currently, Atlantic cod is considered a choke species because its low catch allocation constrains the ability of fishermen to harvest other co-occurring groundfish stocks for which they have ample allocation (e.g., haddock; Huang et al. 2018). In the past, the resolution of the problem of choke species in the multispecies groundfish fishery has been *ad hoc* and included avoidance of stocks with low quotas and selective targeting of those stocks with higher quotas by the fleet, which introduces inefficiencies that are counter-strategic to the intended catch share management system. With the implementation of sector management in 2010, individual fishermen and sectors can lease additional quotas for stocks that are limited (Labaree 2012). Innovations in fishing gear have the potential to mitigate mixed-species interactions as the type of gear and how it is deployed influences the composition of the catch, and changes in gear design can allow non-target fish species to escape while targeted species are retained (Beutel et al. 2008; Dankel et al. 2009). However, the New England groundfish fishery has been shown to be slow to change and adopt new gear technology despite incentives (Eayrs and Pol 2018). Quota holders have utilized transfer of quota (i.e., annual catch entitlement) to cope with this issue. However, the magnitude of the problem and the increasing price of the quota for choke species have constrained the ability of the quota market alone to resolve this issue (Murphy et al. 2018). Presently, the industry has self-selected the approach of bycatch avoidance to deal with this problem, but this approach comes at the cost of limiting the ability of fishermen to capitalize on healthy stocks. If provided to the industry, timely forecasts of species distributions could inform fishing activity and limit mixed-species interactions, potentially improving the execution of this approach (O'Keefe and DeCelles 2013). The inability to catch other species without some bycatch of cod stocks increased its lease value in the catch shares system, occasionally to more than twice its landed value (Murphy et al. 2018). High lease prices have incentivized misreporting, with a notable violation of selling cod as haddock and illegally discarding legal-sized cod on trips with no at-sea observer (NEFMC 2022). Under the Magnuson-Stevens Act, which mandates optimum yield of the fishery (not necessarily individual stocks in a mixed-stock fishery; NOAA 2007), NMFS and the management councils have some ability to ease restrictions on choke stocks to allow for greater access to abundant species. Such logic, supported by economic analysis, was used in 2014 to set the Gulf of Maine cod quotas and in 2022 to set Georges Bank quotas (NEFMC 2022). However, an overall increase in the quota of choke stocks could delay rebuilding already vulnerable stocks, especially in a rapidly changing environment (Pershing et al. 2015).

Understanding whether the current constraints imposed by choke stocks will intensify in the region and identifying solutions to this problem that are robust to climate change is imperative to the persistence of the groundfish fishery. Thus, improved management of Atlantic cod will not only impact the health of the resource but also have broad implications for the groundfish fishery. Due to the choke species issue, there is evidence of some issues with noncompliance where landings or discards may be unreported or misreported for a number of stocks (King and Sutinen 2010; Van Beveren et al. 2019; Holland et al. 2019). However, we do not have an accurate understanding of the magnitude of underreporting and how it has changed over time.

New Paradigm of Atlantic Cod Stock Structure

Recent advances in the application of stock identification methods have revealed inconsistencies between the spatial structure of biological stocks and the definition of management units of Atlantic cod in U.S. waters. In response to scientific uncertainty related to cod stock structure, the Atlantic Cod Stock Structure Working Group was formed in 2018 to inventory and summarize all relevant peer-reviewed information about the stock structure of Atlantic cod in the Gulf of Maine and adjacent areas. The Working Group followed an interdisciplinary review of the case studies approach, forming topical subgroups on fishery management, early life history, genetic markers, life history, natural markers, applied markers and fishermen's ecological knowledge. Disciplinary summaries were synthesized to evaluate the scientific support for alternative biological stock structure scenarios (McBride and Smedbol 2022). This synthesis of complementary stock identification methods identified several mismatches between fishery management units and biological stock structure, which includes both an inshore-offshore separation as well as multiple inshore stocks, including a mixed-stock composition of spring and winter spawners in multiple statistical areas (McBride and Smedbol 2022). Specifically, five biological stocks were identified: (1) Georges Bank, (2) southern New England, (3) western Gulf of

Maine and Cape Cod winter spawning, (4) western Gulf of Maine spring spawning, and (5) eastern Gulf of Maine (Figure 1).

The New England Fishery Management Council and the Northeast Fisheries Science Center (NEFSC) agreed to develop a two-pronged approach to incorporate the stock structure recommendations into science and management. One effort focused on evaluating the data available or needed to support the assessment of the five biological stocks. The other identified the management issues that would need adjustment to implement a different stock structure.

The ongoing research track stock assessment for cod evaluated alternatives for integrating the state of knowledge on cod stock structure into the research track assessment process. Based on a review of the science of Atlantic cod stock structure and the available data to support stock assessment, the stock assessment working group developed a consensus on the most appropriate spatial assessment units. The assessment working group recommended the cod research track proceeds with four spatial units for assessment: (1) eastern Gulf of Maine, (2) western Gulf of Maine (winter and spring spawners combined), (3) Georges Bank, and (4) southern New England (including the Mid-Atlantic Bight; Figure 2). The working group discussed the treatment of the combined winter and spring spawners in the western Gulf of Maine stock unit and suggested short and long-term approaches to addressing this issue. In the short term, the presence of sympatric stocks will be considered in the assembly of input data and interpretation of trends. In the long term, a mixed-stock composition of survey and catch data may be possible to monitor and assess the abundance of winter and spring spawners in the Gulf of Maine. The rationale for the four-unit structure is an improvement in the alignment between the scale of cod stock assessment units and biological stock structure that can be supported with available information.

Currently, the New England Fishery Management Council is considering the appropriate management scale for cod and whether this will align with the new spatial scale of stock assessments or not. There are a variety of reasons why managers may wish to define cod spatial management units that differ from the four assessment units implemented by the stock assessment working group, such as quota allocation or considerations around fishing rights and access. However, a primary concern is conserving population structure to promote rebuilding.

Biology of Atlantic Cod in U.S. Waters

Atlantic cod are highly fecund and can produce 3 to 9 million eggs when they spawn (Fahay 1999). Atlantic cod eggs are pelagic, buoyant, spherical and transparent, with hatching occurring in U.S. waters after 1 to 3 weeks, depending on temperature (Hardy 1978; Thompson and Riley 1981). Larvae occur in surface waters and move deeper as they develop (Lough and Potter 1993). Ichthyoplankton survey data shows several discrete areas of larval production consistent with persistent spawning grounds in the Gulf of Maine, Georges Bank, Cape Cod, and southern New England areas. Juveniles settle in bottom habitats between 3 and 5 cm; however, the time to settlement can vary substantially based on spawning groups based on the influence of temperature on growth (Dean et al. 2022). Recently settled juvenile cod are most abundant at depths < 30 m and where bottom temperatures are < 9°C and exhibit a preference for more complex substrates (e.g., eelgrass, kelp, rock, gravel; Gotceitas and Brown 1993; Linehan et al. 2001; Grabowski et al. 2018). Juvenile settlement areas occur along the coast from New Hampshire to Rhode Island as well as on Georges Bank (Dean et al. 2022). Atlantic cod in U.S. waters can reach a maximum size of around 130 cm (≈ 25 kg) and live as long as 25 years. But today, it is uncommon to see fish older than age 15 in U.S. survey and fishery samples. This is attributed to a reduction in size and age structure through historic overfishing and possibly the influence of recent warming on cod mortality. Currently, cod in the region are reaching sexual maturity at 2 to 3 years old, and there is evidence of a decline in maturity over time (McBride and Smedbol 2022). Atlantic cod exhibit diversity in their timing of spawning, with the majority currently spawning in winter and spring months (Dean et al. 2022). Cod spawning behavior is known for its complexity and strong fidelity to spawning sites and seasons (Robichaud and Rose 2001; Dean et al. 2014; Zemeckis et al. 2014). The predominant production in the region is thought to originate from winter and spring spawners in the western Gulf of Maine and eastern Georges Bank spawners.

Adult cod are typically found associated with coarse sediment and with areas of high relief (i.e., near rocky slopes and ledges). Cod rarely occur deeper than 200 m, with larger individuals tending to remain closer to the bottom in deeper water. Cod can occur in temperatures from near freezing to 20°C and are usually found in temperatures < 10°C, except during fall when they can occur in warmer temperatures. Larger cod are generally found in colder waters (Cohen et al. 1990). Cods tend to move in schools, usually on the bottom, but may also occur in the water column (Fahay 1999). There has been extensive tagging of cod over the last century, and major regional residence and movement patterns have been similar among tagging studies since the early 1900s (Cadrin et al. 2022). Cods in the western Gulf of Maine are relatively sedentary, but there is some movement between the western Gulf of Maine and the Great South Channel. There is extensive movement between eastern Georges Bank and the western Scotian Shelf and historically between Nantucket Shoals and the Mid-Atlantic Bight. In recent years, there has been an indication of cod occurring in more shallow waters, which may relate to changes in the relative abundance of cod populations.

Stock Challenges

Climate and Ecosystem Impacts on Atlantic Cod Stock Dynamics

The Northeast U.S. shelf stands out globally as it has warmed nearly three times faster than the global ocean mean since 1982 and faster than 99.6% of the global ocean since 2004 (Pershing et al. 2015; GMRI 2023). This region has experienced a warming rate that few marine ecosystems have encountered, reshaping the ecosystem in ways that impact the productivity of fishery resources in the region, including Atlantic cod (Drinkwater et al. 2009; Nye et al. 2009; Pinsky et al. 2013; Pershing et al. 2015). Changes in temperature associated with climate change have been found to influence the productivity of Atlantic cod through direct and indirect impacts on key life history processes of fish, including recruitment, growth, and natural mortality (Klein 2016). There have also been observed changes in the distribution of Atlantic cod as well as projected decreases in habitat suitability of U.S. cod stocks at the southern extent of their range in the Northwest Atlantic (Nye et al. 2009; Kleisner et al 2017).

A thorough review of ecosystem and climate influences on cod stock dynamics was undertaken through the ongoing research track stock assessment for cod. This included a review of existing scientific literature, characterization of fishers' ecological knowledge, and exploratory modeling of relationships between climate, ocean and stock variables. Stock assessment models are being developed to test whether the incorporation of environmental covariates on aspects of population dynamics (i.e., recruitment and natural mortality) improves model fit.

Associations between Atlantic cod productivity and temperature have been identified in the region with a trend of weak recruitment and low survival with warming waters identified for certain stocks such as Gulf of Maine cod (Pershing et al. 2015; Dean et al. 2019). Recruitment success has also been linked to retention mechanisms. For example, recruitment in Georges Bank has been associated with the strength of the Georges Bank gyre, which is driven by the position of the north wall Gulf Stream and can determine the degree of egg and larval retention (Canada-U.S. EMFM Workshop Report 2018). Temperature is also known to strongly influence growth, with examples of faster growth and fish reaching smaller asymptotic size with warming (Drinkwater 2005). In recent decades, a decrease in the weight-at-age of cod at older ages is apparent (Cod Research Track Assessment Working group, pers. comm.). Increases in the natural mortality of cod in the Gulf of Maine are viewed as plausible, although an exact mechanism has not been identified (e.g., thermal limits, prey limitation, and predation during various life history stages are all proposed mechanisms) (Pershing et al. 2015; Chen et al. 2022). Furthermore, warmer water temperatures could lead to smaller sizes at age, which could increase susceptibility to predation (Nye et al. 2009; Levangie et al. 2021).

Over the period from 1968 to 2007, Georges Bank Atlantic cod demonstrated a decrease in the area occupied and a northward shift in the center of biomass (1.48 km/year) (Nye et al. 2009). Gulf of Maine cod, however, have exhibited an apparent shift further south which has been associated with the interaction of the ocean warming trend (Nye et al. 2009) as well as localized depletion of populations (Guan et al. 2017). Ocean temperatures in all areas currently occupied by cod in U.S. waters are expected to exceed cod's thermal optimum by 2050, resulting in future decreased habitat suitability (Rogers et al. 2019; Fogarty et al. 2008; Kleisner et al. 2017).

The current management strategies for Atlantic cod do not explicitly consider climate-driven impacts on stock dynamics. Failure to acknowledge the changing nature of relationships between climate and resources can potentially lead to inappropriate management (Mazur et al. 2023). Furthermore, changing dynamics may influence the capacity for Atlantic cod stocks already at low biomass to rebuild and potentially magnify the choke species problem (Pershing et al. 2015).

Challenges to Atlantic Cod Stock Assessment

Initial stock assessments of cod developed in the ICNAF period were descriptive analyses of commercial fishery landings, catch rates and fishery-independent survey indices. An age-structured assessment was developed in the late 1970s for Georges Bank cod and indicated that fishing mortality was relatively high (e.g., greater than the rate expected to produce maximum yield per recruit, F_{max}) (Serchuk and Wigley 1992). Similar methods were applied to the Gulf of Maine cod fishery in the 1980s, and estimated fishing mortality in that area was also greater than F_{max} (NEFC 1989). In 1994, the Georges Bank cod stock was determined to have collapsed (NEFSC 1994); in 2002, the Gulf of Maine cod stock was classified as overfished (NEFSC 2013).

Age-based stock assessments for Georges Bank and Gulf of Maine cod fisheries have been regularly updated since the 1990s. Methods advanced by including commercial discards and recreational catch and transitioning from virtual population analysis and yield per recruit to statistical catch-at-age and spawner-per-recruit models. Assessments indicated that overfishing continued and rebuilding has not been achieved (e.g., NEFSC 2013, 2022). Recent assessments of Georges Bank cod were rejected because of retrospective patterns and replaced with index-based approaches, and similar problems in the Gulf of Maine cod assessment led to the development of multiple models with alternative assumptions of recent natural mortality (NEFSC 2022).

Retrospective inconsistency has been a pervasive problem with New England cod stock assessments (NEFSC 2022). Retrospective patterns in fish stock assessment are directional changes in estimated quantities (e.g., spawning stock biomass) as additional years of data are added to an assessment (Szuwalski et al. 2018). For New England cod assessments, estimates of stock size decrease as more data are added, and estimates of fishing mortality increase (Figure 3). Such patterns suggest that the estimation model is mis-specified. The issues with the Georges Bank stock assessment resulted in a rejection of the analytical assessment in 2015 and retrospective adjustments to the Gulf of Maine assessment being made in recent years (NEFSC 2021). Some of the candidate causes of retrospective patterns include (1) ecosystem change and its impact on population processes (e.g., time-varying natural mortality), (2) changes in fishing behavior and misreporting of catch (e.g., unreported or underestimated discards), or (3) changes in survey or fishery catchability and selectivity. More than one of these processes may occur particularly in the case of climate impacts that simultaneously influence population processes and distribution (Karp et al. 2018). Retrospective inconsistencies in New England cod assessments have been a source of uncertainty for assessment and management, such that precautionary catch limits intended to rebuild stocks and have a low risk of overfishing have resulted in continued overfishing and failure to achieve stock recovery (NEFSC 2022).

The research track stock assessment for New England cod initiated in 2021 confronts these problems. The assessment working group recognized the revised perception of cod population structure (McBride and Smedbol 2022) and re-defined spatial assessment units (Figure 1). More advanced models with the capacity to

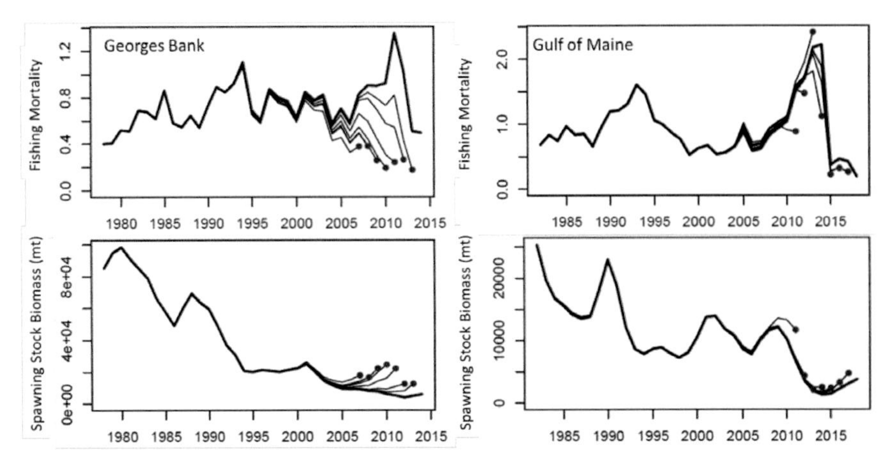

Figure 3. Retrospective patterns in estimates of fishing mortality and spawning stock biomass for New England cod stocks (from NEFSC 2015, 2022). Bold lines indicate the most recent estimates, thin lines indicate retrospective estimates, and circles indicate terminal year estimates from retrospective analyses.

account for time-varying stock dynamics promise to resolve some problems with New England cod assessments. Although analytical assessments were still in development at the time of this writing, survey trends show general stock depletion in each of the revised spatial assessment units, and age composition indicates low recruitment in the last few decades.

Conclusions

New England cod stocks are currently depleted from decades of overfishing, depletion of spawning populations, and climate impacts on productivity. The changing stock dynamics of cod will influence the capacity of these stocks to rebuild and potentially magnify the choke species problem. Our conventional stock assessment and fisheries management approaches do not account for nonstationary productivity imposed by climate change and new information on the biological population of Atlantic cod in U.S. waters. Failure to acknowledge the changing nature of relationships between climate and resources can potentially lead to inappropriate management (Mazur et al. 2023). There is ongoing work to develop approaches that align the spatial scale of assessment and management to account for population structure and climate impacts on cod stock dynamics in assessment and management, which are expected to help rebuild New England cod stocks.

References

Ames, E.P. 2004. Atlantic Cod Stock Structure in the Gulf of Maine. Fisheries, 29: 10–28.

Beutel, D., Skrobe, L., Castro, K., et al. 2008. Bycatch reduction in the Northeast USA directed haddock bottom trawl fishery. Fish. Res., 94(2): 190–198.

Bigelow, H., and Schroeder, W. 1953. Fishes of the Gulf of Maine. Fishery Bulletin of the Fish and Wildlife Service 53.

Cadrin, S.X., Zemeckis, D.R., DeCelles, G.R., et al. 2022. Applied markers. pp. 149–216. *In:* McBride, R.S., and Smedbol, R.K. (eds.). An Interdisciplinary Review of Atlantic Cod (*Gadus morhua*) Stock Structure in the Western North Atlantic Ocean. NOAA Technical Memorandum NMFS-NE-273.

Chen, N., Sun, M., Zhang, C., et al. 2022. Non-stationary Natural Mortality Influencing the Stock Assessment of Atlantic Cod (Gadus morhua) in a Changing Gulf of Maine. Frontiers in Marine Science. 9. https://www.frontiersin.org/articles/10.3389/fmars.2022.845787.

Cohen, D.M., Inada, T., Iwamoto, T., et al. 1990. FAO species catalogue. Vol. 10. Gadiform fishes of the world (order Gadiformes). An annotated and illustrated catalogue of cods, hakes, grenadiers, and other gadiform fishes known to date. FAO Fisheries Synopsis, (125) (10).

Canada-US EMFM Workshop Report 2018.

Cod Research Track Assessment Working group, pers. comm.

Collette, B.B., and Klein-Macphee, G. (eds.). 2002. Bigelow and Schroeder's Fishes of the Gulf of Maine, Third Edition. Smithsonian Books.

Dankel, D.J., Jacobson, N., Georgianna, D., et al. 2009. Can we increase haddock yield within the constraints of the Magnuson-Stevens Act? Fisheries Research, 100(3): 240–247. https://doi.org/10.1016/j.fishres.2009.08.003.

Dean, M.J., DeCelles, G.R., Zemeckis, D.R., et al. 2022. Early life history: spawning to settlement. pp. 19–50. *In:* McBride, R.S., and Smedbol, R.K. (eds.). An Interdisciplinary Review of Atlantic Cod (*Gadus morhua*) Stock Structure in the Western North Atlantic Ocean. NOAA Technical Memorandum NMFS-NE-273.

Dean, M.J., Elzey, S.P., Hoffman, W.S., et al. 2019. The relative importance of sub-populations to the Gulf of Maine stock of Atlantic cod. ICES Journal of Marine Science. 76(6): 1626–1640. https://doi.org/10.1093/icesjms/fsz083.

Dean, M.J., Hoffman, W.S., Zemeckis, D.R., et al. 2014. Fine-scale diel and gender-based patterns in behaviour of Atlantic cod (*Gadus morhua*) on a spawning ground in the Western 46 Gulf of Maine. ICES J. Mar. Sci., 71: 1474–1489. http://dx.doi.org/10.1093/icesjms/fsu040.

Drinkwater, K., Beaugrand, G., Kaeriyama, M., et al. 2009. On the mechanisms linking climate to ecosystem changes.

Drinkwater, K.F. 2005. The response of Atlantic cod (*Gadus morhua*) to future climate change. ICES Journal of Marine Science, 62: 1327–1337.

Eayrs, S., and Pol, M. 2018. The myth of voluntary uptake of proven fishing gear: investigations into the challenges inspiring change in fisheries. ICES Journal of Marine Science, doi:10.1093/icesjms/fsy178.

Fahay, M.P., Berrien, P.L., Johnson, D.L., et al. 1999. Essential fish habitat source document: Atlantic Cod, *Gadus morhua*, life history and habitat characteristics. NOAA Technical Memorandum NMFS-NE-124.

Fogarty, M., Incze, L., Hayhoe, K., et al. 2008. Potential climate change impacts on Atlantic cod (*Gadus morhua*) off the northeastern USA. Mitigation and Adaptation Strategies for Global Change. 13: 453–466. https://doi.org/10.1007/s11027-007-9131-4.

Gulf of Maine Warming Update: 2022 the second-hottest year on record. Gulf Of Maine Research Institute. (2023, February 15). https://www.gmri.org/stories/warming-22/.

Gotceitas, V., and Brown, J.A. 1993. Substrate selection by juvenile Atlantic cod (*Gadus morhua*): effects of predation risk. Oecologia, 93: 31–37.

Guan, L., Chen, Y., Wilson, J.A., et al. 2017. Evaluating spatio-temporal variability in the habitat quality of Atlantic cod (Gadus morhua) in the Gulf of Maine. Fisheries Oceanography, 26(1): 83–96. https://doi.org/10.1111/fog.12188.

Grabowski, J.H., Conroy, C.W., Gittman, R.K., et al. 2018. Habitat associations of juvenile cod in nearshore waters. Rev Fish Sci. Aquac. 26: 1–14. Taylor & Francis. https://doi.org/10.1080/23308249.2017.1328660.

Hardy, J.D., Jr. 1978. Development of fishes of the Mid-Atlantic Bight: An atlas of egg, larval and juvenile stages. Vol. 2: Anguillidae through Syngnathidae. U.S. Fish Wildl. Serv., Biol. Serv. Prog. FWS/OBS-78/12. 458 p.

Hare, J.A., Morrison, W.E., Nelson, M.W., et al. 2016. A vulnerability assessment of fish and invertebrates to climate change on the northeast U.S. continental shelf. PLoS One, 11(2): e0146756. https://doi.org/10.1371/journal.pone.0146756.

He, P., Chopin, F., Suuronen, P., et al. 2021. Classification and illustrated definition of fishing gears. FAO Fisheries and Aquaculture Technical Paper No. 672. Rome, FAO.

Holland, D., Kerr, L., McNamee, J., et al. 2019. New England Fishery Management Council Scientific and Statistical Committee Sub-Panel Peer Review Report for the Groundfish Plan Development Team Analyses of Groundfish Monitoring. Available: https://s3.amazonaws.com/nefmc.org/4.-190513_SSC_Sub_P anel_Peer-Review-Report_OEMethods_FINAL.pdf. Last accesssed date on 11 August 2022.

Huang, L., Ray, S., Segerson, K., et al. 2018. Impact of collective rights-based fisheries management: evidence from the New England Groundfish Fishery. Marine Resource Economics, 33(2): 177–201.

Jensen, A.C. 1967. A brief history of the New England offshore fisheries. US Bureau of Commercial Fisheries, Fishery Leaflet 594.

Karp, M.A., Peterson, J.O., Lynch, P.D., et al. 2019. Accounting for shifting distributions and changing productivity in the development of scientific advice for Fishery Management. ICES Journal of Marine Science. http://doi.org/10.1093/icesjms/fsz048.

Kerr, L., Cadrin, S., Kovach, A., et al. 2014. Consequences of a mismatch between biological and management units on our perception of Atlantic cod off New England. ICES Journal of Marine Science. 71. 1366–1381. 10.1093/icesjms/fsu113.

King, D.M., and Sutinen, J.G. 2010. Rational noncompliance and the liquidation of Northeast Groundfish Resources. Marine Policy, 34(1): 7–21. http://doi.org/10.1016/j.marpol.2009.04.023.

Klein, E.S., Smith, S.L., and Kritzer, J.P. 2016. Effects of climate change on four New England groundfish species. Reviews in Fish Biology and Fisheries, 27(2): 317–338. http://doi.org/10.1007/s11160-016-9444-z.

Kleisner, K.M., Fogarty, M.J., McGee, S., et al. 2017. Marine species distribution shifts on the U.S. Northeast Continental Shelf under continued ocean warming. Progress in Oceanography. 153: 24–36. https://doi.org/10.1016/j.pocean.2017.04.001.

Kulka, D.W. 2012. History and Description of the International Commission for the Northwest Atlantic Fisheries. (https://www.nafo.int/Portals/0/PDFs/icnaf/ICNAF_history-kulka.pdf).

Kurlansky, M. 1997. Cod: A biography of the fish that changed the world. Penguin Books, New York, NY, USA.

Labaree, J.M. 2012. Sector management in New England's groundfish fishery: Dramatic change Spurs Innovation. Retrieved from https://www.cakex.org/sites/default/files/documents/sector_management_in_new_england.pdf.

Legault, C., Restrepo, V. 1999. A Flexible forward age-structured assessment program. ICCAT Coll. Vol. Sci. Pap. 49.

Legault, C.M., Wiedenmann, J., Deroba, J.J., et al. 2023. Data-rich but model-resistant: an evaluation of data-limited methods to manage fisheries with failed age-based stock assessments. Can. J. Fish. Aquat. Sci. 80: 27–42.

Levangie, P.E.L., Blanchfield, P.J., Hutchings, J.A., et al. 2021. The influence of ocean warming on the natural mortality of marine fishes. Environmental Biology of Fishes 105: 1447–1461.

Linehan, J.E., Gregory, R.S., Schneider, D.C., et al. 2001. Predation risk of age-0 cod (*Gadus*) relative to depth and substrate in coastal waters. J. Exp. Mar. Biol. Ecol., 263: 25–44. https://doi.org/10.1016/S0022-0981(01)00287-8.

Lough, R.G., and Potter, D.C. 1993. Vertical distribution patterns and diel migrations of larval and juvenile haddock *Melanogrammus aeglefinus* and Atlantic cod *Gadus morhua* on Georges Bank. Fish Bull. 91: 281–303. https://spo.nmfs.noaa.gov/sites/default/files/pdfcontent/1993/912/lough.pdf.

Massachusetts House of Representatives. 1895. A History of the Emblem of the Codfish in the Hall of the House of Representatives. Compiled by a Committee of the House. Wright and Potter Printing Company. https://archive.org/details/historyofemblemo00mass.

Mazur, M., Jesse, J., Cadrin, S.X., et al. 2023. Impacts of ignoring climate impacts on population dynamics in New England groundfish stock assessments. Fisheries Research, 262: 106652.

McBride and Smedbol, R.K. (eds.). 2022. An interdisciplinary review of Atlantic Cod (*Gadus morhua*) Stock Structure in the Western North Atlantic Ocean. NOAA Technical Memorandum NMFS-NE-273.

Merriman, D. 1982. The history of georges bank. pp. 11–30 *In:* McLeod, G.C., and Prescott, J.H. (eds.). Georges Bank. Taylor and Fancis.

Murawski, S.A., and Finn, J.T. 1988. Biological bases for mixed-species fisheries: Species co-distribution in relation to environmental and biotic variables. Can. J. Fish. Aquat. Sci., 45: 1720–1734.

Murawski, S.A., Maguire, J.J., Mayo, R.K., et al. 1997. Groundfish stocks and the fishing industry. pp. 27–70 *In:* Boreman, J., Nakashima, B.S., Wilson, J.A., et al. (eds). Northwest Atlantic Groundfish; Perspectives on a Fishery Collapse. American Fisheries Society, Bethesda.

Murawski, S.A., Brown, R.W., Cadrin, S.X., et al. 1999. New England groundfish. *In*: Our Living Oceans. NOAA Tech. Mem. NMFS-F/SPO-41: 71–80.

Murphy, T., Ardini, G., Vasta, M., et al. 2018. 2015. Final Report on the Performance of the Northeast Multispecies (Groundfish) Fishery (May 2007–April 2016).

NEFC (Northeast Fisheries Center). 1989. Report of the Seventh NEFC Stock Assessment Workshop (Seventh SAW). NEFC Ref. Doc. 89-04.

NEFMC (New England Fishery Management Council). 2022. Overfishing Limits and Acceptable Biological Catches for Southern New England/Mid-Atlantic Winter Flounder, Georges Bank Yellowtail Flounder and Georges Bank Cod. https://s3.us-east-1.amazonaws.com/nefmc.org/3.-SSC-Report-Aug-25_22-mtg-Memo-09_02_22.pdf.

NEFMC (New England Fishery Management Council. 1985. Fishery Management Plan Environmental Impact Statement Regulatory Impact Review and Initial Regulatory Flexibility Analysis for the Northeast Multi-Species Fishery. Prepared by New England Fishery Management Council in consultation with Mid-Atlantic Fishery Management Council. https://d23h0vhsm26o6d.cloudfront.net/MultiSpecies-FMP.pdf.

NEFMC (New England Fishery Management Council). 2016. Framework Adjustment 55 to the Northeast Multispecies Fishery Management Plan. Newburyport, MA. April 2016.

NEFMC (New England Fishery Management Council). 2022. Northeast Multispecies Fishery, Amendment 23. Newburyport, MA. April 2016.

NEFSC (Northeast Fisheries Science Center). 1994. Report of the 18th Northeast Regional Stock Assessment Workshop (18th SAW) The Plenary. NEFSC Ref Doc 94-23.

NEFSC (Northeast Fisheries Science Center). 2013. 55th Northeast Regional Stock Assessment Workshop (55th SAW) Assessment Report. NEFSC Ref Doc. 13-11.

NEFSC (Northeast Fisheries Science Center). 2015. Operational Assessment of 20 Northeast Groundfish Stocks, Updated Through 2014. NEFSC Ref Doc. 15-24.

NEFSC (Northeast Fisheries Science Center). 2022. Stock Assessment Update of 14 Northeast Groundfish Stocks Through 2018. NEFSC Ref. Doc. 22-06.

NOAA. 2007. Magnuson-Stevens Fishery Conservation and Management Act - 2007 Blue book. Retrieved from https://media.fisheries.noaa.gov/dam-migration/msa-amended-2007.pdf.

Northeast Fisheries Science Center (NEFSC). 2021. Gulf of Maine Atlantic cod - 2021 update assessment report. Northeast Fisheries Science Center, Woods Hole, MA.

Nye, J.A., Link, J.S., Hare, J.A., et al. 2009. Changing spatial distribution of fish stocks in relation to climate and population size on the Northeast United States continental shelf. Mar. Ecol. Prog. Ser., 393: 111–129. 10.3354/meps08220.

O'Keefe, C., and Decelles, G. 2013. Forming a Partnership to Avoid Bycatch. Fisheries., 38. 10.1080/03632415.2013.838122.

Pershing, A.J., Alexander, M.A., Hernandez, C.M., et al. 2015. Slow adaptation in the face of rapid warming leads to collapse of the Gulf of Maine cod fishery. Science, 350: 809–812. http://science.sciencemag.org/content/350/6262/809.full.

Pinsky, Malin, Worm, Boris, Fogarty, Michael, et al. 2013. Marine Taxa Track Local Climate Velocities. Science (New York, N.Y.), 341: 1239–42. 10.1126/science.1239352.

Robichaud, D., and Rose, G.A. 2001. Multiyear homing of Atlantic cod to a spawning ground. Can. J. Fish 49 Aquat. Sci., 58: 2325–2329. https://cdnsciencepub.com/doi/10.1139/f01-190.

Rogers, L.A., Griffin, R., Young, T., et al. 2019. Shifting habitats expose fishing communities to risk under climate change. Nat. Clim. Chang., 9: 512–516. https://doi.org/10.1038/s41558-019-0503-z.

Serchuk, F.M., and S.E. Wigley. 1992. Assessment and Management of the Georges Bank Cod Fishery: An Historical Evaluation. J. Northw. Atl. Fish. Sci. 13: 25–52.

Sinclair, A.F., and S.A. Murawski. 1997. Why have groundfish stocks declined? pp. 71–93. *In*: Boreman, J., Nakashima, B.S., Wilson, J.A., et al. (eds.). Northwest Atlantic Groundfish; Perspectives on a Fishery Collapse. American Fisheries Society, Bethesda.

Szuwalski, C., Ianelli, J., Punt, A., et al. 2018. Reducing retrospective patterns in stock assessment and impacts on management performance. ICES Journal of Marine Science, 75: 596–609. 10.1093/icesjms/fsx159.

Thompson, B.M., and Riley, J.D. 1981. Egg and larval development studies in the North Sea cod (*Gadus morhua* L.). Rapports et Proces Verbaux de Reunions-Conseil Permanent International Pour L'Exploration de la Mer, 178: 553–559.

Van Beveren, E., Duplisea, D.E., Brosset, P., et al. 2019 Assessment modelling approaches for stocks with spawning components, seasonal and spatial dynamics, and limited resources for data collection. PLoS One. 2019 Sep 23; 14(9): e0222472. doi: 10.1371/journal.pone.0222472. PMID: 31545816; PMCID: PMC6756546.

Wang, S.D.H., and A.A. Rosenberg. 1997. US New England Groundfish Management Under the Magnuson-Stevens Fishery Conservation and Management Act. Marine Resource Economics 12: 361–366.

Wang, Y., O'Brien, L., Clark, K., et al. 2011. Assessment of Eastern Georges Bank Atlantic Cod for 2011. Transboundary Resources Assessment Committee. Reference Document 2011/02.

Warner, W.W. 1984. Distant Water: The Fate of the North Atlantic Fisherman. Penguin Publishing Group.

Zemeckis, D.R., Hoffman, W.S., Dean, M.J., et al. 2014. Spawning site fidelity by Atlantic cod (*Gadus morhua*) in the Gulf of Maine: Implications for population structure and rebuilding. ICES J Mar Sci 71: 1356–1365. https://doi.org/10.1093/icesjms/fsu117.

Chapter 4

Greenland Cod Stocks

Anja Retzel,[1,*] *Frank Rigét*[1] and *Rasmus Berg Hedeholm*[1,2]

Introduction

Atlantic cod is an integral part of the Greenland ecosystem and culture. It was subject to the first large-scale commercial fishery in Greenland and continues to be an integral part of it, especially the coastal fishery. Cod fishing, however, remains a precarious endeavor in Greenland, as cod in Greenland exists on the species' northern distribution limit, and its presence and abundance are highly dependent on climatic fluctuations and trends. In the recent warm period, cod have been found as far north as Qaanaaq (77°N) in West Greenland, while previous cold periods have caused near absence in Greenland waters.

The commercial fishery for cod was the first large-scale commercial fishery in Greenland, attracting the attention of European fishing nations following the Second World War. The fishery was intense, and only two decades later, it began to spiral into a collapse, and the cod did not fully recover. The subsistence fishery has, however, always been an important part of Greenland culture, and recent stock increases have seen the re-emergence of an important small-scale inshore fishery. The cod habitat in coastal East Greenland is much smaller than in West Greenland due to the cold-water influence from the Arctic Ocean. Hence, cod inshore abundance has never been as high as in West Greenland, and the few settlements in East Greenland have historically been more oriented toward hunting. The East Greenland offshore area is warmer, and with the modernization of the fishing fleet, the offshore fishing grounds are now accessible year-round and support a > 25,000 t fishery similar to the levels seen during the 1960s and 1970s.

Because of its cultural and commercial importance, cod was the first species in Greenland to be studied in detail by scientists. Particularly, the rapid changes in abundance and, consequently, the rise and fall of the fishery received early attention;

[1] Greenland Institute of Natural Resources, Kivioq 2, 3900 Nuuk, Greenland.

[2] Sustainable Fisheries Greenland, Co/Greenland Business Association, Jens Kreutzmannip Aqq. 3, 3900 Nuuk, Greenland.

* Corresponding author: anre@natur.gl

the link to environmental drivers is still a key aspect of the scientific effort. The migration patterns spurred a series of tagging studies (Hansen 1949; Storr-Paulsen et al. 2004) that continues today (Hedeholm 2018), and in recent years the field of genetics and advanced electronic tags have shed more light on cod movement and behavior (Bonanomi et al. 2016; Nielsen et al. 2023). These aspects of cod life history are supplemented by annual surveys, underlining that cod is a key species both commercially, culturally and scientifically.

Currents

Understanding the current patterns around Greenland is key to understanding cod abundance both historically and currently, as they dictate both the northern limit of cod distribution and, through passive transport of eggs and larvae, have a large influence on stock size fluctuations. The dominating currents are comprised of the Irminger Current, the East Greenland Current and the West Greenland Current (Figure 1). The Irminger current is a branch of the warm and saline North Atlantic current flowing generally westward along the coast of Iceland, where it divides into two currents. One proceeds northwards and eastward around Iceland, and the other flows westward and then southwestward, merging with the East Greenland Current, ultimately forming the West Greenland Current. The East Greenland current originates in the Arctic Ocean and is a cold, low-salinity current that flows southwards along the Greenland continental margin to Cape Farewell in the south (60ºN).

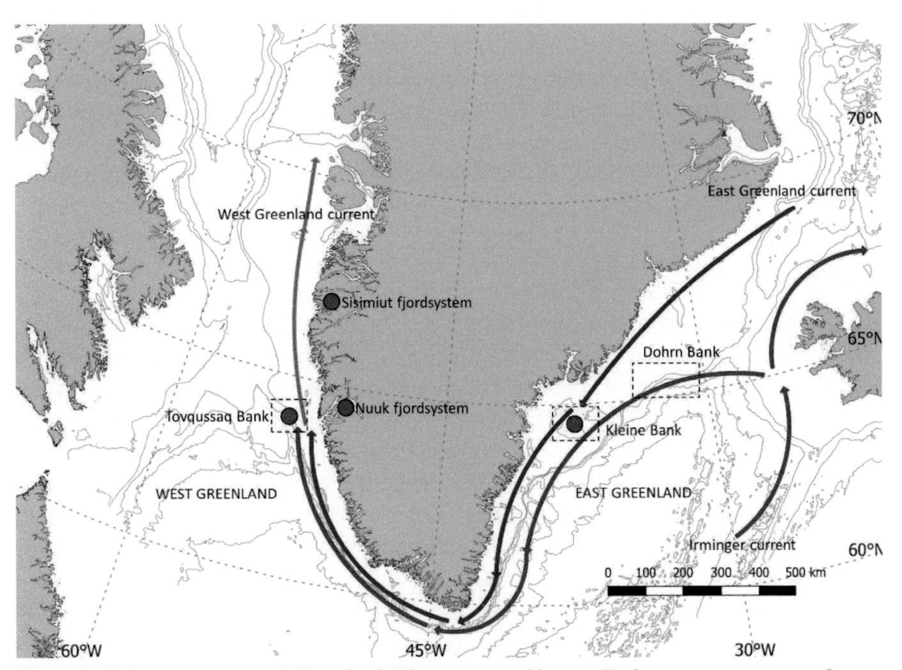

Figure 1. Major currents around Greenland. Blue arrows: cold water. Red arrows: warm water, Orange arrow: mix of cold and warm water. Red dots, spawning hotspots. Boxes: major offshore banks mentioned in the text.

The West Greenland Current flows north along the west coast of Greenland into the Davis Strait, and the heat from the Irminger Current maintains ice-free conditions year-round on the west coast up to approximately 66°N. In general, the near-shore surface waters on the West Greenland banks are comprised of the relatively cold parts of the West Greenland currents. On the outer slopes of the banks, warmer and more saline currents dominated by water from the Irminger Current are found, and it can be traced all the way along West Greenland to Qaanaaq (77°N) (Buch 2000 and 2002). The strength of the Irminger Current has a major impact on the East and West Greenland ecosystems, as it dictates the delineations between warm and cold water; e.g., on the east coast, there is a sharp delineation between the cold waters of the East Greenland current, which runs close to the shelf break, and the warmer waters of the Irminger Current, which runs next to the East Greenland Current further offshore. In West Greenland, the cold Arctic water can extend far offshore in the surface layers, whereas the warmer Irminger water runs deeper and below the cold Arctic water.

West Greenland is characterized by a shallow continental shelf (0–200 m, app. 21,000 km^2) in the south from Cape Farewell (60°N) to Fyllas Bank (64°N). North of Fyllas Bank, the relatively shallow "Store hellefiskebanke" (0–200 m, app. 56,000 km^2) extends far into the Davis Strait, delineating the separation between the Arctic Baffin Bay and the Subarctic Davis Strait. East Greenland is characterized by a shallow continental shelf (0–200 m, app. 12,000 km^2) separated from the deep ocean by a steep slope. The near-shore shelf is dominated by the cold southward moving polar water current—the East Greenland Current—generally considered too cold for cod. This cold-water barrier usually limits cod to latitudes below 69°N, although recently, cod have been found as far north as 77°N (Christiansen et al. 2016). The strength of the Irminger current can have a pronounced effect on the stock size in Greenland; hence, in periods where the Irminger Current is strong, Greenland receives a large inflow of warm water and potentially eggs and larvae from the spawning grounds in Iceland. The warm water promotes successful recruitment and favorable conditions in both East and West Greenland (Stein and Borovkov 2004).

Stocks and Distribution

The challenging environmental conditions in Greenland and strong climatic gradients both north-to-south and east-to-west have promoted local adaptations and the evolution of three distinct cod stocks in Greenland. This distinction was recognized as early as the 1940s, based primarily on tagging experiments and the geographical distribution of eggs and larvae (Hansen 1949). In 2013, genetic analyses of spawning cod supported this stock perception by defining three distinct spawning stocks in Greenland: the West Greenland inshore stock, the West Greenland offshore stock and the East Greenland/offshore Iceland stock (Therkildsen et al. 2013). These stocks maintain separate spawning areas and have stock-specific life-history traits (see below) but are otherwise morphologically indistinguishable. This complicates understanding and describing stock dynamics in Greenland because the stocks mix both on nursery grounds and in association with feeding. This is most pronounced in West Greenland, especially in the inshore area, where stock mixing is ubiquitous along the coast in the coastal zone and the fjords (Christensen et al. 2022).

Currently, there is a latitudinal gradient in the stock proportions along the west coast, with the East Greenland/offshore Iceland stock being dominant in south Greenland (60–63°N), the West Greenland inshore stock dominant in mid-Greenland (63–67°N) and the West Greenland offshore stock dominant in north Greenland (> 67°N) (Figure 2). The low proportions of the West Greenland inshore stock in the south and north Greenland fjords are most likely due to a lack of suitable spawning sites and limited north-south migration of the West Greenland inshore stock (Storr-Paulsen et al. 2004). The West Greenland inshore stock is also virtually absent in offshore waters and only extends its distribution to the coastal zone. The East Greenland/offshore Iceland stock is the only stock in East Greenland, extending its distribution from East Greenland into West Greenland waters. When moving north offshore, the West Greenland offshore stock gradually starts to dominate as the East Greenland/offshore Iceland stock declines, mirroring the pattern seen in the inshore area. This pattern was also prevalent in the large fishery during the middle of the 20th century, so it appears to be a relatively stable overall distribution pattern (Bonanomi et al. 2015).

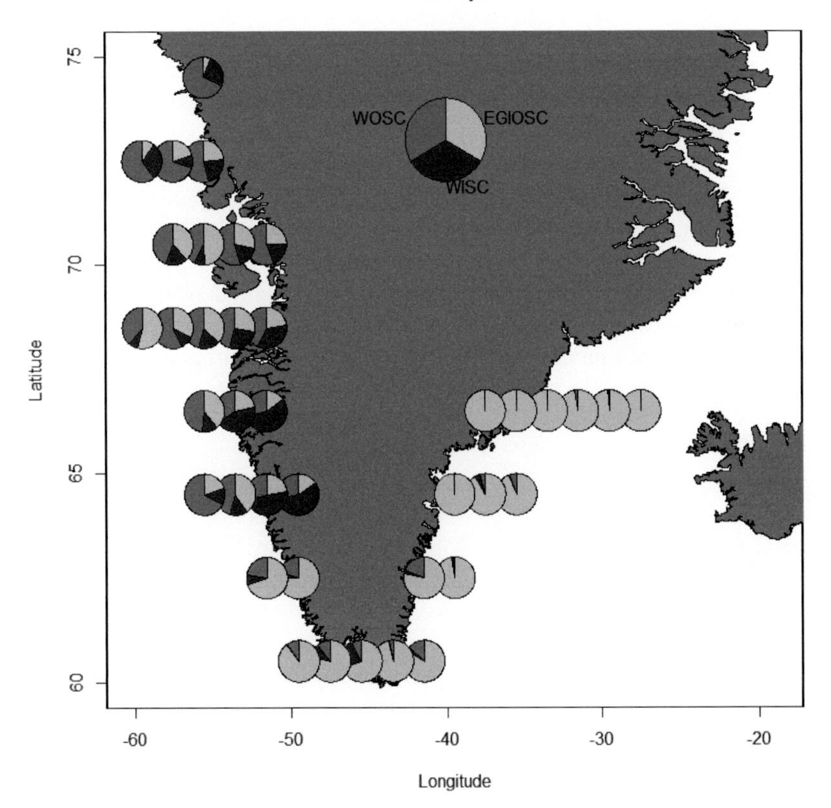

Figure 2. Mix of three stocks around Greenland: EGIOSC: East Greenland/ Iceland offshore spawning cod stock. WISC: West Greenland inshore spawning cod stock. WOSC: West Greenland offshore spawning cod stock. Based on genetic data from 9,509 samples in the period 2000–2022.

Spawning

West Greenland Offshore

Historically, offshore spawning in West Greenland was documented by the presence of eggs and larvae, and the earliest recording of offshore spawning in Greenland was in 1925, where eggs and larvae were found between 63°N and 67°N in the Davis straight (Hansen and Herman 1953). In the 1950s, large quantities of larvae were found as far north as 70°N, and investigations in the 1960s found eggs from 59°N and northwards (Wieland and Hovgård 2002). There is less information on the actual spawning grounds, but spawning has been described regularly on the slope of Fyllas Bank (64°N) at 120 m and a temperature of 1.5°C, and more frequently west of more southern banks below 350 m in the warmer Atlantic water from the Irminger Current. Following the collapse of the West Greenland offshore fishery and the West Greenland offshore stock between 1960 and 1980, significant spawning has only been documented on Tovqusssaq Bank (Figure 1), where spawning takes place in March and April (Retzel 2020).

West Greenland Inshore

Stable and significant spawning occurs in two large fjord systems, the Kapisillit area in the Nuuk fjord system and the Qeqertalik area in the Sisimiut fjord system (Figure 1). Spawning also takes place in the coastal zone and other fjords, especially between Fiskenæsset (63°N) and Sisimiut (67°N), but this is sporadic and less concentrated (Retzel and Hedeholm 2012). In the Kapisillit area, spawning has been documented as far back as the 1920s (Hansen 1949). The timing of spawning is later than in the offshore area, peaking in April/May, and the daily egg production can be as high as 685 eggs per m^2 of the water column (Swalethorp 2013). This highlights the unique character of this area and its importance for inshore cod recruitment, although the actual spawning grounds are small. Hence, egg densities decline with distance from the spawning site, and at the Nuuk fjord entrance 80 km away, cod eggs are rarely found in the annual ichthyoplankton monitoring program (MarineBasis Nuuk).

East Greenland Spawning and Input From Iceland

In East Greenland, spawning was observed between March and June in the 1950s and 1960s along the shelf edge from 62°N to 66°N at depths from 170–400 m and in relatively warm water from 3.2–5.2°C (Wieland and Hovgård 2002). Ichthyoplankton surveys in several years showed that egg densities were substantially higher (1,152 eggs/30 min tow) in Southeast and Southwest Greenland compared to more northern offshore West Greenland (5 eggs/30 min tow). Due to the currents, many of the eggs from East Greenland are passively transported west and northwards, emphasizing the importance of spawning in East Greenland and subsequent drift events in maintaining a high cod abundance in West Greenland (Wieland and Hovgård 2002). Like West Greenland, the stock in East Greenland declined drastically during the 1960s and 1970s. This has resulted in a much-reduced extent of spawning, and recently, significant spawning only takes place in a small area east of Kleine Bank (65°N) (Figure 1) (ICES 2019).

In addition to cod spawned in Greenland waters, Greenland cod abundance is supplemented by an inflow of cod eggs and larvae from Iceland via the Irminger Current, which flows past Southwest Iceland before flowing west towards East Greenland (Figure 1). The Irminger Current overlaps with the distribution of cod eggs spawned on some Icelandic spawning grounds, and a fraction of the eggs drift passively to Greenland. The intensity of these drift events varies with the abundance of eggs and strength of the currents but is equally influenced by the environmental conditions in both Iceland and Greenland (Stein and Borovkov 2004; Brickman et al. 2007). There is most likely an annual westward transport of eggs, but in some years, a considerable quantity of eggs drift across the Denmark Strait to East Greenland (Wieland and Hovgård 2002). In recent decades, such large-scale events occurred in 1973, 1984 and 2003, and the year classes were highly abundant in Greenland as far north as 64°N along the west coast.

Migration

Understanding the age-dependent migration routes of cod in and around Greenland is key to understanding the dynamics and challenges of managing cod in Greenland. When eggs and larvae from the different spawning sites settle, there is already a high degree of stock mixing. In West Greenland, cod from East Greenland and Iceland mix with the two West Greenland stocks and in East Greenland, cod from Iceland mix with locally spawned East Greenland cod. To elucidate the migration patterns, scientists started the first tagging experiments of cod in West Greenland in the mid-1920s with disc tags and continued with traditional plastic T-tags just after World War II. These experiments have been supplemented since, albeit with a break in the years when cod abundance was extremely low. Several thousand recaptures have provided general patterns of migration, and this has, in the recent decade, been supplemented by genetic studies that have put these overall patterns into a stock-specific perspective. At the bottom of the West Greenland fjords, cod are relatively stationary and remain in the area even after maturity. The overall picture from recaptures is that 98% of the cod tagged in the inshore area were recaptured inshore, with only 1% migrating to East Greenland and 1% to Iceland. A recent study using satellite tags on large cod tagged at the Kapisillit spawning site confirmed this stationary behavior over four months, where three large cod remained in the Kapisillit fjord (Nielsen et al. 2023). Cods tagged in the West Greenland coastal region are not similarly stationary. These can either migrate offshore (West, East or Iceland), stay, or move further inshore (Storr-Paulsen et al. 2004), indicating that the coastal zone is a key mixing area, although sections of the coastline have not yet been included in the tagging experiments. Recent genetic studies on the catch compositions in the coastal zone support that the coastal region is a mixing zone for the three cod stocks (Christensen et al. 2022).

The migration patterns of cod tagged in the West Greenland offshore area are a clear indication of the continued influence of the East Greenland/offshore Iceland stock on West Greenland cod abundance. Hence, any cod that migrates away from the West Greenland offshore area swims east (Figure 3). This is most likely linked to a spawning-induced migration, which is clear when comparing the age of recapture and location (Table 1). When cod are recaptured at a maximum age of four before the onset of maturity, they largely remain in West Greenland, but as the age and probability of being mature increases, so does the probability of being recaptured in either East Greenland or Iceland. This pattern is also reflected in the distribution of age groups in surveys, with a clear pattern of increasing age moving east. The gradual shift toward East Greenland starts at age 6, and at age 10 and older, the majority of cod are in the northern part of East Greenland (Figure 4). The same pattern is seen in commercial catch-at-age data, where ages below seven dominate West Greenland, while the older fish are caught in East Greenland.

The combination of surveys, commercial catches and tagging data suggests that East Greenland largely functions as a transit area for cod gradually moving from West Greenland to Iceland. This is confirmed by the East Greenland tagging data, where the probability of being recaptured in Iceland increases with age (Hedeholm 2018). Hence, since 2003, approximately 70% of cod tagged in East Greenland that were recaptured at ages four to five were recaptured in East Greenland. Still, the proportion declined to 19% at ages eight to nine, and all older ages were recaptured in Iceland waters, although the sample size is small. This underlines a key conclusion: albeit from a few examples, the eastward migration from the West Greenland nursery and feeding grounds to East Greenland and, ultimately, Iceland is a unidirectional migration.

By some mechanism, the West Greenland cod, both inshore and offshore, do not follow the East Greenland/offshore Iceland cod when these migrate to their natal spawning area. This has been shown by traditional tagging (Storr-Paulsen et al. 2004)

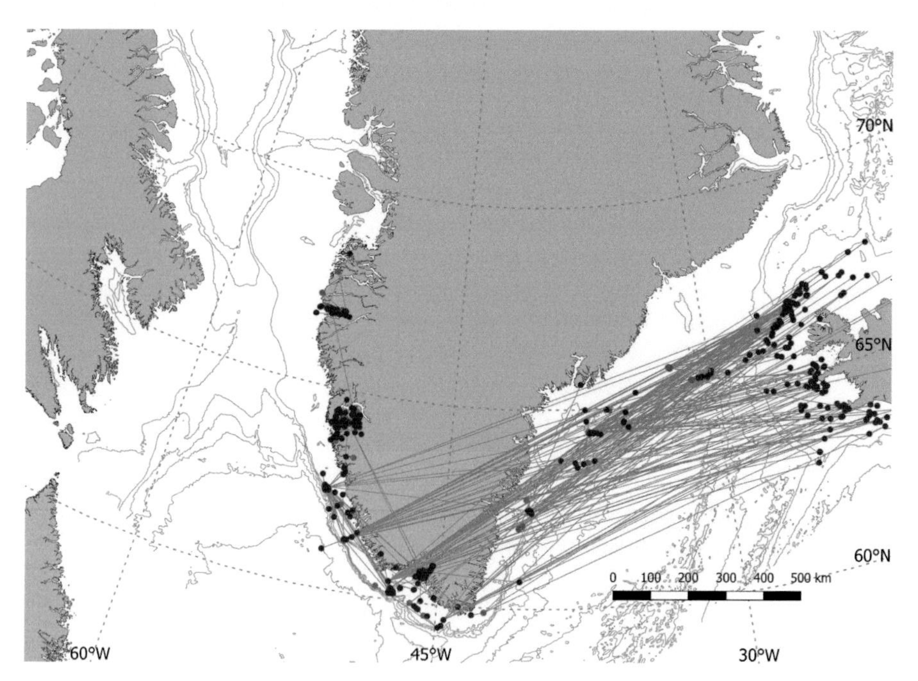

Figure 3. All mark (grey circles) - recapture (black circles) positions from 2003–2016. Lines show the shortest distance between mark and recapture events.

Table 1. Distribution of recaptures of West Greenland offshore marked cod at specific recapture ages.

Age When Recaptured	West Greenland	East Greenland	Iceland
4 (N=7)	86%	14%	0%
5 (N=12)	58%	25%	17%
6 (N=22)	50%	23%	27%
7 (N=18)	39%	17%	44%
8 (N=15)	53%	13%	33%
9 (N=6)	0%	17%	83%
10+ (N=2)	0%	0%	100%

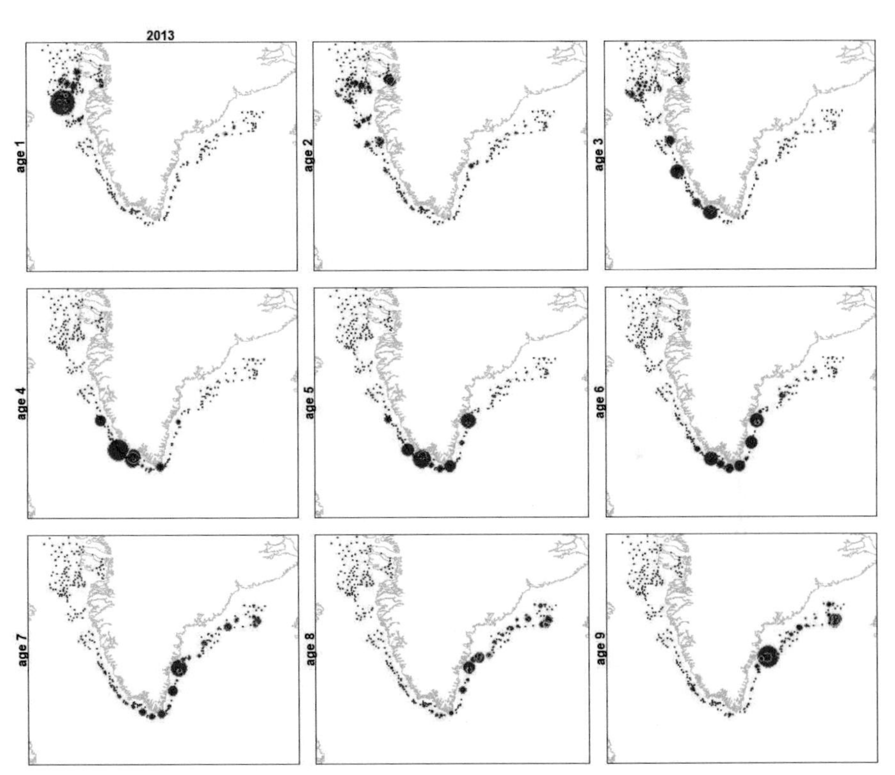

Figure 4. Abundance (%) of ages 1–9 in 2013 from the Greenland survey. Blue circle size denotes the relative distribution (%) of an age group. Each age sums to 100%. Red dots are trawl stations.

and elegantly by extracting ancient DNA from physical tags that were deployed in West Greenland but recaptured in Iceland, thereby determining both the spawning, tagging and recapture locations (Bonanomi et al. 2016). Hence, stock integrity is maintained despite extensive stock mixing, even though stock mixing appears to be present on a small scale, as cod do not seem to move in stock-specific shoals or prefer different habitats.

Inflow of Large Year-Classes

Since the 1970s, when Greenland cod abundance had declined, two year-classes (1973 and 1984) supported a fishery of over 100,000 tons in two short periods: 1977/78 and 1988/89. After 1989, cod disappeared from Greenland waters in what has been called an "overnight event," triggered by a combination of migration and an intense fishery. In Iceland waters, these two year-classes were observed in larger numbers at age six in 1979 and 1990, and these year-classes clearly originated from Iceland and migrated to Iceland "on cue" when approaching maturity at age six. Since the beginning of the 2000s, when cod abundance started to increase in Greenland, only one year-class (2003) has shown the same large-scale migration at age six as the 1973 and 1984 year-classes. The 2003 year class was smaller than the 1973 or 1984 year-classes and supported a fishery of 20,000 tons in Greenland. Although these occasional, large, and rapid migrations of specific year-classes from Greenland to Iceland receive much attention, this pattern is not sporadic but consistent across both years and year-classes. Naturally, the quantity varies with year-class strength (Table 2).

Table 2. Recapture percentages of fish tagged in East Greenland waters by year class.

Year Class	West Greenland (%)	East Greenland (%)	Iceland (%)
2001 (N=10)	0%	80%	20%
2002 (N=31)	0%	13%	87%
2003 (N=43)	0%	33%	67%
2004 (N=23)	0%	43%	57%
2005 (N=3)	0%	33%	67%
2006 (N=6)	0%	50%	50%
2007 (N=54)	4%	48%	48%
2008 (N=25)	8%	28%	64%
2009 (N=11)	0%	9%	91%
2010 (N=4)	0%	50%	50%

Biology

Growth and Maturity

There are distinct regional differences in cod life-history characteristics. Cods from East Greenland are considerably larger at a given age than cod from West Greenland. For instance, cod ages four to six from East Greenland are approximately 10 cm larger than their West Greenland inshore and offshore counterparts (Figure 5). The difference is seen in most years, but it has increased in recent years.

In addition, part of the difference can be explained by the earlier onset of maturation in West Greenland. Here, cod mature at age four to five, which is a full year earlier than in East Greenland (Figure 6). This allocation of resources toward reproduction rather than growth could drive the size-at-age differences between West and East Greenland. In the 1980s and 1990s, cod in the West Greenland offshore area

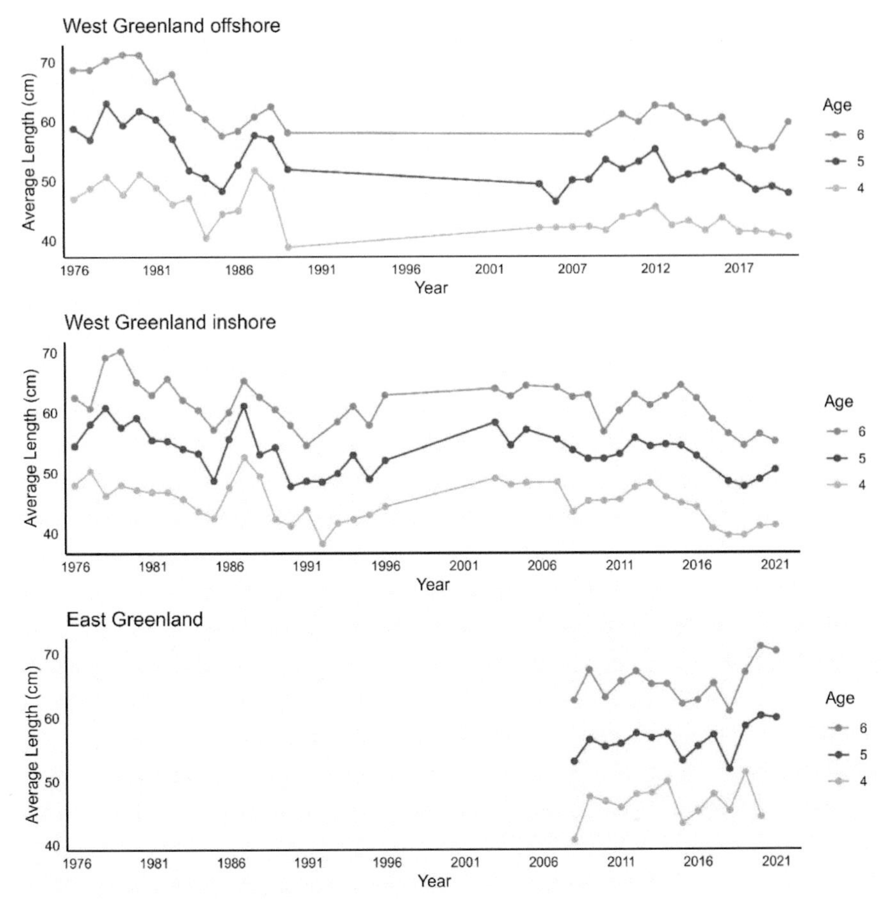

Figure 5. Mean length-at-age of the four, five and six-year-old cod from offshore West Greenland (between 62°30N–68°50N, limited data available in the period 1990–2004 and 2021), inshore West Greenland (between 62°30N-68°50N, limited data available in the period 1997–2003) and East Greenland East of 44°00 W (limited data available prior to 2008).

was larger than today and similar in size to the East Greenland cod. However, this coincided with the last large year-classes that drifted to West Greenland from East Greenland/Iceland, and when these year-classes migrated back to East Greenland/Iceland for spawning, the size at age dropped drastically in West Greenland. The temperature also dropped in West Greenland at that time (Figure 7), which could contribute to the observed reduction in size-at-age. However, recent temperatures are high in Greenland, but the size-at-age in West Greenland has not increased to the historical level. Hence, the growth differences between the stocks are apparently also driven by genetic differences, with the East Greenland cod growing the fastest. This also supports the documented difference in tolerance to fishing as the higher productivity inherent in the East Greenland stock can support a more intense fishery (see section "Historical Perspective—Changes in Abundance"). The effect temperature has on cod growth was pronounced from 1950 to 1990.

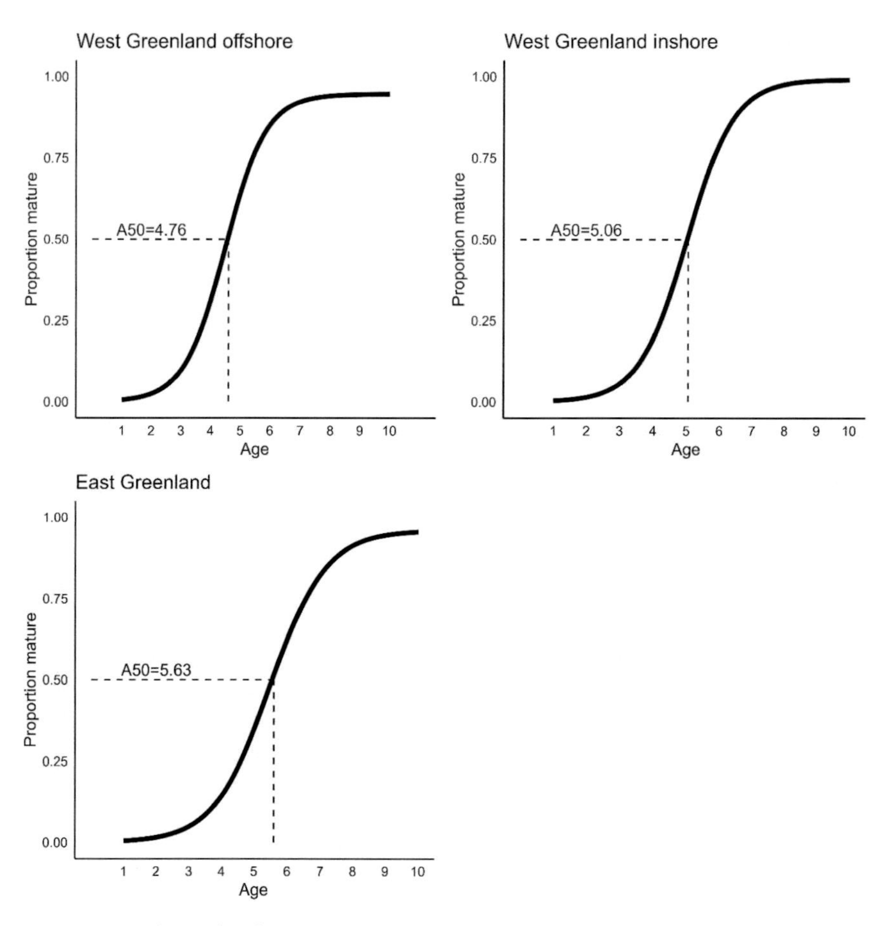

Figure 6. Maturity Ogive for the three stocks. Years 2010–2022. Based on genetically assigned individuals.

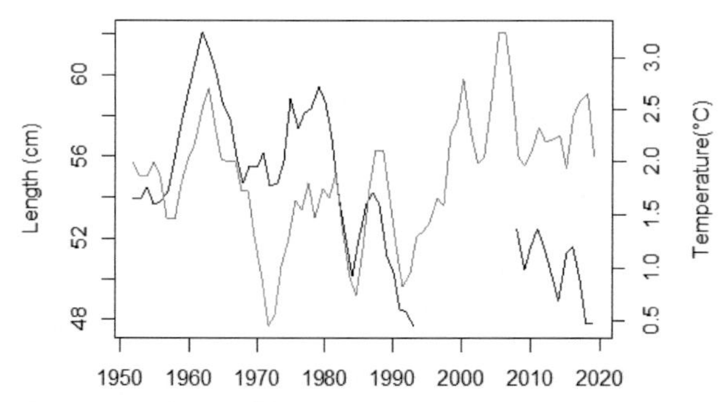

Figure 7. Size and temperature in the offshore West Greenland area are between 62°30N and 64°15N. Black line: Three-year running mean of mean length of age five in the second quarter of the year (April–June), limited data available from 1992–2006. Grey line: Three-year running mean of mid-June temperature index at the surface layer of the Fyllas Bank (64°N).

During this period, the mean size of 5-year-old cod was highly correlated with temperature fluctuations (Figure 7). However, after the stock collapse around 1990, the temperature increased and remains high today, but cod growth remains low. A combination of factors can contribute to this, including a difference in input from the faster-growing East Greenland/offshore Iceland stock, but it could also indicate a fundamental shift in the environmental conditions that are in some way less favorable for cod. There is insufficient knowledge on trophic interactions and environmental influences on cod, especially at younger life stages, to conclude on possible effects, but the lack of significant stock rebuilding coincides with this decoupling of growth and temperature, which could suggest that the cod population will not automatically rebound to historical levels even in the absence of fishing.

Condition

The better growth conditions in East Greenland and the apparently worsened West Greenland offshore conditions are also reflected in conditional indices (Fulton's Condition factor K and Hepatosomatic index) (Figure 8). Cods from the West Greenland offshore area are in poorer condition than cod from the West Greenland inshore area and East Greenland. Cods on Dohrn Bank in the northern part of East Greenland are particularly well-conditioned compared to the rest of Greenland. The simplest explanation for the difference is a seasonal westward shift in part of the Iceland cod stock, which is traditionally in a better condition than the Greenland cod (ICES 2021).

The West Greenland inshore area is currently considered a key feeding area, which is also reflected in high conditional indices. The West Greenland inshore area has a north-south gradient of cod distribution of approximately 1,500 km. Along this gradient, temperatures gradually decline and ultimately dictate the limits of cod spawning (68°N) and common cod presence (72°N). Expectedly, cod condition and growth, as a measure of habitat suitability, should gradually deteriorate along the gradient, but this is not the case. However, contrary to this expectation, there is a tendency for increased growth at the northern range of the gradient (Figure 9). This underlines that temperature is not the sole driver of habitat suitability in Greenland. This is also seen in local capelin (*Mallotus villosus*) fjord stocks that grow larger, moving north along the same gradient before reaching their northern limit around the same latitude as cod (Hedeholm et al. 2010). One complicating factor could be the gradual shift in the different stock's contribution to the overall population along the same gradient. Furthest north, cod are mainly from the West Greenland offshore stock, and this stock is probably the stock best adapted to colder conditions compared to the distribution of the other stocks (see Figure 2) and could, therefore, maintain a high growth rate in colder conditions.

The appeal of the West Greenland inshore area as feeding grounds for all stocks is most likely linked to the presence of capelin, which is a key prey during the early summer months when they approach their beach spawning sites, but also at other times of the year as capelin can form dense aggregations in the fjords (Friis-Rødel and Kanneworff 2002). Capelin is extremely energy-rich (Hedeholm et al. 2011), and feeding on this high-quality prey could be reflected in a higher condition factor of

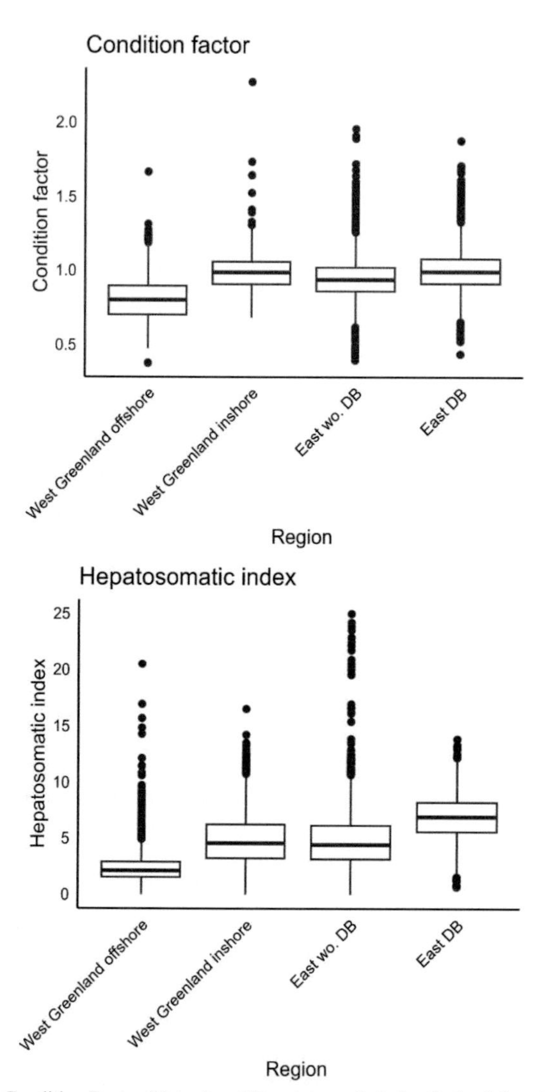

Figure 8. Fulton's Condition Factor (K, top) and Hepatosomatic index (below) from 20–100 cm cod from 1st quarter (January–March), 2010–2021. Inshore: West Greenland (between 62°30N–68°50N), Offshore: West Greenland (between 60°45N–68°50N). East without DB: East of 44°W and west of 32°W. East DB: East of 32°W, corresponding to the Dohrn Bank area.

cod in the inshore area compared to offshore areas in both East and West Greenland (Figure 8).

Trophic Interactions

Greenland bottom fish assemblages are relatively species-poor, and cod constitutes an abundant and integral part of the ecosystem, whereas other gadoids such as haddock (*Melanogrammus aeglefinus*) and saithe (*Pollachius virens*) are absent

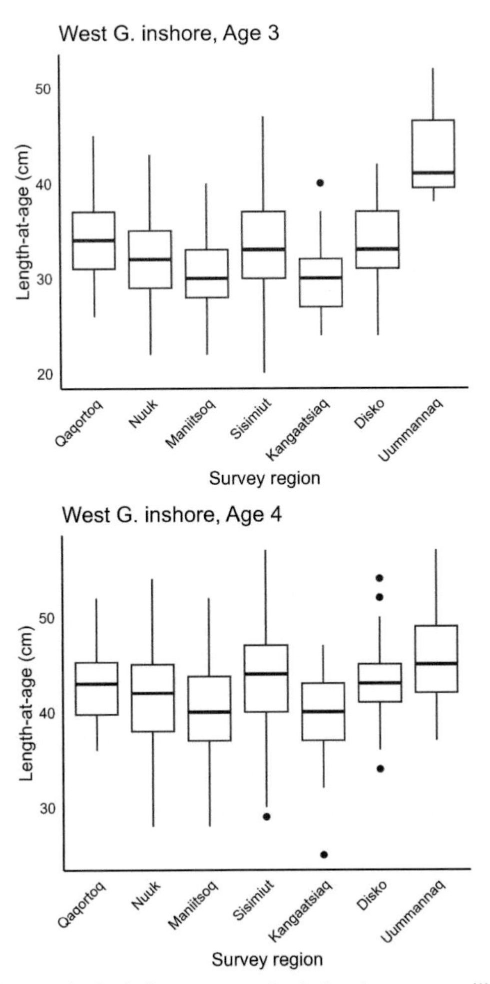

Figure 9. Mean length-at-age in the inshore area caught during the summer gillnet survey. QA: South Greenland (60ºN); NU: Nuuk area (64ºN); MA: Maniitsoq area (65º30N); SI: Sisimiut area (67ºN); KA: Kangaatsiaq area (68º30N); Disko: Disko Bay (69º30N) and Uummannaq area (71ºN).

in this region. Hence, cod plays a key role as a predator of other commercial and highly abundant species, such as northern shrimp (*Pandalus borealis*) and capelin. Cods are opportunistic feeders, and their diet in Greenland naturally includes the most abundant prey. Hence, northern shrimp is a key prey, especially for smaller cod (Nielsen and Andersen 2001; Hedeholm et al. 2017; Werner et al. 2019). Because of this, cod is included in the assessment of northern shrimp, where the estimated removed biomass is incorporated specifically as part of the natural mortality. In recent years, the removed biomass was estimated at around 5,000 t. However, with shrimp mainly distributed north of 66°N and cod mainly south of 66°N in the last decades, the overlap in distribution areas is currently limited. In earlier periods, shrimp were more abundant further south, which coincided with the period of low cod abundance. Hence, as in other North Atlantic regions, there appears to be an inverse relationship between shrimp and cod (Worm and Myers 2003). Other important cod crustacean prey are krill and euphausiids, which is the case across the entire Greenland area.

Fish feeding preferences are generally size dependent, and larger cod have an increasingly larger input from fish in the diet (Hedeholm et al. 2017). The species involved seem to be area-dependent, reflecting cod's generalist feeding; the main prey species are redfish, capelin and cod. Not all prey is equally important in terms of energy, ranging from low in brittle stars and jellyfish to high in krill and capelin. Because of this, certain prey items may be pivotal; for instance, in the inshore Kapisillit area, where cod spawn, intense post-spawning feeding on capelin accounts for more than 30% of the annual cod growth in just a matter of weeks (Grønkjær et al. 2020). Capelin is also important in offshore waters, especially in northern East Greenland, where the large capelin stock that spawns in Iceland spends part of its juvenile life stage (Pálsson and Bjórnsson 2011; Hedeholm et al. 2017; Werner et al. 2019).

The feeding of cod larval stages is less well documented in Greenland. In the key inshore spawning area, Kapisillit, the early life stages feed on different organisms, dependent on size. The earliest life stages prefer calanoid nauplii, but with increasing larval size, *Pseudocalanus*, larger calanoids and cladocerans increase in importance (Swalethorp et al. 2014). No comparable studies have been conducted in Greenland offshore waters.

Adult cod are known to be preyed upon by Greenland sharks (*Somniosus microcephalus*) (Nielsen et al. 2019), and cod cannibalism is common in all cod populations, including Greenland (Hedeholm et al. 2017). Juvenile stages of cod are naturally prey to larger fish such as Atlantic halibut (*Hippoglossus hippoglossus*), Greenland halibut (*Reinhardtius hippoglossoides*), redfish (*Sebastes norvegicus*) and seals, but there are no studies that describe juvenile or adult cod mortality in detail.

Historical Perspective

During the 19th century, there were two periods of relatively high cod abundance in West Greenland; the first was in the 1820s, and the second was in the 1840s (Hansen and Herman 1953). This was followed by a large stock decline caused by declining temperatures, and the fishery could only be sustained in local fjords in southwest Greenland. This lasted until the 1920s when increasing temperatures provided cod with more favorable conditions, and over the next decade, the abundance increased in West Greenland, driven by local spawning supplemented by East Greenland/offshore Iceland cod drifting to the area. This resulted in record-high cod abundance in the 1920s and 1930s, and the distribution expanded as far north as Uummannaq (71°N). Consequently, a large offshore fishery developed after World War II with annual catches above 400,000 tons during the 1950s and 1960s taken by foreign vessels primarily from Portugal, Germany and the Faroe Islands (Figure 10 and 11) (Horsted 2000). During this period in the 1950s, biomass was estimated as high as 4 million tons (Hovgård and Wieland 2008) (Figure 12). The abundance had already started to decline in the 1950s, probably caused by overfishing, as temperatures were still favorable for cod. In the 1970s, declining temperatures, homing of East Greenland/offshore Iceland cod and, not least, a sustained high fishing intensity combined to cause a rapid stock decline and ultimately a collapse. By 1975, the biomass estimate had declined to 110,000 t and by 1990, cod had virtually disappeared from West Greenland's offshore waters, and the area remains depleted today. This coincided

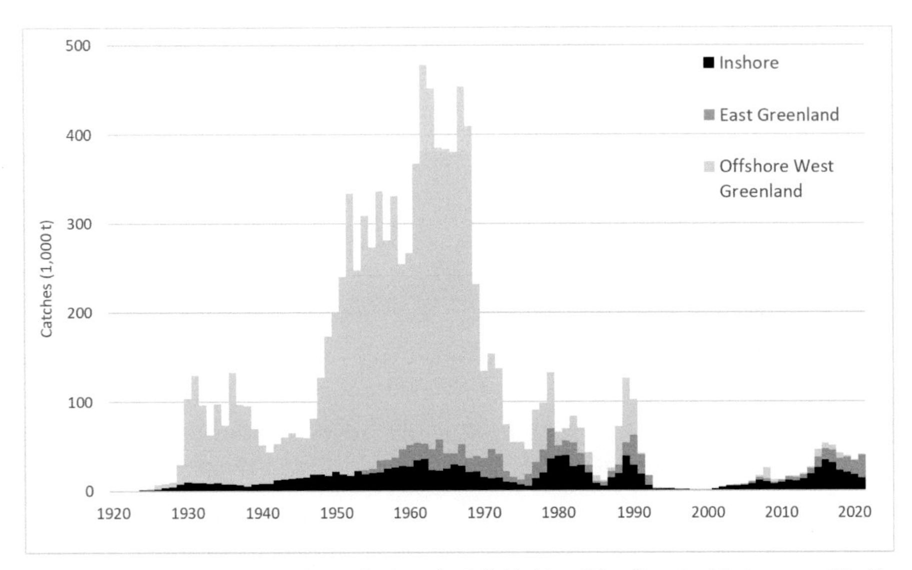

Figure 10. Cod catches (thousand tons) in Greenland divided into West Greenland inshore area (black), East Greenland (grey) and West Greenland offshore (light grey).

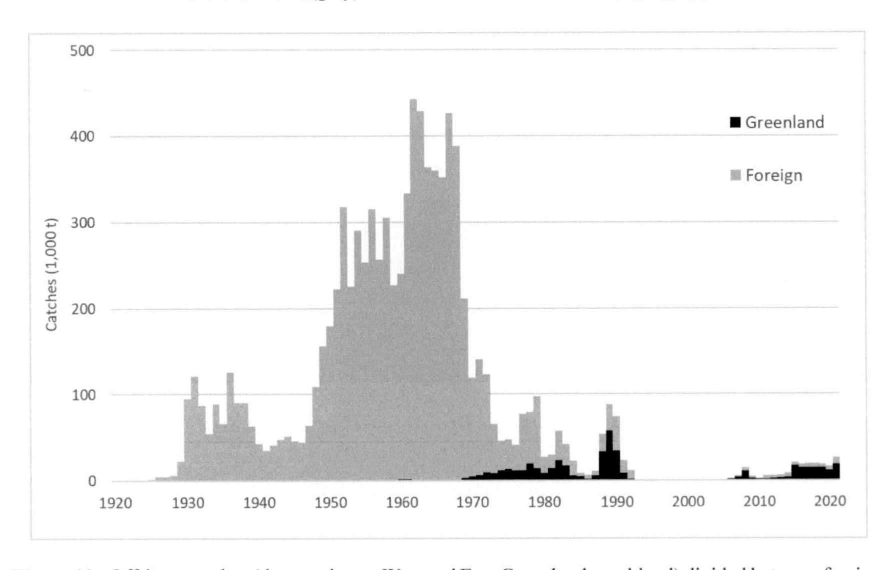

Figure 11. Offshore catches (thousand tons, West and East Greenland combined) divided between foreign (grey) and Greenland (black) fleet.

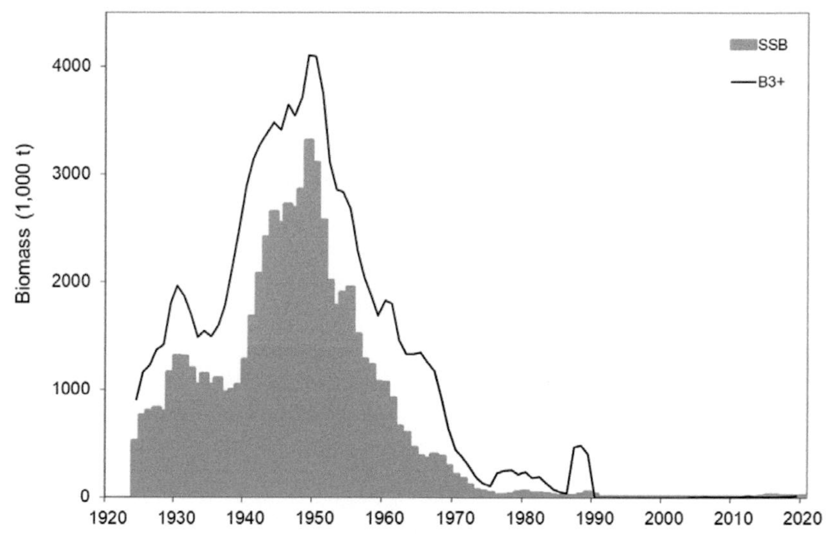

Figure 12. Estimates of stock biomass (thousand tonnes, age 3+) and spawning stock biomass (SSB) at West Greenland.

with the famed collapse of the Canadian cod fishery of Newfoundland both in terms of factors (overfishing and temperature decline) and volume, with only a few percentages of the peak biomass remaining after the collapse.

The fishery in West Greenland during this intense fishing period was a mixed fishery in terms of cod stocks, with both the West Greenland offshore stock and the East Greenland/offshore Iceland stock contributing to the catches (Bonanomi et al. 2015). Hence, the stock collapse was a combination of two collapses, with the West Greenland offshore stock supporting the early catches, and as that stock declined and collapsed, the fishery gradually shifted toward the East Greenland/offshore Iceland stock. The difference in the timing of the declines was probably caused by the two cod stocks having different tolerances for fishing. Bonanomi et al. (2015) estimated that the West Greenland offshore stock needed 5.6 kg of spawning stock biomass to produce one recruit compared to 1.1 kg of the East Greenland/offshore Iceland stock, and Hovgård and Wieland (2008) found similar values of 6.2 kg and 0.94 kg, respectively. Similarly, Hovgård and Wieland (2008) estimated an equilibrium fishing mortality of 0.14 for the West Greenland offshore stock, and Bonanomi et al. (2015) estimated it to be 0.82 for the East Greenland/offshore Iceland stock.

In the early part of the 1950–1970 fishery, the West Greenland offshore stock constituted the main part of the catch (> 60%). During this period, the fishing mortality was well above 0.14, and the West Greenland offshore stock quickly declined by the late 1960s; it constituted < 10% of the catches and the East Greenland/offshore Iceland stock dominated. This pattern persisted until the final collapse. In the 1970s and 1980, there were two periods of increased abundance after what seemed a one-way decline during the 1960s and early 1970s. These increases were caused by the 1973 and 1984 year-classes, which egg surveys and drift modelling studies strongly suggest were spawned primarily in Iceland waters (Ribergaard and Sandø 2005).

These year-classes gave rise to a large fishery that caught 100,000 t in a few years, but with the onset of maturation at age six, both year-classes returned to Iceland. Because the West Greenland offshore stock had collapsed, the East Greenland/offshore Iceland cod was the only stock in the area. The migration event in 1989–90 is probably the clearest example of the speed and consequence of these migration events on regional cod abundance, and it clearly underlines the need to consider such events by managers. Following the disappearance of the 1984 year-class, the West Greenland offshore biomass finally collapsed completely and remains so today, highlighting the sporadic occurrence of these large drift events that have not happened since in any significant quantity.

Following the collapse, the West Greenland offshore area could not sustain the fishery, and from 1992–2006, there were no catches, while a small inshore fishery below 1,000 tons annually persisted during the 1990s. In East Greenland, cod in the 1990s was mainly caught as bycatch in redfish and Greenland halibut fisheries, and annual catches were below 500 tons. In all areas, catches started to increase at the beginning of the 2000s, first in the inshore area in 2002, followed by the offshore area in East Greenland in 2006. The first major fishery took place when the 2003 year-class supported a fishery of 25,000 tons in 2008 in South Greenland, both East and West. The 2003 year-class most likely drifted to Greenland from Iceland, and it migrated back to Iceland at age five to seven, but new year-classes have since supported a steadily increasing fishery (ICES 2021). The West Greenland inshore fishery currently averages 17,000 t annually, peaking at 34,000 tons in 2016. Genetic analyses have shown this to be a mixed fishery, with approximately 50% of the catches being locally spawned cod, while the remaining cod are from the West Greenland offshore stock (30%) and the East Greenland/offshore Iceland stock (20%). While the West Greenland inshore fishery has developed positively in recent decades, the West Greenland offshore fishery continues to struggle with annual catches averaging only approximately 4,000 tons from 2015–2018 and consistently remaining below 500 tons in all other years since 2007.

In East Greenland, catch trends have developed similarly to the West Greenland inshore area, and recent catches are close to the levels observed during the 1950s and 1960s. Hence, recent annual catches are between 25,000 t and 30,000 t. These high catches have since 2018 primarily (70%) taken place on the southern continental slope of Dohrn Bank just west of Icelandic waters (66°N) (Figure 1), and catches are dominated by large (> 80 cm) cod. This northern shift in the fishery footprint is most likely caused by a combination of warmer waters, favourable feeding conditions and a large Icelandic cod stock.

Stock and Management Challenges

History of Management

From 1993, when ICES provided the first scientific advice, until 2012, the Greenland stock complex was treated as a single stock, with advice given for all of Greenland. The advice was "Zero Catch" in all years, as the large fishery prior to the collapse served as the baseline. In 2012, a new advice procedure distinguished between an inshore (West Greenland) and an offshore area (offshore West Greenland and East Greenland). The advice remained Zero Catch for the offshore area but was increased to 8,000 tons for the inshore area. In 2016, the offshore area advice was split into a West Greenland advice (north of 60°45N in West Greenland) and an East Greenland advice (South of 60°45N in West Greenland and east of 44°W in East Greenland) to reflect stock definitions (Figure 1). The advice remained Zero Catch for the West Greenland offshore area until 2023 as surveys continued to show no sign of stock recovery, and in East Greenland, the advice has fluctuated between 3,000 and 12,000 t in the period 2016–2023.In 2023, the stock assessment framework underwent a major revision (ICES 2023). Based on a genetic sampling program over several years, it is now possible to split surveys and catch data by stock rather than geographic delineations back to 2000. This resulted in a major revision of the catch data, particularly in the West Greenland inshore region, where a large part of the catch is from the West Greenland offshore and the East Greenland/Iceland offshore stocks (Figure 13). With the new assessment framework, the advice for the West Greenland offshore stock was changed from a decade-long moratorium to approximately 2,500 tons. The relatively high residency of cod in the inshore area, reflected by distinct regional stock dynamics and separate spawning sites, is now reflected in the stock advice given for a northern and a southern stock component, respectively (Figure 1). The northern stock is in a depleted state, and the advice for 2024 was a fishery of less than 1,000 tons, whereas the southern stock is in a better condition, and the advice was approximately 3,000 tons.

The Greenland authorities set individual quotas annually for each of the management areas (Inshore West Greenland, Offshore West Greenland and East Greenland). Offshore in both West and East Greenland, the quotas are allocated to individual vessels—trawlers and longliners—and inshore; the quota sets the limit for an Olympic (first come, first serve) fishery for a fleet of small vessels/boats. Inshore, the quotas have in all years since 2012 been set well above the scientific advice, and there has been a de facto free fishery, limited only by catch and processing capacity. The East Greenland fishery has been managed according to different management plans, the first implemented in 2011. The current from 2021 includes a harvest control rule, and accordingly, the annual quota is based on the scientific advice in the southern part of the management area (south of 60°45 N in West Greenland (from 2024 moved to east of 44°00 W) and west of 35°15 W in East Greenland) (Figure 1). This quota is supplemented by a quota for the northern part of the management area (Dohrn Bank) under the rationale that cod fished here are the result of periodic migrations from Iceland that are separate from the stock assessment for the East Greenland area. The new assessment procedure includes the Dohrn Banke area, and the advice for 2024 has therefore increased to more than 23,000 t. In the assessment models, ongoing work is to reflect better the apparent migration from the Icelandic part of the stock's distribution area into Greenland waters. The main spawning area around Kleine Bank (Figure 1) has been closed for fishing in general but recently only during spawning in March-May. The fishing industry will most likely close the key spawning sites in West Greenland for fishing during the spawning period starting in 2024. In 2025, a cod management plan will most likely be implemented

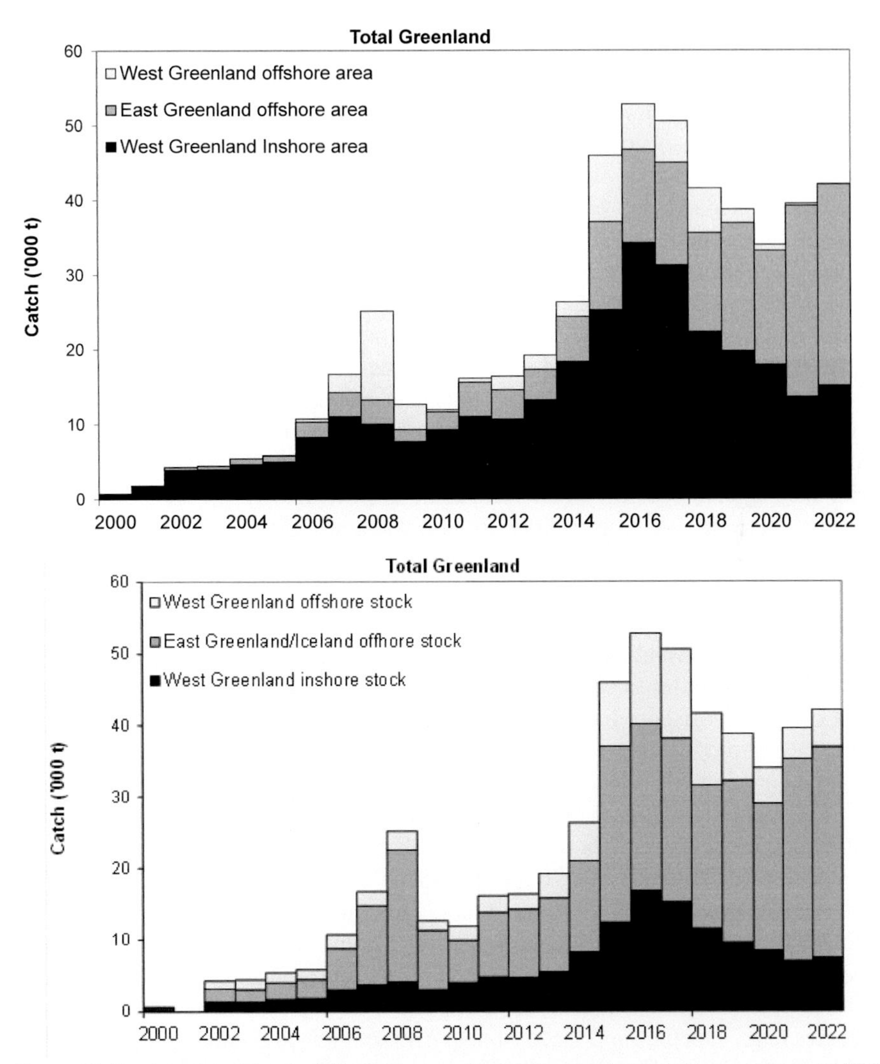

Figure 13. Total catch in Greenland divided by area (top, delineation between East and West at 44°W) and genetic assignment to stocks (below). 2001, not possible to split by stock.

for the entire Greenland area. Because the different stocks can be fished in different areas at different times by different fleet components due to the migration patterns, the management regime will be a balancing act that must consider the fishing from a more holistic approach. The cod can only be fished once, and the complexity is a national dilemma more often associated with highly migratory international stocks such as mackerel and herring. The details of the 2025 management plan are currently unknown, but it will have to be an innovative approach to management that must include both "stock management" and "area management."

Management Challenges

The complexity of the Greenland stock complex presents managers with a range of challenges when trying to optimize the fishery while maintaining long-term sustainability. The overarching issue in this Gordian knot of stock complexity is that genetically distinct stocks migrate across management delineations, mix in the West Greenland area and are morphologically indistinguishable. This causes a series of challenges that vary with area. In general, the stock mixing adds uncertainty to the input parameters for stock assessment purposes, and even though the new assessment procedure uses genetic information, there is still uncertainty regarding seasonal and spatial variability. Hence, managers are still presented with the difficult task of balancing ecological and socioeconomical considerations.

The key challenge in the West Greenland inshore area is managing a fishery with multiple stocks. The current objective for all the stocks is to fish them at the maximum sustainable yield, and the managers can explore management strategies with the regional-specific catch compositions provided by the genetic sampling scheme. Currently, fishing inshore without catching offshore cod in considerable amounts is impossible. The different stocks are in different states and require different approaches to management. For instance, the unlimited fishery inshore has led to an unsustainable catch of the West Greenland offshore stock for decades and may be a contributing factor to the depression of the West Greenland offshore stock (Christensen et al. 2022). Another challenge is the spawning migration of East Greenland/offshore Iceland cod. When these leave the West Greenland inshore area, the inshore biomass can drop drastically within 1–2 years, and managers must respond quickly to deal with such sporadic and rapid events sustainably.

The main challenge in the West Greenland offshore area is determining if the current regime will allow for rebuilding the abundance levels of the West Greenland offshore stock observed 50–70 years ago or if conditions have changed such that realistic stock levels are lower today. This, in turn, outlines the management strategy needed to reach a sustainable harvest. Recent ocean temperatures have been favorable for cod in Greenland and the North Atlantic and should be conducive for cod recruitment and stock growth (Campana et al. 2020). Furthermore, the small pelagics (e.g., capelin, sand lance (*Ammodytes dubius*), and polar cod (*Boreogadus saida*) typically preferred in cod feeding are not harvested commercially in Greenland and, together with an abundant northern shrimp stock should provide favorable feeding conditions. Nevertheless, the stock remains at a relatively low level. This leaves managers with three options: (1) minimize the known mortality of West Greenland offshore cod in the inshore fishery; (2) hope the current strategy will eventually result in a stock increase; or (3) adopt a new baseline based on a shift in the regime. The first option is the only one that can be influenced directly, but reducing the inshore catches has large socioeconomic consequences and holds no guarantee of a stock increase. Hence, a combination of the latter two options is the more likely, but also highly uncertain, approach currently pursued by managers. The new advice procedure operates with a shift in the regime, suggesting that the West Greenland offshore stock will not reach historical levels, and reference points have been set accordingly. Genetic sampling, spatially explicit management and a new management plan are prerequisites for an optimal strategy and harvest.

From a management perspective, the occasional inflow of cod from East Greenland/Iceland can result in a tempting pulse fishery in West Greenland, but again, stock mixing is a complicating factor. A fishery of these "visiting" cod will inevitably result in bycatch of locally spawned cod, and from an exploitation point of view, it would be better to harvest these fish when they have migrated back to East Greenland/Iceland. Otherwise, the collapse scenario of the 1950s–70s could be repeated. In East Greenland, the main challenge is also related to the migration events. The key aspect is that quotas must be designed to allow for the harvest of locally spawned cod as well as cod migrating to and through the area while at the same time maintaining a fishing intensity that allows for sustained spawning in East Greenland. If not, the fishery depends on sporadic drift events from Iceland and migration from West Greenland. The spawning migration from West Greenland to East Greenland at ages five to six cannot be quantified in advance. Similarly, the migration from East Greenland to Iceland is difficult to quantify, although tagging data are useful (Hedeholm 2018). Finally, there is uncertainty about the seasonal migration to, and especially from, Iceland in the Dohrn Bank area. Jointly, this presents stock assessors with a challenging dynamic that assessment models have so far been unable to handle satisfactorily. Hence, the advice is uncertain, and managers have generally been reluctant to base the East Greenland quota solely on scientific advice. The best approach to understanding these migrations is a combination of tagging and monitoring. However, tagging is a long process, better at describing overall patterns and not seasonal fluctuations, and the resources needed to track seasonal movement through surveys are substantial. Clearly, the connectivity of cod in Greenland and Iceland calls for a joint approach to management.

Environmental Challenges

The interaction between climatic variability and the distribution of fish in Greenland has been documented in several studies (e.g., Wisz et al. 2015). Ecosystem changes associated with the warm period during the 1920s and 1930s included the expansion northwards of boreal species, such as cod, haddock, and herring (*Clupea harengus*), while colder water species, such as capelin, retreated northwards. During a cold period in 1960–1970, cod abundance declined (Horsted 2000; Drinkwater 2006 and 2009) while cold-water species such as northern shrimp and Greenland halibut increased, and the shrimp fishery replaced cod as the dominant industry in Greenland (Hamilton et al. 2003). Hence, historically, cod and other species have responded swiftly to temperature changes and the recent warm period was expected to produce another cod stock increase (Wisz et al. 2015); although there have been positive trends, the increase does not match historical levels. The effects of ocean warming on cod can influence migration and egg drift events in Greenland, and both the scientific community and managers must develop frameworks to handle this variability. This is pivotal if the continued harvest of cod is to be sustainable while also supporting relatively stable conditions for the fishery.

Contaminant levels of, e.g., persistent organic pollutants (POP), in the Greenland marine ecosystem are relatively low compared to more southern latitudes. However, relatively high concentrations of some contaminants are found in the higher trophic levels due to biomagnification in the food web. These compounds are of concern because marine mammals and seabirds constitute a significant part of the human diet across Greenland. However, contaminants are not considered to constitute a challenge for the cod stocks of Greenland.

The biggest concern about oil and gas activities in Greenland is related to a potential catastrophic oil spill in the marine environment. The probability of such an incident is low as there are currently no indications of an oil industry developing in Greenland waters. Nevertheless, the risk is apparent, and the environmental impacts from a large spill can be severe and long-lasting. Tankers are potential sources for a spill, and ship traffic in the Arctic is increasing with the retreatment of sea ice because of the warming climate.

References

Bonanomi, S., Pellissier, L., Therkildsen, N.O., et al. 2015. Archived DNA reveals fisheries and climate-induced collapse of a major fishery. Scientific Reports, 5: 15395.

Bonanomi, S., Therkildsen, N.O., Retzel, A., et al. 2016. Historical DNA documents long-distance natal homing in marine fish. Molecular Ecology. doi: 10.1111/mec.13580.

Brickman, D., Marteinsdottir, G., Logemann, K., et al. 2007. Drift probabilities for Icelandic cod larvae. ICES Journal of Marine Science, 64: 49–59.

Buch, E. 2000. A monograph on the physical oceanography of the Greenland waters. Danish Meteorological Institute Scientific Report, 00-12: 405 p.

Buch, Erik. 2002. Present oceanographic conditions in Greenland Waters. DMI Scientific Report 2.

Campana, S.E., Stefánsdóttir, R.B., Jakobsdóttir, K., et al. 2020. Shifting fish distributions in warming subArctic oceans. 10: 16448. doi.org/10.1038/s41598-020-73444-y.

Christensen, H.T., Rigét, F., Retzel, A., et al. 2022. Year-round genetic monitoring of mixed stocks in an Atlantic cod (*Gadus morhua*) fishery; implications for management. Journal: ICES Journal of Marine Science, 0: 1–5. DOI: 10.1093/icesjms/fsac076.

Christiansen, J.S., Bonsdorff, E., Byrkjedal, I., et al. 2016. Novel biodiversity baselines outpace models of fish distribution in Arctic water. Sci. Nat., 103–8.

Drinkwater, K.F. 2006. The regime shift of the 1920s and 1930s in the North Atlantic. Progress in Oceanography, 68: 134–151.

Drinkwater, K.F. 2009. Comparison of the response of Atlantic cod (*Gadus morhua*) in the high-latitude regions of the North Atlantic during the warm periods of the 1920s–1960s and the 1990s–2000s. Deep Sea Research II, 56: 2087–2096.

Friis-Rødel, E, and Kanneworff, P. 2002. A review of capelin (*Mallotus villosus*) in Greenland waters. ICES Journal of Marine Science, 59: 890–896.

Grønkjær, P., Ottosen, R., Joensen, T., et al. 2020. Intra-annual variation in feeding of Atlantic cod *Gadus morhua*: the importance of ephemeral prey bursts. Journal of Fish Biology, 97: 1507–519. DOI: 10.1111/jfb.14520.

Hamilton, L.C., Brown, B., and Rasmussen, O.R. 2003. West Greenland's cod-to shrimp transition: local dimensions of climate change. Arctic, 56: 271–282.

Hansen, P.M. 1949. Studies on the biology of the cod in Greenland waters. ICES Rapp. ~roc.~Verb., 73: 1–85.

Hansen, P.M., and Hermann, F. 1953. Fisken og havet ved Grønland. Skrifter fra Danmarks Fiskeri- og Havundersøgelser, 15: 128p. (In Danish).

Hedeholm, R., Grønkjær, P., Rosing-Asvid, A., et al. 2010. Variation in size and growth of West Greenland capelin (*Mallotus villosus*) along latitudinal gradients. ICES Journal of Marine Science, 67: 1128–1137.

Hedeholm, R., Grønkjær, P., Rysgaard, S. 2011. Energy content and fecundity of capelin (*Mallotus villosus*) along a 1,500-km latitudinal gradient. Marine Biology, 158: 1319–1330. DOI 10.1007/s00227-011-1651-5.

Hedeholm, R.B., Mikkelsen, J.H., Svendsen, S.M., et al. 2017. Atlantic cod (*Gadus morhua*) diet and the interaction with northern shrimp (*Pandalus borealis*) in Greenland waters. Polar Biol., 40: 1335–1346. DOI 10.1007/s00300-016-2056-1.

Hedeholm, R. 2018. Analysis of 2003–2016 tagging data from Greenland waters as it relates to assessment of the East Greenland offshore stock and the West Greenland inshore stock. Benchmark 2018, IBPGCod (WD 03).

Horsted, S.A. 2000. A review of the cod fisheries at Greenland, 1910–1995. Journal of Northwater Atlantic Fish Science, 28: 1–112.

Hovgård, H., and Wieland, K. 2008. Fishery and environmental aspects relevant for the emergence and decline of Atlantic cod (*Gadus morhua*) in West Greenland waters. pp. 89–110. *In:* Kruse, G.H., Drinkwater, K., Ianelli, J.N., et al. (eds.). Resiliency of Gadid Stocks to Fishing and Climate Change. University of Alaska Sea Grant.

ICES 2019. Report on experimental fishery in East Greenland in April 2018. ICES North Western Working Group, 25 April – 1 May 2019, Working doc.: 08.

ICES 2021. Northwestern Working Group (NWWG). ICES Scientific Reports. 3: 52. 766pp.https://doi.org/10.17895/ices.pub.8186.

ICES 2023. Benchmark workshop on Greenland cod stocks (WKBGREENCOD). ICES Scientific Reports. 5: 42. 287 pp. https://doi.org/10.17895/ices.pub.22683151.

MarineBasis Nuuk, ttps://gcrc.gl/research-programs/marine-monitoring-programs/).

Nielsen, J.R., and Andersen, M. 2001. Feeding habits and density peatterns of Greenland cod, *Gadus ogac* (Richardson 1836), at West Greenland compared to those of the coexisting Atlantic cod, *Gadus morhua* L. J. Northw. Atl. Fish. Sci., 29: 1–22.

Nielsen, J., Christiansen, J.S., Grønkjær, P., et al. 2019. Greenland Shark (*Somniosus microcephalus*) Stomach Contents and Stable Isotope Values Reveal an Ontogenetic Dietary Shift. Frontier in Marine Science, 6: 125. doi: 10.3389/fmars.2019.00125.

Nielsen, J., Estévez-Barcia, D., Post, S., et al. 2023. Validation of pop-up satellite tags (PSATs) on Atlantic cod (*Gadus morhua*) in a Greenland fjord. Fisheries Research, 266: 106782.

Pálsson, Ó.K., and Björnsson, H. 2011. Long-term changes in trophic patterns of Iceland cod and linkages to main prey stock sizes. ICES journal of Marine Science, 68(7): 1488–1499. doi:10.1093/icesjms/fsr057.

Retzel, A., and Hedeholm, R. 2012. Greenland commercial data for Atlantic cod in Greenland waters for 2011. ICES North Western Working Group, 26 April- 3 May 2012, Working doc 22.

Retzel, A., 2020. Greenland Shrimp and Fish survey results for Atlantic cod in NAFO subareas 1A-1E (West Greenland) and results from survey on spawning cod in NAFO subarea 1C in 2019. ICES North Western Working Group (NWWG) April 23-28, 2020, WD 05.

Ribergaard, M.H., and Sandø, A.B. 2005. Modelling transport of cod eggs and larvae from Iceland to Greenland waters for the period 1948–2001. Environmental Science.

Stein, M., and Borovkov V.A. 2004. Greenland cod (*Gadus morhua*): modeling recruitment variation during the second half of the 20th century. Fish. Oceanogr., 13(2): 111–120.

Storr-Paulsen, M., Wieland, K., Hovgård, Rätz, et al. 2004. Stock structure of Atlantic cod (*Gadus morhua*) in West Greenland waters: implications of transport and migration. ICES Journal of Marine Science, 61: 972–982.

Swalethorp, R. 2013. Early life history of inshore fishes in Greenland with emphasis on Atlantic cod (*Gadus morhua*). PhD Thesis. DTU Aqua, Copenhagen.

Swalethorp, R., Kjellerup, S., Malanski, E., et al. 2014. Feeding opportunities of larval and juvenile cod (*Gadus morhua*) in a Greenlandic fjord: temporal and spatial linkages between cod and their preferred prey. Mar Biol. 161: 2831–2846. DOI 10.1007/s00227-014-2549-9.

Therkildsen, N.O., Hemmer-Hansen, J., Hedeholm, R.B., et al. 2013. Spatiotemporal SNP analysis reveals pronounced biocomplexity at the northern range margin of Atlantic cod *Gadus morhua*. Evolutionary Applications, 6: 690–705.

Werner, K., Taylor, M.H., Diekmann, R., et al. 2019. Evidence for limited adaptive responsiveness to large-scale spatial variation of habitat quality. Marine Ecology Progress Series, 629: 179–191. https://doi.org/10.3354/meps13120.

Wieland, K., and Hovgård, H. 2002. Distribution and drift of Atlantic cod (*Gadus morhua*) eggs and larvae in Greenland waters. Journal of Northwest Atlantic Fishery Science, 30: 61–76.

Wisz, M.S., Broennimann, O., Grønkjær, P., et al. 2015. Arctic warming will promote Atlantic–Pacific fish interchange. Nature Climate Change, 5: 261–265.

Worm, B., and Myers, R.A. 2003. Meta-analysis of cod-shrimp interactions reveals top-down control in oceanic food webs. Ecology, 84(1): 162–173.

Chapter 5

Icelandic Cod Stock

Ingibjörg G. Jónsdóttir, Christophe Pampoulie, Einar Hjörleifsson* and *Jón Sólmundsson*

Stock Description

Distribution

Atlantic cod is a commercially important groundfish species, which Icelanders have exploited since settlement (Jónsson 1947) and has constituted a major resource for the nation for over a century (Jónsson 1990). In Icelandic waters, cod is close to the centre of its distribution range, both in relation to geographical location and temperature. However, cold currents of Arctic origins flowing north, east and west of Iceland and great depths outside the Icelandic continental shelf limit cod distribution in this area. Cod is widely distributed around Iceland, mainly at depths < 400 m, but can migrate to 600 m or deeper during the feeding season. It is mainly found at temperatures ranging from –1 to 9°C (Sólmundsson et al. 2021), with observed minimums and maximums of –1.5 and 13.4°C (Righton et al. 2010). Seasonal changes in the distribution of adult cod are linked to migrations between spawning and feeding grounds.

Based on survey data in March and October (see Text Box), the distribution of cod shows slight changes in recent decades (Figure 1). In March, highest abundance of cod is found northwest, southeast and north of Iceland. Cod biomass south of Iceland was higher in 2006–2021 compared to 1985–2005. The increase occurred during the same time as that of the total stock biomass (MFRI 2021a) and increasing spawning stock biomass (SSB) (Bogason et al. 2021). In October, the highest abundance of cod is found from the northwest to east of Iceland, but abundance is lowest in the south and west. At this time of year, mature cod have left spawning grounds in the south and west.

Marine and Freshwater Research Institute, Fornubúðum 5, 220 Hafnarfjörður, Iceland.
* Corresponding author: ingibjorg.g.jonsdottir@hafogvatn.is

TEXT BOX

Cod is the main (or one of the main) target species of three ongoing standardised groundfish surveys carried out in Icelandic waters. In addition, cod is part of the sampling programme in more localised shrimp surveys in nearshore and offshore waters.

The Icelandic groundfish survey (spring survey) is a trawl survey initiated in 1985 and carried out annually since then (MRI 2010; Sólmundsson et al. 2020). The main goal is to monitor changes in abundance, condition, and distribution of demersal fish in Icelandic waters, thereby improving the precision of stock assessments. The spring survey is important for stock assessment and fishery advice for cod and many other groundfish species and provides information on the distribution and condition of these species, as well as on sea-water temperature. The spring survey covers the continental shelf of Iceland to depths of 500 m (Figure 1). Stratification in the survey and the allocation of stations was based on pre-estimated cod density patterns, highlighting that cod was the main target species when the study was designed. For each statistical square, fishery scientists selected random single positions for half of the stations, and the towpath was determined by experienced fishermen. The other half of the stations were selected by fishermen based on their fishing experience. The number of stations in the spring survey was originally around 590, but in the first years, some tows were dropped due to difficulties in towing certain areas. The annual number of stations have varied from 510 to 600; 438 stations have been sampled each year, and 549 stations have been sampled more than 27 times in the 37-year study period.

The Icelandic autumn groundfish survey (autumn survey) is a trawl survey conducted since 1996 (MRI 2010). The main objective is to gather fishery-independent information on the biology, distribution, and biomass of demersal fish species in Icelandic waters, with particular emphasis on Greenland halibut (*Reinhardtius hippoglossoides*), demersal beaked redfish (*Sebastes mentella*), golden redfish (*Sebastes marinus*) and cod. Data from the autumn survey are used in annual stock assessments for these species, and the survey further provides information on the distribution and condition of both shallow-water and deep-water species, as well as on sea-water temperature. The study area is the Icelandic continental shelf and slopes and is divided into a shelf area (10–400 m) with about 180 stations and a slope area (400–1,500 m) with about 200 stations (Figure 1). Initially, shelf stations in the autumn survey (mainly targeting cod, golden redfish, and haddock (*Melanogrammus aeglefinus*)) were randomly selected from the spring survey station list. For the slope area (mainly targeting Greenland halibut and demersal beaked redfish), half the stations were randomly positioned in the area, but the other half was randomly chosen from logbooks of commercial trawlers fishing for Greenland halibut and demersal beaked redfish in 1991–1995. The autumn survey was not fully carried out in 2011, but for the rest of the 24 years, a total of 241 stations have been collected annually, and 354 stations more than 17 times.

The Icelandic gillnet survey, with around 300 stations targeting spawning cod mainly at depths of 10–250 m, has been carried out in April each year since 1996, using commercial gillnetters (Bogason et al. 2018, 2021). The purpose of the gillnet survey is to monitor spawning, general conditions and the abundance of mature cod in various spawning areas around Iceland. Initially, the survey covered only the south and southwest coasts but was expanded to the north coast in 2002. At each station, 12 nets with different mesh sizes, made of either monofilament or multifilament material, are linked together in a standardised manner (24 nets at the south slope where nets are set as deep as 500 m). Half of the stations in the gillnet survey have fixed positions and were initially set out by captains with long experience of gillnet fishing on cod spawning grounds. For each fixed station (except in the deeper south area), captains select a non-fixed station at 0.5–4.0 nm distance from the fixed station annually.

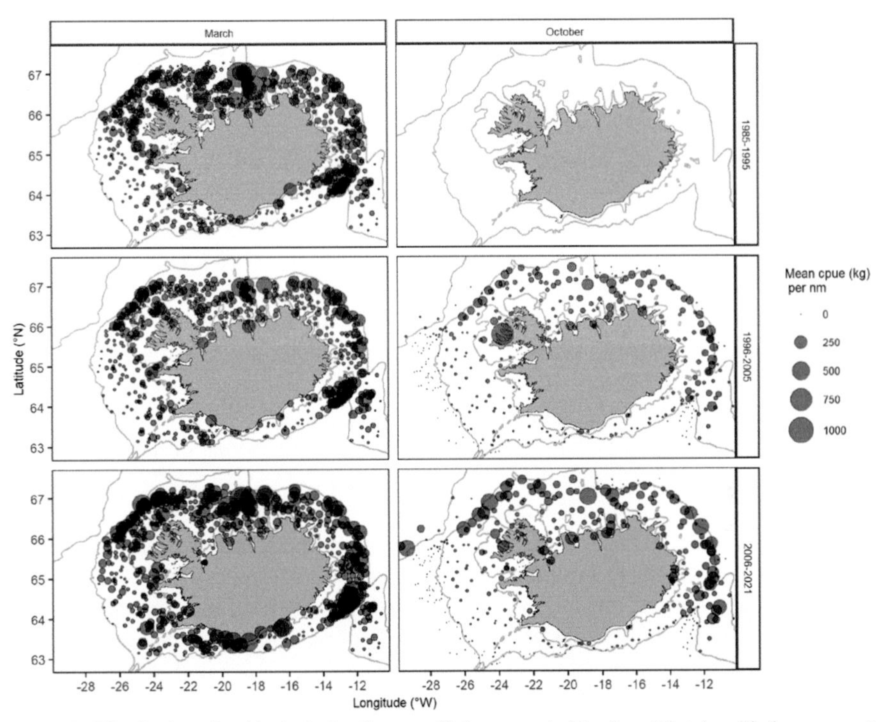

Figure 1. Distribution of cod in the Icelandic groundfish surveys in March and October. Circles represent mean catches per unit effort (CPUE) of cod at each station in 1985–1995, 1996–2005 and 2006–2021. No data is available for October before 1996.

Spawning

The main spawning area is southwest and west of Iceland, where spawning lasts usually from the middle of March until the middle of May (Sæmundsson 1926; Jónsson 1947, 1982; Marteinsdottir and Björnsson 1999). Around 1940, the density of cod at the main spawning area was low, and it was therefore assumed that cod did not need to migrate all the way to the main spawning area to find suitable spawning conditions (Friðriksson 1949). Substantial spawning has also occurred in fjords west of Iceland (Sæmundsson 1906; Bogason et al. 2021), and regional spawning aggregations are found along the coast from the north to the southeast of the country (Figure 2A). Spawning events in the north seemed to occur slightly later than in the south and southwest, from late April through May (Sæmundsson 1906).

The observed diversity in spawning components has been proposed to explain the relatively stable recruitment of three-year-old cod around Iceland compared to other cod stocks in the North Atlantic (Marteinsdottir et al. 2000b). More widespread spawning locations may provide greater variability of environmental opportunities to match hatching times and larval food production and increase the probability of larval settlement in favourable areas. The number of juveniles originating from the southwest spawning area is variable and, when low, may have a large negative effect on the overall recruitment of three-year-old cod. In 1995 and 1996, relatively few 0-group cod were estimated to have originated from the southwest spawning area

(Marteinsdottir et al. 2000b), and the recruitment index for three-year-old cod of the 1999 year class was exceptionally low (MFRI 2021a).

Migration

Icelandic cod conduct seasonal migrations between spawning and feeding areas (Figure 2). After spawning, cod migrate to feeding areas (feeding migrations) and return to the spawning areas in early spring (spawning migrations) the following year. The oldest description of spawning migrations of cod in this area dates back to 1785 (Magnússon 1785 in Jónsson (1996)). Magnússon described two main routes: westwards along the southeast and south coast and a route southward along the west coast. Spawning and feeding migrations within Icelandic waters were later demonstrated by extensive tagging experiments conducted from 1948 to 1986 (Jónsson 1996). The area along the southwest coast is at the centre of those migrations, where tagging experiments have demonstrated return migrations of repeat spawners (Jónsson 1996), suggesting spawning site fidelity.

After spawning, cod disperse from the spawning areas towards feeding areas, but the migration distance varies. During warm periods (after 1920), spawning occurred mainly along the south and southwest coast of Iceland and feeding areas were located west, north and east of the country. During this period, feeding migrations towards Greenland were common (Friðriksson 1949). Since the 1940s, cod spawning at the main spawning area in the southwest migrated towards feeding areas west and northwest of Iceland or east along the south coast, while cod spawning in the southeast migrated towards the east (Jónsson 1996; Sólmundsson et al. 2015). Tagging data collected from 1991–2008 suggested that for some spawning areas,

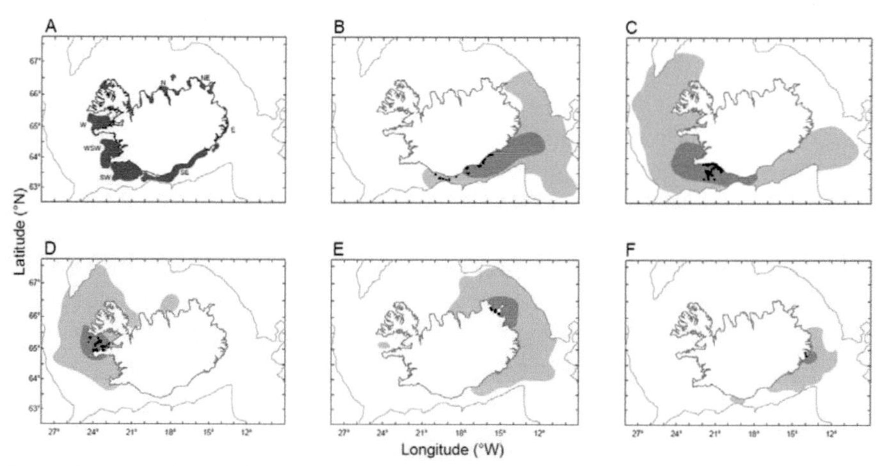

Figure 2. (A) Main spawning areas of Icelandic cod (W=West, N=North, NE=Northeast, E=East, SE=Southeast, SW=Southwest, WSW=West Southwest). (B-F) Main feeding areas based on recaptures of cod tagged within five spawning areas where light-grey areas denote an estimated 95% kernel home ranges during the feeding season (June-February) and dark-grey areas denote 50% core areas. Black dots show tagging sites within each spawning area. The 800 m depth contour is shown. Based on Sólmundsson et al. (2015).

there was limited overlap of home ranges during the feeding season (Figure 2), i.e., cod spawning in the southeast exhibited a large home range from the southwest to the northeast of the country (Figure 2B). Similarly, cod spawning in the southwest exhibited a large home range from the southeast to the northwest but did not visit the northern or northeastern areas of the country (Figure 2C). Cod spawning in the west and northeast migrated to nearby areas but were not found on the opposite side of the country (Figures 2D, E). Cod spawning in the east had the smallest home ranges and were found in the east and southeast during the feeding season (Figure 2F). All these tagging data therefore suggested that for some of the spawning areas there is a limited overlapping of home range of cod, whereas the feeding areas of cod from other areas overlap considerably.

Eggs and larvae drift towards the main nursery areas northwest, north and east of Iceland, where the juveniles settle. Juvenile cod are relatively stationary during the first years, showing high fidelity to their nursery area, but when they reach maturity, they migrate towards the spawning areas (Jónsson 1954; Saemundsson et al. 2020). The distance of the migration greatly varies. For example, juveniles from the north migrated towards the main spawning area in the southwest, while juveniles from the west made shorter spawning migrations and showed little affiliation with the main southwest spawning areas (Saemundsson et al. 2020).

Tagging studies suggest that adult cod rarely migrate out of Icelandic waters. From studies performed off Iceland in 1924–1939, a total of 19 out of just over 1,000 recaptures were taken off Greenland (Hansen 1941). In 1933, a cod tagged in Icelandic waters was recaptured off Newfoundland and two off the Faroe Islands (Tåning 1934). Furthermore, one cod tagged south of Iceland in 1931 was recaptured off north Norway (Sæmundsson 1935). From 1948–1986, 18 out of approximately 11,000 recaptures were from East Greenland and Dohrnbank, four from West Greenland, six from the Faroe Islands, three from the North Sea and seven from Norway and the Barents Sea (Jónsson 1996). In tagging experiments from 1991–2021, a few individuals have been recaptured at the Iceland-Faroe Ridge and one recaptured off the Faroe Islands (Jónsdóttir et al. 2021). Tagging in 1991–2019 also showed a number of recaptures on the Iceland-Faroe Ridge during the feeding season (Jónsdóttir et al. 2021). As very few cod tagged in Icelandic waters have been recaptured off the Faroe Islands, these cod likely migrate back to the Icelandic shelf to spawn. In 2018, spawning cod were caught off Jan Mayen and there were indications that some of them originated from Iceland (Fiskifréttir 2019; Havforskningsinstituttet 2019).

Cod migrations into Icelandic waters have been more common, mostly from Greenland. The first observation was a cod tagged in southwest Greenland in August 1924, which was recaptured southwest of Iceland in March 1927 (Sæmundsson 1933). Migrations of mature cod from Greenland to Iceland are common, but these individuals likely originate from spawning areas in southwest and west Iceland, from which they drifted with currents as eggs/larvae to Greenland. When they reach maturity, they migrate back to their natal spawning grounds (Jónsson 1947). Cod catches can increase substantially when large year classes migrate back to Iceland, as noted with year classes 1922, 1924 and later 1943, 1945 and 1950 (Jónsson 1996).

The observed lower migration rate from Greenland in 1960–2000 was likely due to different environmental conditions and/or small spawning stock size, as it is expected that larger SSBs enhance larvae drift to Greenland.

Migration from areas other than Greenlandic waters is very rare. Three individuals tagged off Jan Mayen in 1930 were recaptured in Icelandic waters (Sæmundsson 1933).

Cod Diet

Cod is an omnivorous top predator and the most abundant demersal fish in Icelandic waters. Therefore, dietary studies on cod are important in understanding trophic interactions. The diet of Icelandic cod has been studied for decades (see Pálsson 1983) and is monitored in two annual groundfish surveys covering cod distribution (see Text Box). Additionally, large-scale studies on spatio-temporal and ontogenetic patterns in cod diets and trophic linkages have been conducted as part of a multispecies diet programme (MRI 1997). Furthermore, there are several studies on trophic ecology of certain age groups or areas, on diet of commercially caught cod around the country, and on predator-prey interactions with individual prey species, mainly capelin (*Mallotus villosus*) and northern shrimp (*Pandalus borealis*) (Jaworski and Ragnarsson 2006; Björnsson et al. 2011; Pálsson and Björnsson 2011; Jónsdóttir et al. 2012; Jónsdóttir 2017).

Diet studies on larval and newly settled cod in Icelandic waters are scarce. In 1985, larvae and juveniles were sampled monthly, as cod drifted clockwise from spawning areas in the southwest towards the northwest (Thorisson 1989). In May, the smallest larvae (4–5 mm) primarily fed on nauplii and some copepod eggs, but the larger larvae fed on copepods and copepodites (mostly *Calanus finmarchicus*). Thorisson (1989) highlighted the importance of copepod eggs as the first feeding item of Icelandic cod larvae. In June, the cod juveniles were 10–20 mm, feeding mainly on the copepods *Temora* and *Acartia,* although *C. finmarchicus* was most abundant by weight. Only a few stomachs were analysed in July (size range 10–28 mm), but *Pseudocalanus* copepods were the most dominant prey. In August, cod juveniles (45–65 mm) were split between a shallow station where the juveniles mainly consumed fish (capelin) larvae and euphausiids and a deeper station where copepods were more important (41% compared to 12%). An earlier study found the main diet of 3–6 months pelagic cod (35–59 mm) to be *Acartia* and *C. finmarchicus* west of Iceland, but euphausiids were the most important prey north of Iceland, followed by *Temora* and pteropods (Pálsson 1974).

A change in diet composition of juvenile cod in Northwest Iceland, from pelagic copepods to polychaetes, amphipods and fish, occurred when 0-group cod gradually shifted towards the benthic niche (Ólafsdóttir et al. 2015), indicating a wide prey spectrum during this life stage. The diet of newly settled cod (about 3–6 cm) caught with a beach seine in the same fjord system was analysed from late August to early November, revealing a varying reliance on benthic (mainly Harpacticoida), epibenthic (mainly Cyclopoida), and pelagic (mainly Calanoida) copepods (Nickel 2016).

The diet of cod > 20 cm (1 year and older) has been well documented. Using a multivariate approach, Jaworski and Ragnarsson (2006) found that fish size was the most important variable explaining the diet of cod sampled throughout the year in 1992, followed by season, latitude, and depth. Cod < 30 cm preyed mainly on amphipods, polychaetes, euphausiids, and capelin, while 30–80 cm cod preyed mainly on capelin, northern shrimp, and other decapods (Figure 3). Finally, cod > 80 cm mainly preyed on fish (capelin, sand eel, redfish, and gadids) (Jaworski and Ragnarsson 2006). This is in line with a study on long-term (1981–2010) changes in trophic patterns of Icelandic cod, highlighting the importance of capelin as the main diet of cod > 20 cm, especially early in the year, and showing that capelin, northern shrimp and euphausiids dominate the diet and may be classified as the stable food source for cod (Pálsson and Björnsson 2011). In autumn, the feeding intensity on capelin is not as apparent as late in the winter, and the diet composition is more diverse, with a higher reliance on other fish species (e.g., sand eel, blue whiting, and herring) (Figure 4).

Cod is frequently found in inshore northern areas where shrimp fishing occurs. Stomach content data show that shrimp is an important diet for cod in these areas

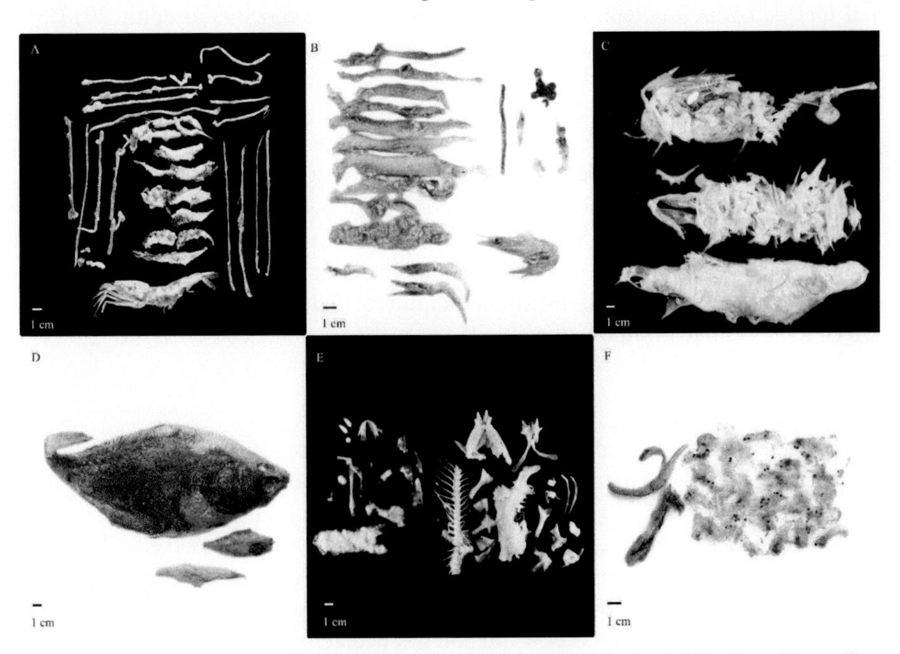

Figure 3. Various prey items from cod stomachs sampled north of Iceland. (A) About 23 capelin, a *Pasiphaea shrimp*, and a euphausiid from an 86 cm cod, (B) Nine capelin, three northern shrimp, three euphausiids, a polychaete, and an ophiuroid from a 100 cm cod, (C) Two golden redfish from a 123 cm cod, (D) Two long rough dab and one unidentified flatfish from a 107 cm cod, (E) A gadoid and an Atlantic wolffish from a 130 cm cod, (F) Two capelin and about 23 euphausiids from a 91 cm cod. Photo: Svanhildur Egilsdóttir, MFRI.

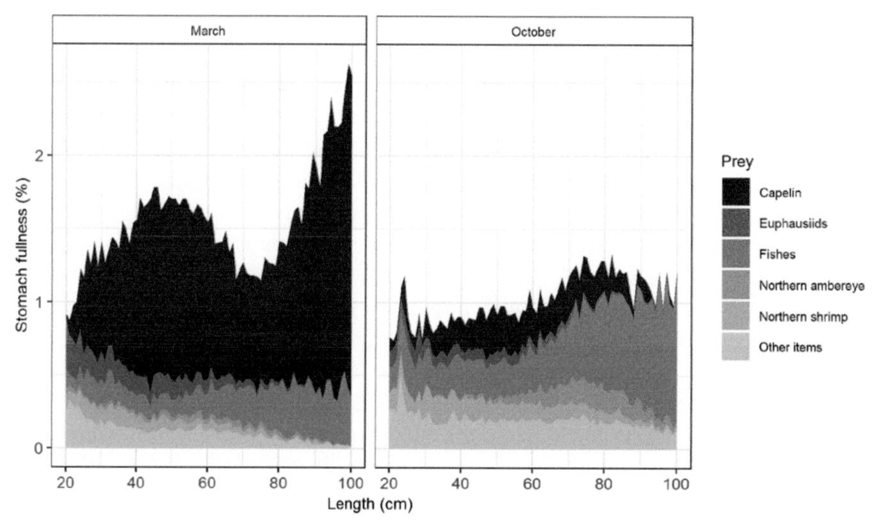

Figure 4. Stomach fullness (%) by length (stomach content weight as a percentage of predator body weight, calculated according to Lilly and Rice (1987)) of cod in March and October from 1996–2021.

and that cod was a potential regulator of shrimp abundance (Jónsdóttir 2017). In Arnarfjörður, one of the inshore shrimp areas in the northwest, the diet of cod in 2005–2006 consisted mainly of fishes, northern shrimp and euphausiids, with some seasonal and annual differences, the most striking one being a shift to almost entirely euphausiids for all length classes of cod in June 2006 (Björnsson et al. 2011). In Breiðafjörður, West Iceland, the diet of cod was monitored for three years, revealing clear seasonal changes in food composition where sand eel was the main prey late in the year while capelin was the main prey in March-April (Karlsson et al. 2005). However, capelin was not found in cod stomachs in 2004, indicating that capelin did not migrate to spawn in Breiðafjörður that year, which has apparently affected the stomach fullness of cod in the area (Karlsson et al. 2005).

Life History

Mean weight-at-age is one of the factors used to estimate the biological state of cod. In 1955–1975, the mean weight of three-year-old cod was below the long-term average of 1955–2021 (Figure 5). Mean weight increased sharply from 1972 to 1979, fluctuated without a trend between 1975 and 1995, and decreased steadily between 1995 and 2008. During this time, the stock size increased steadily and mean weight of many age groups has remained below average since 2001–2004 (MFRI 2021a). Mean weight of three-year-old cod has been correlated to capelin stock size as capelin is a main component of cod diet throughout the year and it seems that cod can only partially substitute capelin with other prey items (Pálsson and Björnsson 2011). Since 2000, the capelin stock has decreased, which can explain the decrease in the cod mean weight. In addition, northern shrimp stocks at the main nursery areas north of Iceland have decreased, and shrimp is an important prey item for juvenile cod (Pálsson and Björnsson 2011).

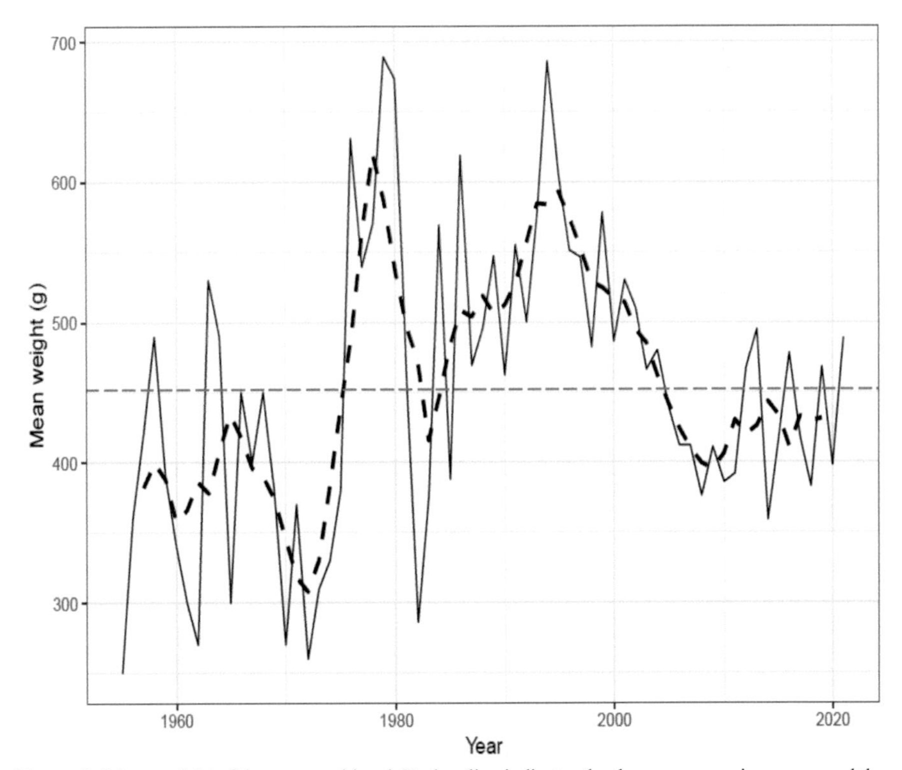

Figure 5. Mean weight of three-year-old cod. Broken line indicates the three-year running mean, and the broken horizontal line indicates long-term average.

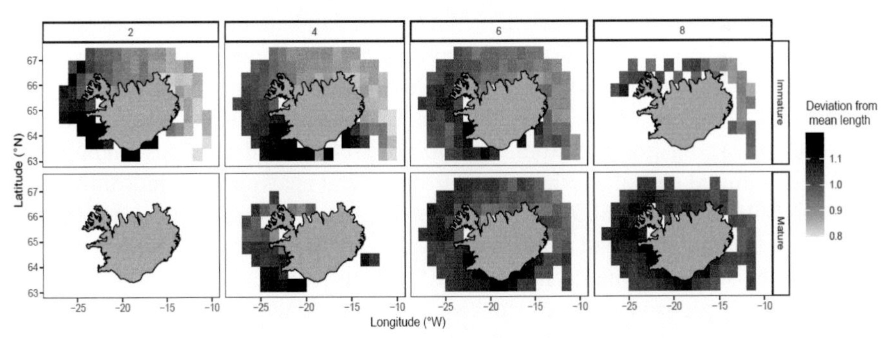

Figure 6. Mean length (as deviation from mean length 1) of immature and mature cod at four different ages (two, four, six and eight years old).

Immature cod is found widely in Icelandic waters (Figure 6). Immature cod of eight-year-olds is found from the northwest to east of Iceland, but at this age, no immature cod is found in the south and west. The mean length of immature cod is highest in the south and west of Iceland and lowest in the east. This clockwise size gradient is probably temperature-induced as temperatures on the south and west coasts are higher than on the north and east coasts. Mature and spawning cod are found all around Iceland, largely distributed within a temperature range of 2–8°C (Begg and Marteinsdottir 2002a). The cod of a four-year-old is mature in the south and west of Iceland, but at this age, it is not mature in the north and east (Figure 6). The clockwise size gradient seen for immature cod was also observed for mature cod, but the gradient is less pronounced for eight-year-old cod than the younger ages.

Age diversity is often overlooked in stock-recruitment models, and two of the largest year classes in Icelandic waters after 1950 (the 1983 and 1984 year classes) were generated during a period of high age diversity (Marteinsdottir and Thorarinsson 1998). Decreased age diversity may reduce the duration of the spawning period and the extent of areas over which spawning occurs, which may lead to reduced spatial distribution of fish eggs and larvae and a more homogeneous ambient environment within each year class. This may lead to reduced buffering against negative environmental influences and result in lower recruitment. Furthermore, total egg production is higher for older and larger females (Stige et al. 2017), producing more viable eggs and larvae (Marteinsdottir and Steinarsson 1998). In the past years, the emphasis was on increasing the spawning stock biomass and, hence, the proportion of older females, a successful strategy reflected in current high biomass estimates (MFRI 2021a). With decreasing fishing efforts, the spawning stock and age diversity have increased (MFRI 2021a).

Stock Structure

General Stock Structure

Historically, the Icelandic cod has been thought to belong to a single stock (Schopka 1994) and early research using the mitochondrial DNA method confirmed the presence of a homogeneous cod stock (Árnason and Rand 1992; Árnason et al. 1992). In the past decades, the stock structure of Icelandic cod has been investigated with several stock identification methods, from variability in life-history traits to otolith shape and genetic markers. The first potential indication of the stock structure of Icelandic cod emerged from tagging experiments conducted as early as the 1900s (Schmidt 1907; Sæmundsson 1913) and from successive tagging experiments (1948 to 1986; see Jónsson 1996). Analyses of the recapture positions of cod conducted on seven spawning grounds from 1997 to 2008 confirmed previous findings and revealed a limited spatial overlap of cod from several spawning grounds throughout the year (Sólmundsson et al. 2015). These results, spanning many years, were the first to suggest that Icelandic cod exhibited spawning site fidelity and homing behaviour, both prerequisites to limited gene flow and stock structure.

The most extensive study of stock structure in Icelandic waters was conducted during an EU project (METACOD) carried out in the early 2000s. In order to investigate the stock structure, more than 2,500 Icelandic cod were collected at 22 different spawning grounds all around Iceland in two consecutive years (2002 and 2003) (Figure 7). These cod were analysed for otolith shape, otolith chemistry and genetic markers (the pantophysin locus, *Pan* I and nine microsatellite makers). Concurrently, an analysis of conventional tagging was also performed to assess potential homing and migration patterns. Overall, all these methods indicate that the

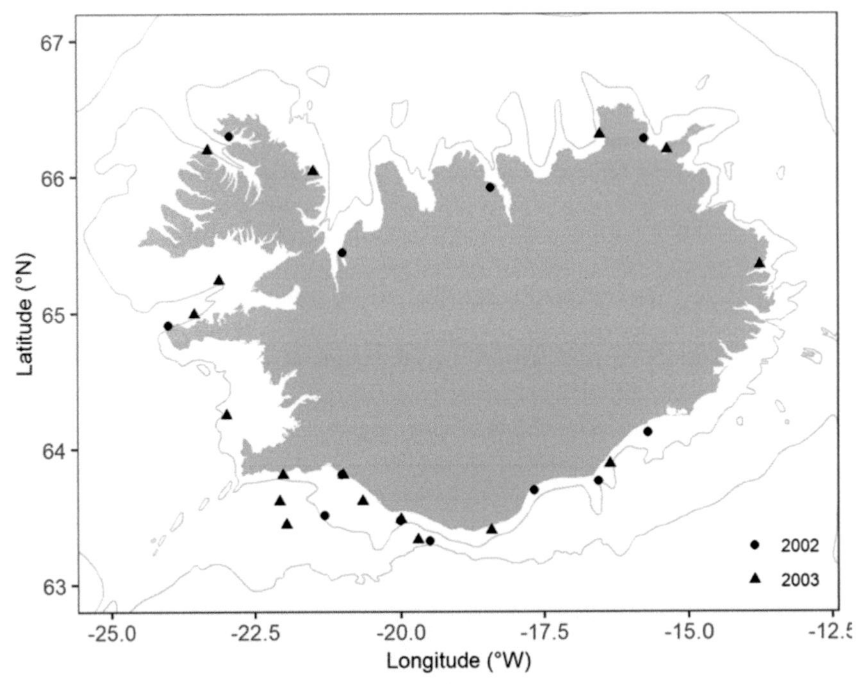

Figure 7. Sampling locations in the METACOD project in spring 2002 (circles) and 2003 (triangles). Contours display the 100 and 500 m depth limits.

spawning stock of Icelandic cod was composed of three regional groups: the north, south and deep south regions (Jónsdóttir et al. 2006a,b; Pampoulie et al. 2006).

North Versus South

A substantial difference was observed in fish length, weight and growth rate, where cod spawning in the north of Iceland was smaller, lighter and grew more slowly than cod spawning in the south (Jónsdóttir et al. 2006a). Considerable temperature differences are found between these two areas (Malmberg and Valdimarsson 2003), which could, to some extent, explain the observed growth differences since the influence of temperature on cod growth has been well established (Björnsson et al. 2001). It could also indicate that cod in the north and south of Iceland use different nursery and/or feeding areas, as environments outside the spawning areas influence the growth. Most Icelandic cod only remain in the spawning areas, after which they migrate to the main feeding areas northwest and east of Iceland (Jónsson 1996; Thorsteinsson and Marteinsdottir 1998).

The large variation in the environment experienced by cod that spawn in the areas north and south of Iceland may also have contributed to the differences in otolith chemistry and shape. Only small differences in otolith shape and chemistry were found among adjacent spawning locations within each area off Iceland (Jónsdóttir et al. 2006a, b). Mixing of cod among adjacent spawning locations may occur, but its magnitude will likely vary among areas.

Genetic studies revealed the presence of stock structure in Icelandic waters as early as the 1980s, when Jamieson and Birley (1989) confirmed the presence of a relatively high *HbI* polymorphism among cod in Icelandic waters. More interestingly, they were the first to mention a clear difference between cod from the North and the South of Iceland, owing to a shift of the HbI^1 allele frequency from 0.61 in the northeast to 0.09–0.32 in the southwest. Based on a sampling of more than 2,500 cod at spawning grounds located all around the country and using the pantophysin locus (*Pan* I) and nine microsatellite loci, Pampoulie et al. (2006) suggested that the Icelandic cod stock was geographically partitioned into two groups, the Northeastern and the Southwestern components. While the differentiation level at the microsatellite loci was low, the one observed at the *Pan* I locus was 80-fold higher, a result interpreted as evidence of local adaptation in the Icelandic cod stock.

Inshore Versus Offshore

At the main spawning area south of Iceland, inshore and offshore populations were first separated based on the number of vertebrae (Schmidt 1931). Discrimination of cod spawning from different depths was later achieved using otolith shape (Jónsdóttir et al. 2006a; Petursdottir et al. 2006) and otolith chemistry (Jónsdóttir et al. 2006b). Differences between inshore and offshore cod were also described at the *Pan* I locus at relatively small geographical scales in the main spawning ground southwest of Iceland (Jónsdóttir et al. 1999, 2001; Imsland et al. 2004). Icelandic cod populations in deep water exhibited a higher proportion of *Pan* I^{BB} genotypes than their coastal counterparts, increasing *Pan* I^B allele frequency from the coastal to the deep environment. Pampoulie et al. (2006) also described an increase of *Pan* I^B allele frequencies with depth, i.e., *Pan* I^B frequencies being constantly higher in deeper water. This pattern was also detected at microsatellite loci, with both observed and expected heterozygosities increasing with depth (Pampoulie et al. 2006). Árnason et al. (2009) also found a decrease in the *Pan* I^A frequencies in deeper water using spring spawning samples collected during three consecutive years (2005–2007) by the MFRI. Overall, these studies highlighted the limited gene flow among cod spawning in inshore and offshore habitats of the main spawning area in southwest Iceland. This limited gene flow, probably associated with selection (*Pan* I locus results), contributes to the observed genetic differentiation between inshore and offshore spawning cod in the southwest spawning areas.

Behavioural Ecotypes

Pálsson and Thorsteinsson (2003) initiated tagging experiments of Icelandic cod with Data Storage Tags (DSTs), small devices which are implanted in the abdominal cavity of cod and can register temperature and pressure (later converted to depth) at regular time intervals. Using DSTs during spawning time (April 1996–1999) showed that cod tagged at the main spawning area in the southwest exhibited different migration patterns. The depth and ambient temperature profiles suggested that cod exhibited two alternatives in feeding strategies characterised by deep- and shallow-

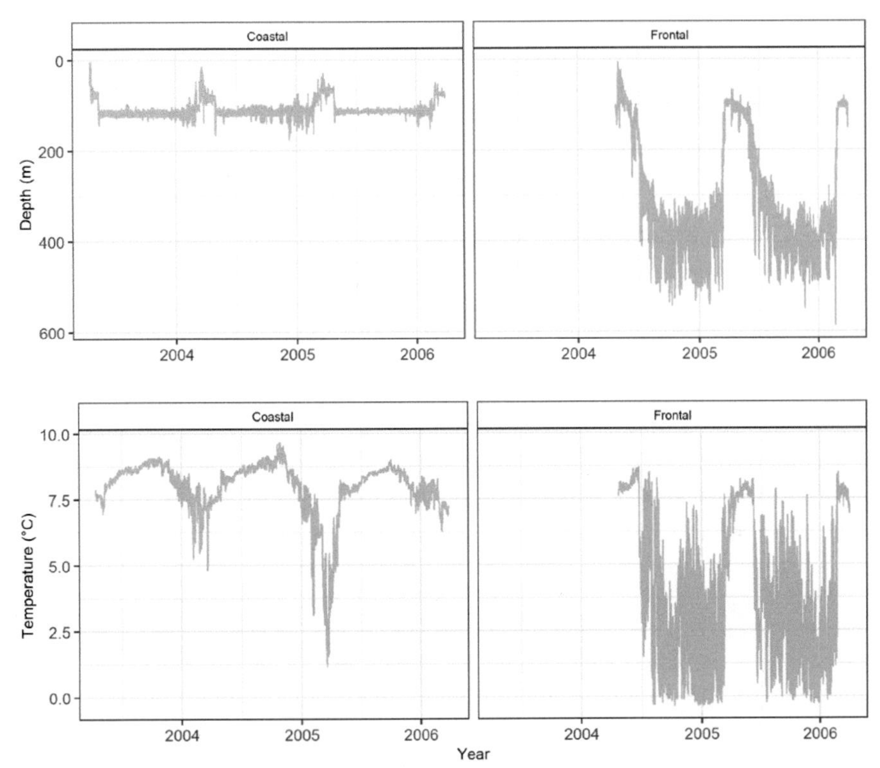

Figure 8. Depth and temperature profiles from data storage tags (DSTs) for coastal and frontal cod.

water migrations (later referred to as frontal and coastal ecotypes) (Figure 8). Coastal cod followed the seasonal trend in temperature characteristic for the shelf waters while the frontal cod tended to migrate to colder and deeper water during the feeding season. Several years later, the availability of more DST data revealed that the migration pattern of both coastal and frontal cod was consistent from year-to-year, indicating that the migration timing was synchronous and that both ecotypes tend to migrate in groups or shoals (Thorsteinsson et al. 2012).

As more DSTs were deployed, geneticists investigated whether the two behavioural ecotypes exhibited genetic/genomic differences. Pampoulie et al. (2008) investigated variation at the *Pan* I locus of coastal and frontal cod in Icelandic waters. Using data from 69 recaptured individuals, they found that 97% of *Pan* I^{AA} genotypes exhibited a typical coastal migration behaviour, while 88% of *Pan* I^{BB} genotypes exhibited a frontal behaviour. However, 50% of the heterozygotes exhibited coastal behaviour, while the other 50% exhibited a frontal one, suggesting that the *Pan* I locus was not a perfect candidate for assessing the behaviour of cod in Icelandic waters. Another gene polymorphism was investigated on more DSTs tagged cod (n = 148) a few years later, the *RH1* opsin gene polymorphism (Pampoulie et al. 2015). This approach was based on the fact that amino acid replacements at tuning sites of the *RH1* gene had been associated with altered spectral sensitivity, which is suspected to provide a mechanism to adapt to ambient light conditions and adaptation to depth. The coastal and frontal ecotypes diverged at the *RH1* gene, exhibiting a synonymous single nucleotide polymorphism in the protein-coding region and the 3' untranslated region (3'-UTR). Both the *Pan* I locus and the *RH1* opsin gene were later shown to be part of a chromosomal inversion in the linkage group 1 (LG1), which contains hundreds of genes maintained by selection processes and was suggested to facilitate coevolution of genes underlying complex traits of the behavioural ecotypes (Berg et al. 2016, 2017). A total of four chromosomal inversions were also identified within samples collected across the North Atlantic Ocean (Berg et al. 2017). Population sequencing efforts, therefore, identified genome-wide patterns of divergence, shedding light on local adaptation processes. Interestingly, the whole-genome sequence approach also revealed that, while the Icelandic cod stock exhibits coastal and frontal behavioural ecotypes similar to the coastal Norwegian and the North East Arctic Cod, respectively, the divergence between behavioural ecotypes in Icelandic waters is of much more recent origin than the one in Norway (Matschiner et al. 2022; Berg et al. 2017), making it more difficult to assess the genomic structure of the Icelandic cod population.

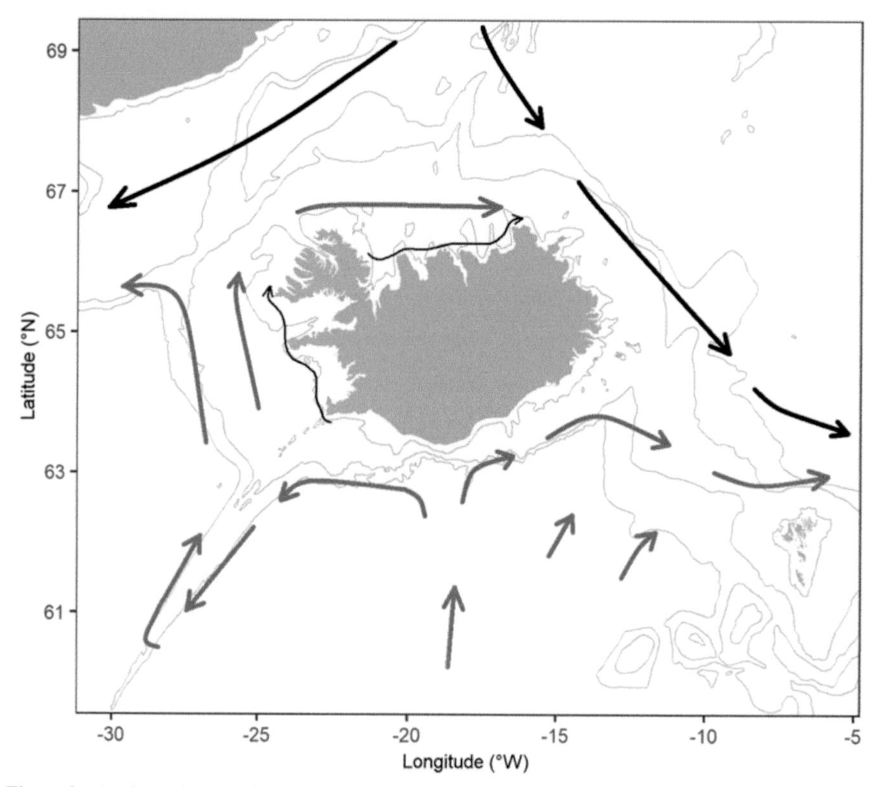

Figure 9. A schematic map of the main near-surface circulation around Iceland. Grey arrows indicate the flow of warm Atlantic water, thick black arrows flow of colder water of polar origin and thin black arrows along the coast of the coastal current. Based on Ólafsdóttir et al. (2020).

Environment

Oceanic Currents

The biology of Icelandic cod is influenced by the highly variable environment around Iceland, located at the boundary between warm Atlantic waters and waters of polar origin (Figure 9); south of Iceland, the Irminger Current branches off the North Atlantic Current (NAC), carrying warm and saline Atlantic water along the southwest coast of Iceland, continuing as North Icelandic Irminger Current (NIIC) into the Denmark Strait. While a large part recirculates south, some Atlantic water continues onto the shelf north of Iceland. The strength of the inflow onto the northern areas varies greatly between years and seasons (Stefánsson 1962; Jónsson and Valdimarsson 2005, 2012). On the Greenland side of Denmark Strait, the East Greenland Current (EGC) conveys cold and fresh waters of polar origin southward (Våge et al. 2013). Finally, the cold East Icelandic Current (EIC), also originating in the north, flows along the country's northeast coast (Macrander et al. 2014). Sharp fronts are created between the NIIC and EGC in Denmark Strait, and the NAC and EIC in the east. Close to Iceland, salinity is reduced due to freshwater runoff; this coastal water is generally flowing clockwise along the coastline.

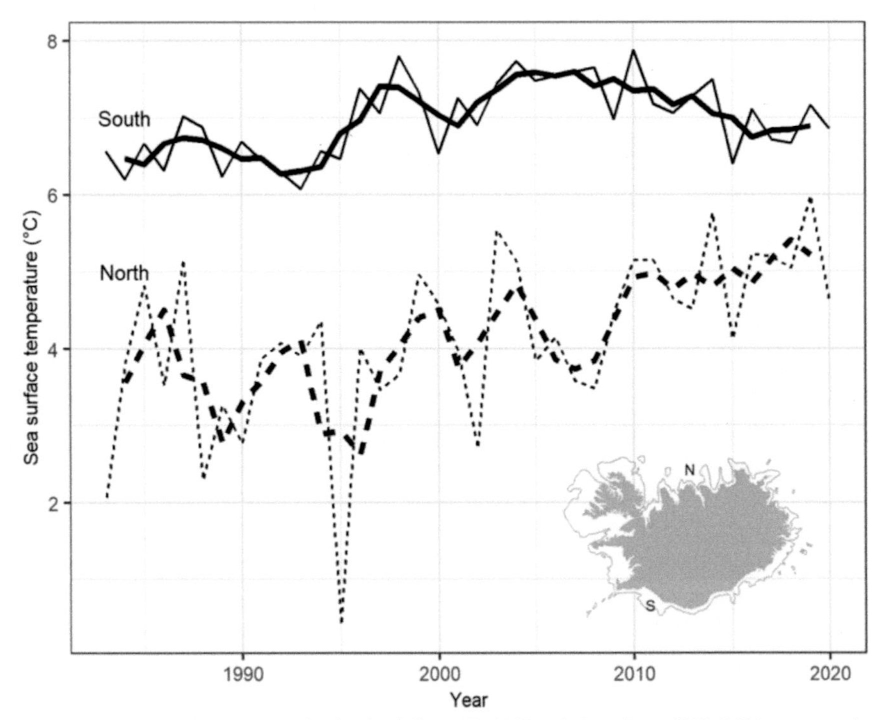

Figure 10. Mean sea temperature for the depth layer "0–100" m in May/June 1983–2020 at one station south (S) of Iceland and one station north (N) of Iceland. Bold lines indicate three-year running means (Source: sjora.hafro.is).

Temperature

Mean seawater temperature at 0–100 m south of Iceland in early summer (May/June) fluctuated between 6.1–7.9°C from 1983–2020, with about 1°C warmer waters between 1996–2014 compared with 1984–1995. However, mean seawater temperature in the same period and season ranged between 0.4 and 6.0°C north of Iceland, increasing since 1996 (Figure 10). In recent years, the lower difference in mean seawater temperature between the southern and northern areas is due to increased Atlantic influence, leading to increasing temperature in the north (Figure 10). Temperature differences between these two areas affect cod characteristics in many ways, such as different timing of spawning, growth and size at maturity (Jónsdóttir et al. 2008; Pardoe et al. 2009).

Stock Challenges

Climate Change

A recent study has been performed to understand how climate change will affect the distribution of commercial species such as Atlantic cod across its distribution range (Morato et al. 2020). According to Morato et al. (2020), habitat suitability for Atlantic cod will remain quite stable in the future in Icelandic waters, while it will decrease in the southern range of the species (North Sea and USA) and increase in the northern range (Greenlandic waters). However, the assessment and prediction of climate change's impact on the distribution of cod in Icelandic waters will remain a challenge and might have a societal impact, which is necessary to assess for the future. The monitoring data of Atlantic cod in Icelandic waters has so far not suggested any big changes in habitat suitability and cod distribution (Sólmundsson et al. 2021). With increasing sea temperature, it is nevertheless likely that cod distribution will shift towards more northern distribution in the future (Mason et al. 2021).

Larval Drift

Ocean physics plays an important role in the distribution of juveniles through the retention and/or dispersal of eggs and larvae to favourable nursery grounds. The strength of the coastal current is a dominant factor affecting pelagic dispersal and, hence, cod recruitment (Begg and Marteinsdottir 2002b). The main spawning grounds are located in the warmer areas in the south and southwest, and larvae from these areas experience higher temperatures compared to those spawned in the north. This favours fast growth during larval stages. A clockwise size gradient, with smaller juveniles in the northern part compared to the western part, is also age-related and due to cod spawning later in the northern areas (Marteinsdottir et al. 2000b). The freshwater runoff facilitates stratification by forming density layers and provides essential nutrients for the high primary production that characterises this area in early spring (Thordardottir 1986). Modelling results of Brickman et al. (2007) suggest that Icelandic cod larvae from the main spawning grounds drift with coastal currents towards the main nursery areas north of Iceland, but juveniles were shown to settle in fjords and the continental shelf northwest, north and northeast of Iceland (Astthorsson et al. 1994; Marteinsdottir et al. 2000b; Brickman et al. 2007;

Jonasson et al. 2009). However, cod larvae from the northern spawning locations are likely to remain within northern areas (Begg and Marteinsdottir 2000; Marteinsdottir et al. 2000a) or drift further out onto the continental shelf (Brickman et al. 2007). Therefore, larvae drifting from the south are exposed to warmer waters than those originating from the north.

Spawning success in the southerly spawning areas partly depends on how effectively the flow of the Irminger Current transports eggs and larvae to more northerly nursery areas (Begg and Marteinsdottir 2000). The drift varies over the years, and in 1997, an increased influx of Atlantic water to the northern shelf enhanced the drift of larvae and growth, leading to an increased abundance of juveniles (Jónsson and Valdimarsson 2005). In any given year, more than half of the juveniles found in the northern region were likely to have originated from the main spawning area (Marteinsdottir et al. 2000a; Begg and Marteinsdottir 2002a; Pampoulie et al. 2012). In years of high 0-group abundance (1973, 1976, 1984 and 1985), large aggregations of pelagic larvae were found in East Greenland waters (Begg and Marteinsdottir 2000), most likely originating from Icelandic spawning grounds.

Juvenile Habitat

After the pelagic phase, juveniles settle into nursery areas within shallower waters, i.e., the coastal shelf and fjords, in late summer and autumn. In the first year, juveniles are exposed to high predation pressure, and growth and survival are highly dependent on food availability and suitable habitat (Fjøsne and Gjøsæter 1996). Preferred nursery areas are structured habitats where juveniles can seek shelter and where food is also highly available (Gotceitas et al. 1995; Fraser et al. 1996).

Due to proximity to land, these are the marine habitats that are most likely to be influenced by human activity, and various coastal developments and utilisation may cause coastal habitat loss. Increased aquaculture in Icelandic fjords may be one of the greatest threats for juvenile cod as considerable expansion is planned in fjords along both the western and eastern coastlines. Influx of organic waste, treatment with pesticides and usage of anaesthetics and disinfectants can have harmful effects on the marine community. Pesticides have negative effect on invertebrates like zooplankton (Willis and Ling 2003; Van Geest et al. 2014; Macken et al. 2015), which becomes an important prey for juvenile cod when they change from pelagic to benthic habitats (Nickel 2016). Exploitation of maerl and macro-algae, mainly at the west and north-west coast of Iceland, may reduce the availability of habitat, with negative effect on growth and survival (Fjøsne and Gjøsæter 1996). Furthermore, in the past 60 years construction across fjords have been performed at 13 locations (Gunnarsson and Björnsson 2019). These constructions can limit flow into the fjords, change sediment and limit access of juveniles to these habitats. It may be difficult to estimate beforehand the magnitude of coastal development on the ecosystem and the impact on cod juvenile habitats. Therefore, two important challenges remain of concern for cod juvenile's habitat, e.g., to identify all juvenile habitat through mapping and distribution modelling and examine the impacts of habitat changes. Therefore, it is of utmost importance that areas are well studied before they will be changed or utilised to ensure that the impact on habitats and organisms will be as little as possible.

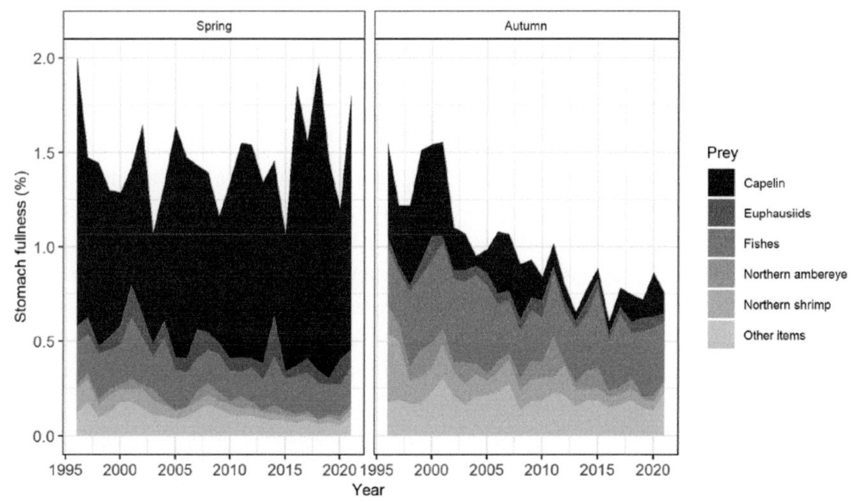

Figure 11. Stomach fullness (%) (stomach content weight as a percentage of predator body weight, calculated .according to Lilly and Rice (1987)) of cod (20–100 cm) in March and October from 1996–2021.

Trophic Interaction

The most important long-term trends in the diet composition of cod involve those of capelin and northern shrimp (Pálsson and Björnsson 2011). Changes in the abundance and distribution of these two species greatly impact food availability for cod. Concurrent with increasing sea temperature, capelin distribution changed around 2000. Juvenile capelin, common north of Iceland, moved towards the west and north and is now mainly found on the Greenlandic shelf (Bárðarson et al. 2021). After the mid-1990s, the consumption of capelin declined relative to the preceding decade (Figure 11). There was also a continuous decline for northern shrimp over the same period (Figure 11). The biomass of northern shrimp decreased drastically from 1996 to 2004, and biomass levels have remained low since then (MFRI 2021b). A link was found between cod consumption and the capelin and shrimp stock sizes. The relationship between capelin stock size and cod consumption of capelin was strong (but deteriorating) for medium-sized (30–89 cm) cod in March, but not in autumn suggesting lower predator-prey overlap at that time of year. Cod were not fully able to compensate for reduced availability of capelin by increasing predation on other species, and this may have affected growth rate of cod in the early 2000s (Pálsson and Björnsson 2011).

Fisheries Induced Selection

In Icelandic waters, two main studies based on the *Pan* I locus have been performed to assess the potential effect of fisheries on the genetic diversity of Atlantic cod. Covering space and time over a relatively short period, Árnason et al. (2009) detected an intense fishery-induced selection against the *Pan* I^{AA} genotype, i.e., the shallow-water fish. The second study assessed genetic variability both at the *Pan* I locus and microsatellite loci over a period of more than 50 years using archived otoliths

(Jakobsdóttir et al. 2011). This temporal study did not reveal any temporal changes at the microsatellite loci, but the *Pan* I locus clearly showed a constant decrease of the *Pan* I^{BB} genotypes from 1940 to 2002. The contradictory results produced by these two studies could be challenged further if catches in Icelandic waters were genetically analysed to assess the proportion of the different *Pan* I genotype captured per gears and geographical areas. So far, no genetic catch data have been produced in Icelandic waters.

Fisheries, Assessment and Management

Fisheries

Landings

The Icelandic waters ecoregion covers the shelf and surrounding waters inside the Icelandic Exclusive Economic Zone (EEZ, Figure 12). The zone was established after the extension to 200 nautical miles in 1975. Systematic annual landing statistics of cod by nations fishing in Icelandic waters are available since 1905 (Figure 13), but fisheries had been taking place by Icelanders for centuries prior to that (Jónsson 1947). Reported landings were on the order of 100 to 150 thousand tonnes prior to World War I (WWI). Following a decline in landing in WWI, when only limited foreign fleet fisheries took place, there was an explosion in the cod fisheries, reaching

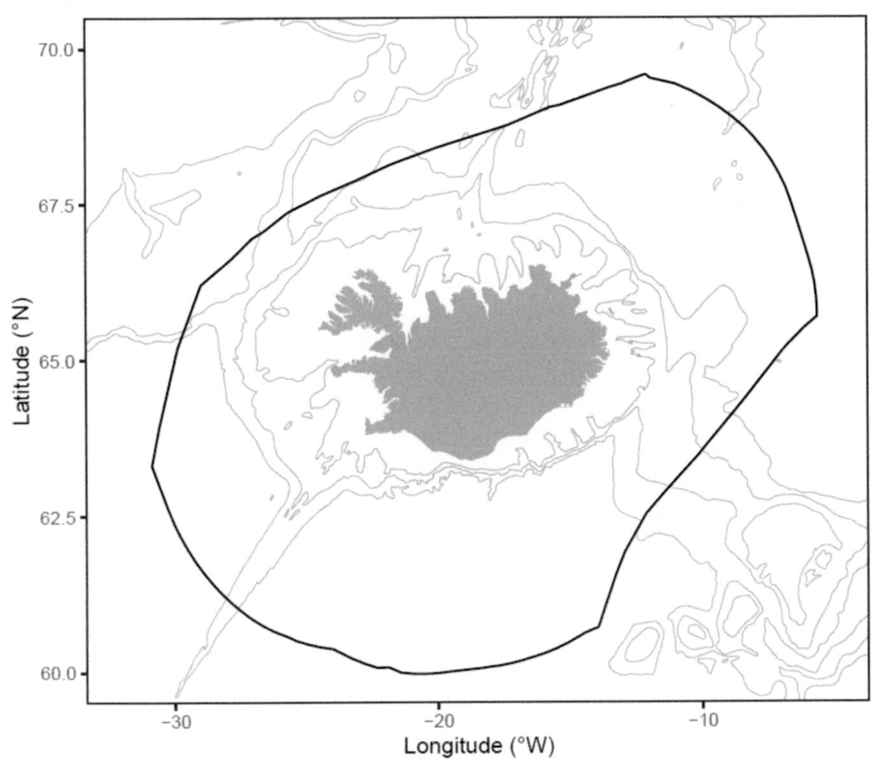

Figure 12. The Icelandic waters ecoregion.

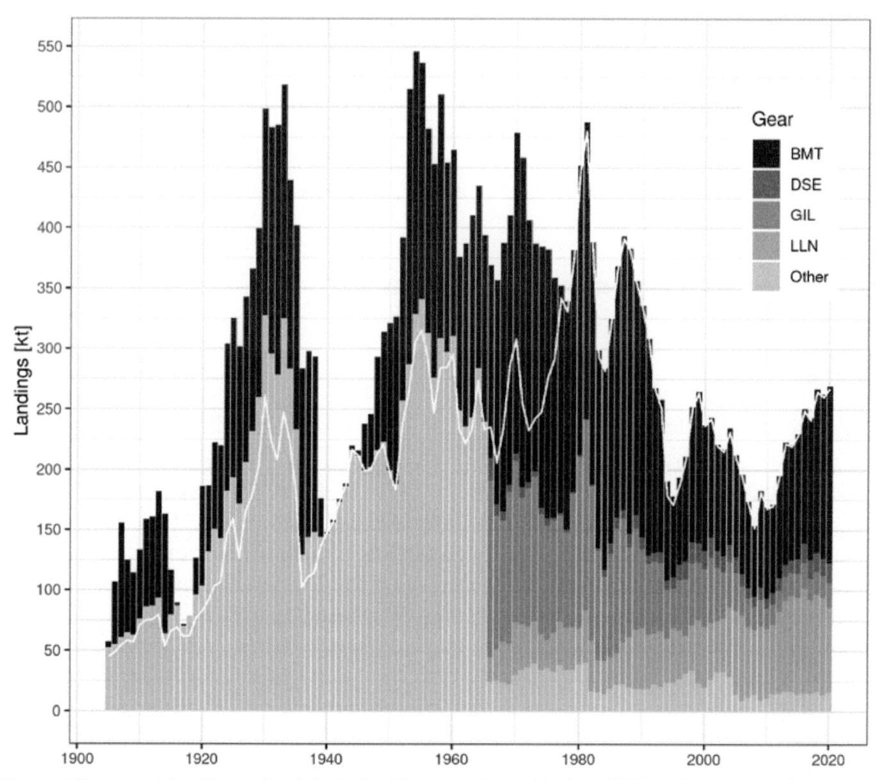

Figure 13. Annual landings of cod in Icelandic waters from 1905 to 2020, split by gear where data is available. BMT=Bottom trawl, DSE=Demersal seine, GIL=Gillnets, LLN=Longline. The white line shows the total landings of the Icelandic fleet. (Source: ICES and national database).

a peak of around 550 thousand tonnes from 1930 to 1935. This was followed by a rapid decline down to 300 thousand tonnes prior to World War II (WWII). Similar to WWI, cod fisheries were largely limited to the Icelandic fleet in WWII. After the war, the fisheries quickly resumed their former quantities, with catch levels returning to a peak of around 500 thousand tonnes in the 1950s, followed by a gradual decline, reaching a record low landing of just under 155 thousand tonnes in 2008. The fisheries have recovered over the last ten years, reaching over 250 thousand tonnes in 2020.

Four nations, Iceland, the United Kingdom, Germany, and the Faroe Islands, account for around 99% of the total landings since the start of the 20th century. Iceland and the United Kingdom accounted for around 3/4 or more of the annual landings. The foreign fleet accounted for around 50% of the annual landings prior to WWII. After WWII, the influence of the foreign fleet in the cod fisheries waned gradually from 40% of the total catch in the 1950s to less than 2–3% of the annual catch after the extension of the EEZ in 1975.

Spatial and Seasonal-Catch Distribution

Accurate quantitative data of annual catch by fleet and/or gear were not compiled before 1966 (Figure 13). The fisheries of the United Kingdom (with some exception

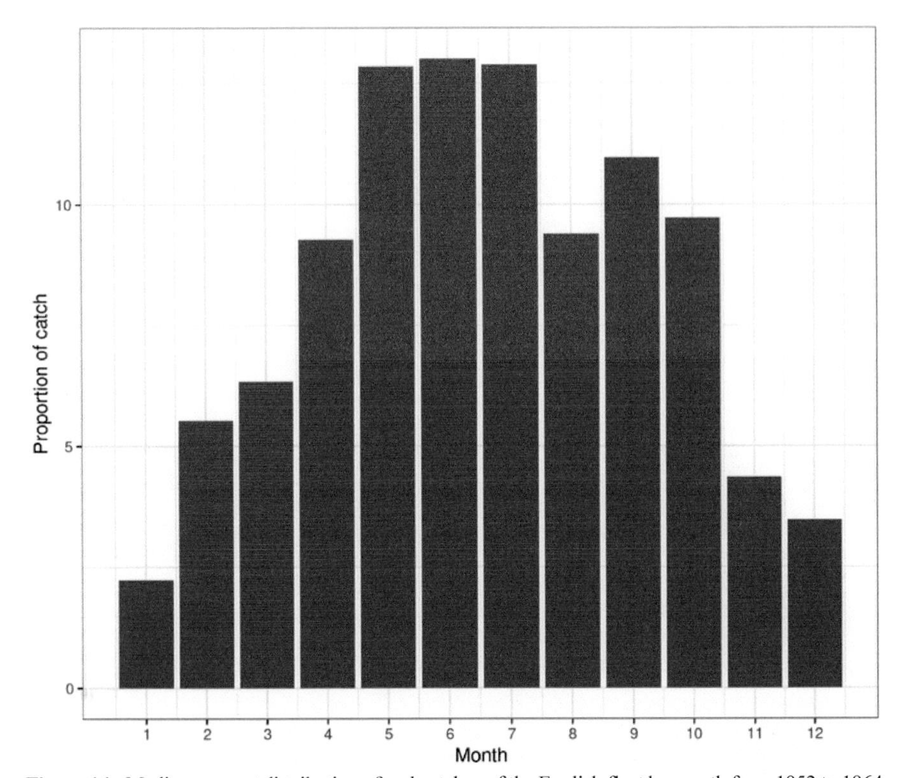

Figure 14. Median per cent distribution of cod catches of the English fleet by month from 1952 to 1964.

of the small Scottish fleet) and Germany were largely limited to trawling over the whole period. Before 1905, the Icelandic fisheries were solely confined to small boats using longlines and handlines. Trawling by Icelanders commenced in 1905, with the share of that gear increasing to around 25% in 1915, peaking at 50% around 1925 but declining to under 30% of the Icelandic catch at the start of the 1940s (Jónsson and Magnússon 1997). A large part of the Icelandic fisheries prior to WWII was on the spawning grounds and season, with the bulk of the catches taken in shallow coastal waters in the south and southwestern areas of the country. In contrast, the fisheries of the United Kingdom prior to WWII were mostly outside the spawning season and distributed in more offshore waters (Engelhard 2005).

Information on the catch distribution of the English vessels (Figure 14) showed that the bulk of the catches in the 1950s and early 1960s occurred in May–October, outside the spawning season, and that the fishing was distributed around Iceland (Figure 15), with some hotspots of higher catches in areas along the eastern and western side of the island.

Seasonal data on the Icelandic cod fleet are available from 1966 onwards, with information on the spatial distribution of the fisheries only available after the 1970s, first from part of the trawling fleet but a complete record of all the fleets since the start of this century. The seasonal nature of the Icelandic fisheries changed

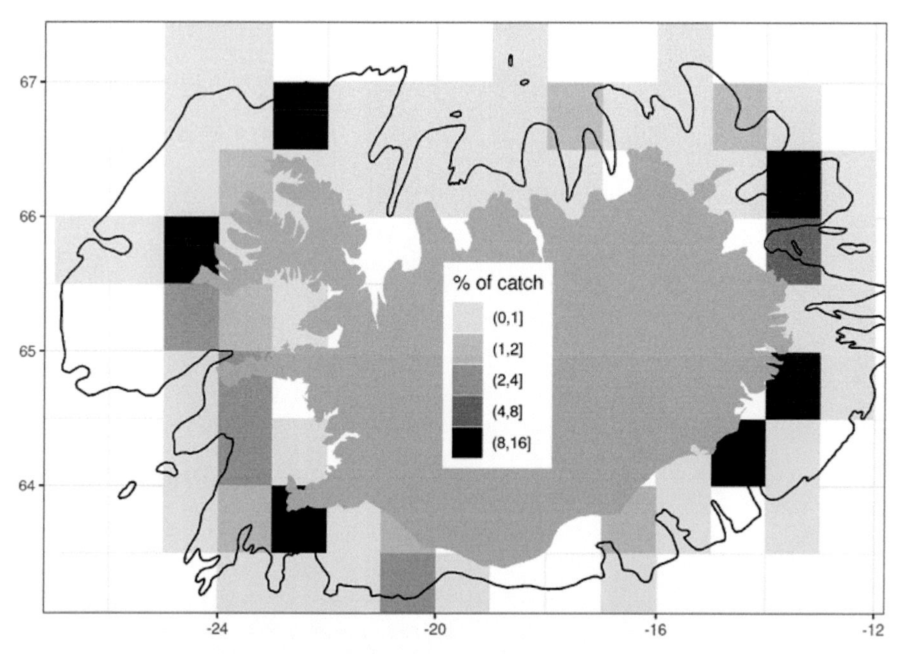

Figure 15. Per cent distribution of cod catches of the English fleet in 1952–1953 by statistical rectangles (Source: UK SeaCharts).

drastically in the period from 1966 to 1996, from the bulk of the catches being taken during the spawning season in March and April largely by gillnets (around 65% of the annual Icelandic catches and around 40% of the total catches), towards more continuous fisheries throughout the year (Figure 16) with almost no fishing with gillnets during spawning. This pattern change was partly driven by the diminishing spawning stock (see next section), expansion of the trawling fleet and later longline fleet (see catch development in Figure 13) and management action. The seasonal pattern in the Icelandic fisheries in 1966 is likely a reasonable representation of the fisheries prior to WWII, except that during that period, handline and longline were used in the spawning fisheries, gillnets being more or less absent (Schopka 1994). At present, the dominant gears are bottom trawl (55%) and longline (26%), with gillnets, demersal seine and jiggers each representing only 6–7% of the catches (Figure 13) (ICES 2020).

As indicated earlier, accurate spatial information on the Icelandic fisheries is limited only to recent decades. However, it is likely that prior to the 1970s, when the spawning fisheries dominated the Icelandic catches, the bulk of the catches were taken in shallow coastal waters off the south and west coasts of Iceland. Currently, the fishing is distributed around the whole shelf (Figure 17), with some aggregation of fisheries along the continental slope in the northwest, southeast, east and shallower waters of the western part of the shelf.

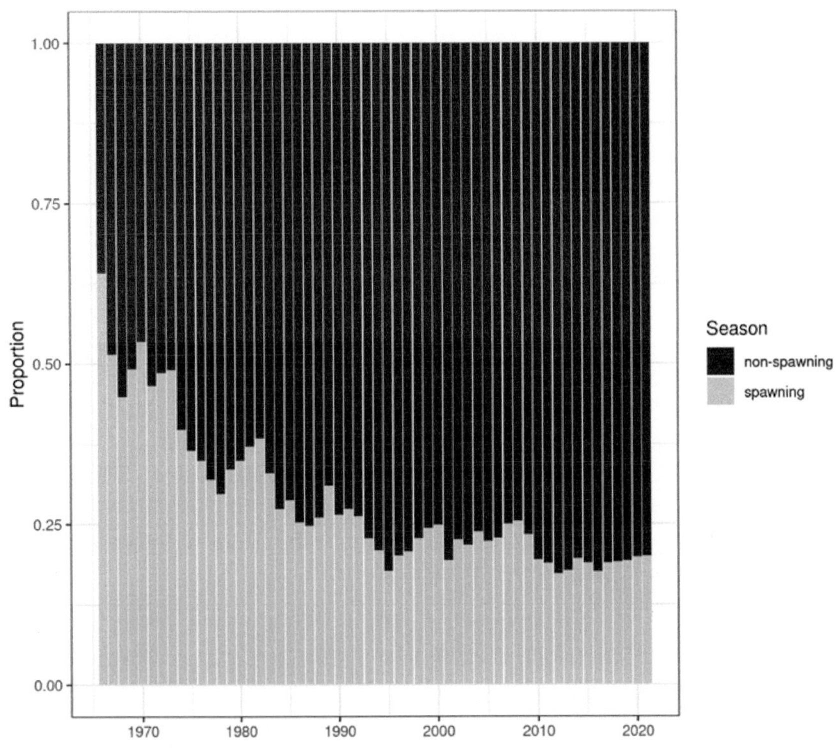

Figure 16. Development in the proportion of the Icelandic cod catches taken during the spawning season (months of April and May) vs. non-spawning season (months of May to January) (Source: National landings statistics).

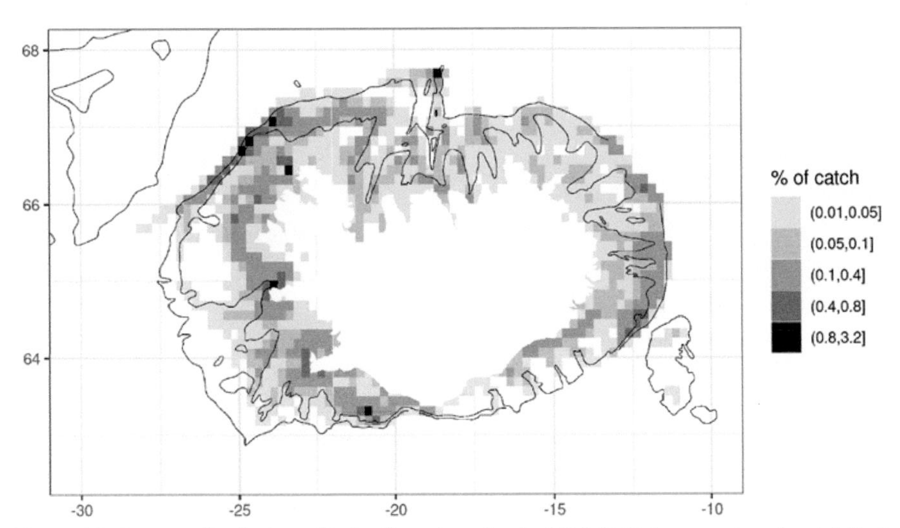

Figure 17. Per cent distribution of Icelandic cod catches in 2016–2020 by a quarter of a statistical rectangle. Contour lines indicate 200 and 400 m depths.

Assessment

The earliest analytical assessment of the Icelandic cod is that of Jónsson (1954), who estimated the relative size and total mortality of the spawning stock over 1930–1954 based on catch-at-age from the spawning fisheries of the Icelandic fleet. In the 1960s, the focus of analysis based on catch-at-age was related to changes in mesh size and, in particular, the effect of those measures on the yield of the main national fleet that participated in the fisheries at the time (ICES 1966, 1969). The first VPA assessment of cod was done in 1970 (ICES 1971), covering the years 1960–1969, this being extended back to 1955 in the 1976 assessment (ICES 1976), then to 1949, 1941 and 1928 by Cushing (1982) and Schopka (1993, 1994), respectively. Annual assessment upon which the annual advice is based have been generally limited to the years 1955 onwards (ICES 2021).

The reconstruction of the long-term stock dynamics of the Icelandic cod stock presented here was based on combining the catch- and weight-at-age (ages 3–14) matrices for the years 1928–1954 as compiled by Schopka (1994) with the input used in current annual ICES assessment that goes from 1955 forwards (ICES 2021). The assessment model is a statistical catch-at-age tuned with the fall and the spring survey, details of additional inputs provided in Table 1.

Table 1. Input to the stock assessment model.

Separable periods	1928–1937, 1938–1949, 1950–1975, 1976–1993, 1994–2003, 2004–2021
M (for all age groups)	0.2
Estimation of immigration from Greenland (years and ages)	1930–8, 1933–9, 1953–8, 1958–9, 1959–9, 1960–10, 1962–9, 1964–10, 1969–8, 1970–8, 1972–9, 1980–7, 1981–8, 1990–6 and 2009–6
Spawning stock weights 1928-1984	Predicted from catch weights, based on catch and stock weight relationship from 1985 onwards.
Maturity	1928–1954 Mean of the period 1955–1984
Age groups Spring survey	1–14
Age groups Autumn survey	3–13

Stock Dynamics

The fishing pressure was relatively low until the end of WWII, with reference fishing mortality being generally below 0.25 (Figure 18). After WWII, fishing mortality increased at a more or less steady rate, reaching above 0.9 at the beginning of the 1990s. Since the early 1990s, fishing mortality has been declining to approximately 0.3 in recent years, mortality not observed since before the 1950s.

The biomass is estimated to have been quite high before the mid-1950s; the reference biomass (B4+) rarely being below 2,000 tonnes and SSB rarely below 800 thousand tonnes (Figure 18). Over the long term, there is, however, a general decline in reference biomass and SSB throughout the last century, the mean of the estimates in the early 1990s being around 640 and 160 thousand tonnes, respectively. Since 2000, there has been some gradual recovery. The reference biomass and SSB in 2021 were around 1,000 and 360 thousand tonnes, respectively, and the SSB approaching values that were observed in the early 1960's.

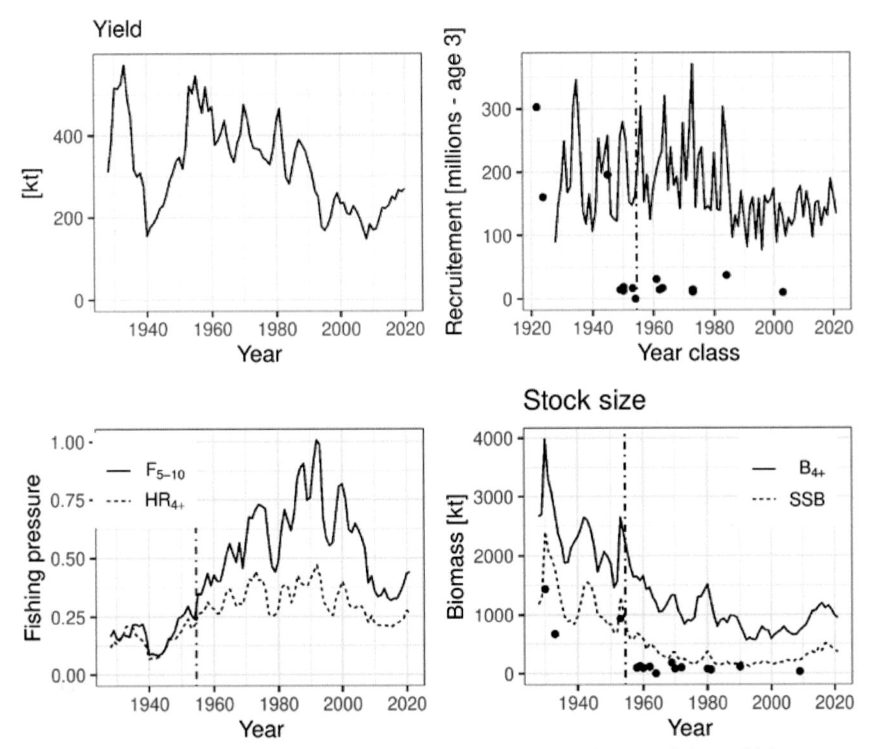

Figure 18. Summary of the stock assessment: yield, recruitment at the age of three, fishing pressure, and stock size (reference biomass (B_{4+}) and spawning stock biomass (SSB)). The black points indicate the contribution of the immigrants from Greenland to the reference biomass at the year of the migration.

The biomass dynamics in the early 1930s were to some extent influenced by the large immigration from Greenland; the 1922-year class (estimation of 300 million fish, around 1.5 million tonnes at age 8 in 1930) and the 1924-year class (160 million fish, 700 thousand tonnes at age 9 in 1933). Similarly, the peak in the 1950s was influenced by the large immigration of the 1945-year class from Greenland (196 million fish, 900 thousand tonnes at age eight in 1953). Immigrations of other year classes were generally estimated to be much smaller, although they had some short-term influence on the stock dynamics and fisheries because of generally low stock size.

The productivity of the stock (measured in terms of recruitment at age three) shows two very distinctive periods (Figure 18): the period 1928–1984, where the geometric mean recruitment was around 190 million fish, and 1985–2019, where the mean recruitment was around 135 million fish (a reduction of 30%). The recruitment variability has been relatively low within those two periods, particularly in the latter period (cv = 0.32 and 0.26, respectively). A remarkable feature is that in the latter period, recruitment has rarely been higher than the average recruitment in the former period. In the former period, there was no strong indication of a decline in recruitment over the observed spawning stock size (Figure 19). Although the mean SSB was generally much larger in 1928–1954 than in 1955–1984 (1,234 vs 352 kt),

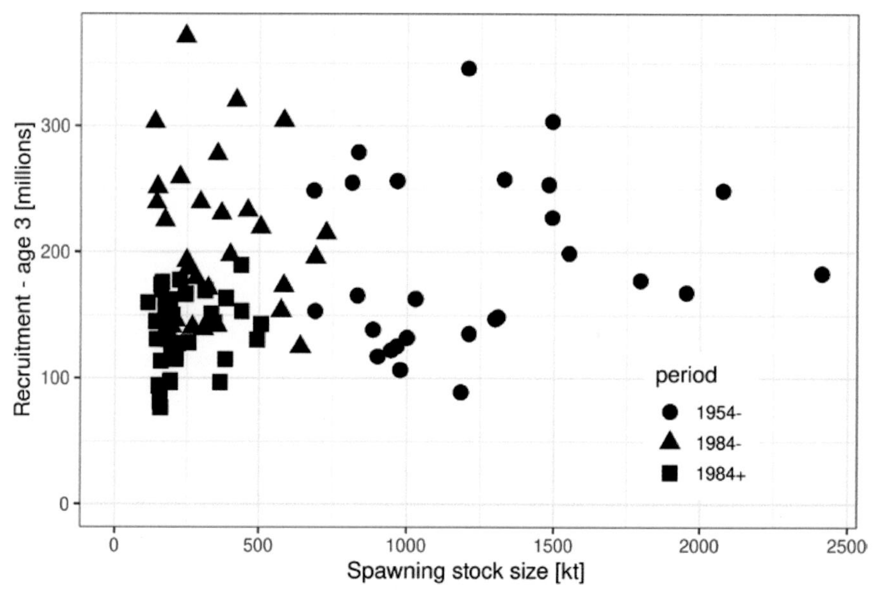

Figure 19. The relationship between spawning stock size and recruitment at age three.

mean recruitment was very similar (180 vs 200 million) with little indication that the frequency of poor recruitment increased at smaller stock size. As noted above, the mean recruitment pattern changed abruptly after 1984, this being despite the spawning stock size having already reached low levels by the early 1970s. There are some indications that after 1984 the frequency of year-class size under the long-term mean (167 million), particularly very poor year class (under 100 million), occurs more frequently at spawning stock sizes smaller than around 200 thousand tonnes (Figure 19).

Fisheries Management

By the mid-1970s, fisheries scientists started to raise concerns that SSB was at a historical minimum and that the fishing mortality was five times higher than an advocated reference fishing mortality rate of 0.22 (F0.1) (ICES 1976). The government was slow to respond to this concern but set the first total allowable catch (TAC) in place in 1984 to try to cap the ever-increasing fishing mortality (Figure 18). The set TAC was in the beginning normally set higher than that recommended by fisheries scientists and the total catches were normally higher than the set TAC because of various loopholes in the management system.

By the early 1990s, it was clear that more effective measures were needed. A commission was established by the Ministry of Fisheries in 1992, which recommended in 1994 that the TAC would be set as 0.20 of the reference biomass. Based on this work, the government established a formal harvest control rule to cap the TAC, first effective for the quota year 1995/96, but based on a multiplier of 0.25. In addition, an individual transfer quota system had been set in place in 1993, covering all vessels except the smallest. These measures initially resulted in a sharp reduction in fishing mortality (Figure 18), which increased again around the 2000s partly because of the overestimation of stock size. In the early 2000s, fisheries scientists started to advocate for a reduction in the multiplier in the harvest control rule (HCR) to that originally proposed in the early 1990s, in part driven by a lack of response in the spawning stock size. In 2007, the government followed this guideline and set the TAC for the first time based on the 0.20 HCR rule for the fishing year 2007/08; this rule has been in place since.

Acknowledgements

We thank Andreas Macrander and Guðmundur Þórðarson for valuable comments on the manuscript.

References

Árnason, E., and Rand, D.M. 1992. Heteroplasmy of short tandem repeats in mitochondrial DNA of Atlantic cod, *Gadus morhua*. Genetics, 132: 211–220.

Árnason, E., Pálsson, S., and Arason, A. 1992. Gene flow and lack of population differentiation in Atlantic cod, *Gadus morhua* L., from Iceland, and comparison of cod from Norway and Newfoundland. Journal of Fish Biology, 40: 751–770.

Árnason, E., Hernandez, U.B., and Kristinsson, K. 2009. Intense habitat-specific fisheries-induced selection at the molecular *Pan* I locus predicts imminent collapse of a major cod fishery. PLoS ONE 4: e5529. doi: 10.1371/journal.pone.0005529.

Astthorsson, O.S., Gislason, A., and Gudmundsdottir, A. 1994. Distribution, abundance, and length of pelagic juvenile cod in Icelandic waters in relation to environmental conditions. ICES Journal of Marine Science, 198: 529–541.

Bárðarson, B., Guðnason, K., Singh, W., et al. 2021. Loðna (*Mallotus villosus*). *In*: Óskarsson, G.J. (ed.). Staða umhverfis og vistkerfa í hafinu við Ísland og horfur næstu áratuga. Haf- og vatnarannsóknir, HV 2021-14: 31–34.

Begg, G.A., and Marteinsdottir, G. 2000. Spawning origins of pelagic juvenile cod *Gadus morhua* inferred from spatially explicit age distributions: potential influences on year-class strength and recruitment. Marine Ecology Progress Series, 202: 193–217.

Begg, G.A., and Marteinsdottir, G. 2002a. Environmental and stock effects on spatial distribution and abundance of mature cod *Gadus morhua*. Marine Ecology Progress Series, 229: 245–262.

Begg, G.A., and Marteinsdottir, G. 2002b. Environmental and stock effects on spawning origins and recruitment of cod *Gadus morhua*. Marine Ecology Progress Series, 229: 263–277.

Berg, P.R., Star, B., Pampoulie, C., et al. 2016. Three chromosomal rearrangements promote genomic divergence between migratory and stationary ecotypes of Atlantic cod. Scientific Reports 6: 23246: 1–12. doi: 10.1038/srep23246.

Berg, P.R., Star, B., Pampoulie, C., et al. 2017. Trans-oceanic genomic divergence of Atlantic cod ecotypes is associated with large inversions. Heredity, 119: 418–428. doi: 10.1038/hdy.2017.54.

Björnsson, B., Steinarsson, A., and Oddgeirsson, M. 2001. Optimal temperature for growth and feed conversion of immature cod (*Gadus morhua* L.). ICES Journal of Marine Science, 58: 29–38.

Björnsson, B., Reynisson, P., Solmundsson, J., et al. 2011. Seasonal changes in migratory and predatory activity of two species of gadoid preying on inshore northern shrimp *Pandalus borealis*. Journal of Fish Biology, 78: 1110–1131. doi: 10.1111/j.1095-8649.2011.02923.x.

Bogason, V., Björnsson, H., and Sólmundsson, J. 2018. Stofnmæling hrygningarþorsks með þorskanetum (SMN) 1996–2018. Haf- og vatnarannsóknir HV 2018-30: 1–42.

Bogason, V., Sólmundsson, J., Björnsson, H., et al. 2021. Stofnmæling hrygningarþorsks með þorskanetum (SMN) 2021 - framkvæmd og helstu niðurstöður. Haf- og vatnarannsóknir HV 2021-31: 1–25.

Brickman, D., Marteinsdottir, G., Logemann, K., et al. 2007. Drift probabilities for Icelandic cod larvae. ICES Journal of Marine Science, 64: 49–59.

Cushing, D.H. 1982. A simulacrum of the Iceland cod stock. Journal du Conseil/conseil Permanent International pour. Expllration de la Mer, 40: 27–36.

Engelhard, G.H. 2005. Catalogue of Defra historical catch and effort charts: six decades of detailed spatial statistics for British fisheries. Science Series Technical Report, Cefas Lowestoft, 128: 42.

Fiskifréttir. 2019. Tveir bátar kanna þorskgengd við Jan Mayen. Fiskifréttir 23. maí.

Fjøsne, K., and Gjøsæter, J. 1996. Dietary composition and the potential of food competition between 0-group cod (*Gadus morhua* L.) and some other fish species in the littoral zone. ICES Journal of Marine Science, 53: 757–770.

Fraser, S., Gotceitas, V., and Brown, J.A. 1996. Interactions between age-classes of Atlantic cod and their distribution among bottom substrates. Canadian Journal of Fisheries and Aquatic Sciences, 53: 305–314.

Friðriksson, Á. 1949. Boreo-tended changes in the marine vertebrate fauna of Iceland during the last 25 years. Rapports et Procés-Verbaux des Réunions du Conseil International pour l'Exploration de la Mer., 125: 30–32.

Gotceitas, V., Fraser, S., and Brown, J.A. 1995. Habitat use by juvenile Atlantic cod (*Gadus morhua*) in the presence of an actively foraging and non-foraging predator. Marine Biology, 123: 421–430.

Gunnarsson, B., and Björnsson, H. 2019. Grunnsævið - firðir og flóar eru vagga margra helstu nytjastofna við Ísland. Kver Hafrannsóknastofnunar KV 2019-01: 8.

Hansen, P.M. 1941. Studies on the biology of cod in Greenland waters. Rapports et Procés-Verbaux des Réunions du Conseil International pour l'Exploration de la Mer, 123: 2–77.

Havforskningsinstituttet. 2019. Torskemysteriet ved Jan Mayen er löst. https://www.hi.no/hi/nyheter/2019/januar/torske-mysteriet-ved-jan-mayen-er-lost.

ICES. 1966. Report of the North-Western Working Group 1965.

ICES. 1969. Reprt of the North-Western Working Group 1968.

ICES. 1971. Report of the North-Western Working Group.

ICES. 1976. Report of the North-Western Working Group.

ICES. 2021. Northwestern Working Group (NWWG). ICES Scientific Reports, 3: 52: 766.

ICES. 2020. North Western Working Group (NWWG). ICES Scientific Reports. 2:5. 670 pp.

Imsland, A.K., Jónsdóttir, Ó.D.B., and Daníelsdóttir, A.K. 2004. Nuclear DNA RFLP variation among Atlantic cod in south and south-east Icelandic waters. Fisheries Research, 64: 227–233.

Jakobsdóttir, K.B., Pardoe, H., Magnússon, Á., et al. 2011. Historical changes in genotypic frequencies at the Pantophysin locus in Atlantic cod (*Gadus morhua*) in Icelandic waters: evidence of fisheries-induced selection? Evolutionary Applications, 4: 562–573. doi: 10.1111/j.1752-4571.2010.00176.x.

Jamieson, A., and Birley, A.J. 1989. The demography of a haemoglobin polymorphism in the Atlantic cod, *Gadus morhua* L. Journal of Fish Biology, 35: 193–215.

Jaworski, A., and Ragnarsson, S.A. 2006. Feeding habits of demersal fish in Icelandic waters: a multivariate approach. ICES Journal of Marine Science, 63: 1682–1694. doi: 10.1016/j.icesjms.2006.07.003.

Jonasson, J.P., Gunnarsson, B., and Marteinsdottir, G. 2009. Abundance and growth of larval and early juvenile cod (*Gadus morhua*) in relation to variable environmental conditions west of Iceland. Deep Sea Research Part II: Tropcal Studies in Oceanography, 56: 1992–2000. doi: 10.1016/j.dsr2.2008.11.010.

Jónsdóttir, I.G. 2017. Predation on northern shrimp (*Pandalus borealis*) by three gadoid species. Marine Biology Research, 13: 447–455. doi: 10.1080/17451000.2016.1272697.

Jónsdóttir, I.G., Campana, S.E., and Marteinsdottir, G. 2006a. Otolith shape and temporal stability of spawning groups of Icelandic cod (*Gadus morhua* L.). ICES Journal of Marine Science, 63: 1501–1512.

Jónsdóttir, I.G., Campana, S.E., and Marteinsdottir, G. 2006b. Stock structure of Icelandic cod *Gadus morhua* L. based on otolith chemistry. Journal of Fish Biology, 69: 136–150.

Jónsdóttir, I.G., Marteinsdóttir, G., and Pampoulie, C. 2008. Relation of growth and condition with the Pan I locus in Atlantic cod (*Gadus morhua* L.) around Iceland. Marine Biology, 154: 867–874.

Jónsdóttir, I.G., Björnsson, H., and Skúladóttir, U. 2012. Predation by Atlantic cod *Gadus morhua* on northern shrimp *Pandalus borealis* in inshore and offshore areas of Iceland. Marine Ecology Progress Series, 469: 223–232. doi: 10.3354/meps09977.

Jónsdóttir, I.G., Sólmundsson, J., Hjörleifsson, E., et al. 2021. Göngur og atferli þorsks: Þorskmerkingar við Ísland í rúma öld. Náttúrufræðingurinn, 91: 5–15.

Jónsdóttir, Ó.D.B., Imsland, A.K., Daníelsdóttir, A.K., et al. 1999. Genetic differentiation among Atlantic cod in south and south-east Icelandic waters: synaptophysin (*Syp* I) and haemoglobin (*HbI*) variation. Journal of Fish Biology, 54: 1259–1274.

Jónsdóttir, Ó.D.B., Daníelsdóttir, A.K., and Naedval, G. 2001. Genetic differentiation among Atlantic cod (*Gadus morhua* L.) in Icelandic waters: temporal stability. ICES Journal of Marine Science, 58: 114–122.

Jónsson, E. 1982. A survey of spawning and reproduction of the Icelandic cod. Rit Fiskideildar, 6: 1–45.

Jónsson, J. 1947. Þorskveiðar og þorskrannsóknir við Ísland. Náttúrufræðingurinn, 1: 7–16.

Jónsson, J. 1954. Göngur íslenzka þorsksins. Ægir, 47: 2–9.

Jónsson, J. 1990. Hafrannsóknir við Ísland II. Eftir 1937. Bókaútgáfa Menningarsjóðs, Reykjavík.

Jónsson, J. 1996. Tagging of cod (*Gadus morhua*) in Icelandic waters 1948-1986. Rit Fiskideildar 14: 1–82.

Jónsson, G., and Magnússon, M.S. 1997. Hagskinna: Icelandic historical statistics. Statistics Iceland, Reykjavík, Iceland.

Jónsson, S., and Valdimarsson, H. 2005. The flow of Atlantic water to the North Icelandic Shelf and its relation to the drift of cod larvae. ICES Journal of Marine Science, 62: 1350–1359.

Jónsson, S., and Valdimarsson, H. 2012. Water mass transport variability to the North Icelandic shelf, 1994–2010. ICES Journal of Marine Science, 69: 809–815. doi: 10.1093/icesjms/fss024.

Karlsson, H., Ármannsson, H., Pétursson, H., et al. 2005. Fæða þorsks á Breiðafjarðarsvæðinu. Ægir 7: 12–13.

Lilly, G.R., and Rice, J.C. 1987. Food of Atlantic cod (*Gadus morhua*) on the northern Grand Bank in spring. NAFO Scientific Counsil Research Document 83/IX/87.

Macken, A., Lillicrap, A., and Langford, K. 2015. Benzoylurea pesticides used as veterinary medicines in aquaculture: Risks and developmental effects on nontarget crustaceans. Environmental Toxicology and Chemistry, 34: 1533–1542.

Macrander, A., Valdimarsson, H., and Jónsson, S. 2014. Improved transport estimate of the East Icelandic Current 2002–2012. Journal of Geophysical Research: Oceans 119. doi: 10.1002/2013JC009517.

Malmberg, S.-A., and Valdimarsson, H. 2003. Hydrographic conditions in Icelandic waters, 1990–1999. ICES Marine Science Symposia, 219: 50–60.

Marteinsdottir, G., and Björnsson, H. 1999. Time and duration of spawning of cod in Icelandic waters. ICES CM 1999/Y:34.

Marteinsdottir, G., and Steinarsson, A. 1998. Maternal influence on the size and viability of Iceland cod *Gadus morhua* eggs and larvae. Journal of Fish Biology, 52: 1241–1258.

Marteinsdottir, G., and Thorarinsson, K. 1998. Improving the stock-recruitment relationship in Icelandic cod (*Gadus morhua*) by including age diversity of spawners. Canadian Journal of Fisheries and Aquatic Sciences, 55: 1372–1377.

Marteinsdottir, G., Gudmundsdottir, A., and Thorsteinsson, V. 2000a. Spatial variation in abundance, size composition and viable egg production of spawning cod (*Gadus morhua* L.) in Icelandic waters. ICES Journal of Marine Science, 57: 824–830.

Marteinsdottir, G., Gunnarsson, B., and Suthers, I.M. 2000b. Spatial variation in hatch date distributions and origin of pelagic juvenile cod in Icelandic waters. ICES Journal of Marine Science, 57: 1182–1195.

Mason, J., Woods, P.J., Thorlacius, M., et al. 2021. Projecting climate-driven shifts in demersal fish habitat in Iceland's waters. ICES Journal of Marine Science, 78: 3793–3804.

Matschiner, M., Barth, J.M.I., Torresen, O.K., et al. 2022. Supergene origin and maintenance in Atlantic cod. Nature Ecology and Evolution, 6: 469–481.

MFRI. 2021a. Cod - *Gadus morhua*. State of Marine Stocks and Advice 2020. MFRI Assessment Reports.

MFRI. 2021b. Offshore shrimp - *Pandalus borealis*. State of Marine Stocks and Advice 2020. MFRI Assessment Reports.

Morato, T., González-Irusta, J.M., Dominguez-Carrió, C., et al. 2020. Climate-induced changes in the suitable habitat of cold-water corals and commercially important deep-sea fishes in the North Atlantic. Global Change Biology, 26: 2181–2202. doi: 10.1111/gcb.14996.

MRI. 1997. Fjölstofnarannsóknir 1992-1995. Hafrannsóknastofnun Fjölrit 57.

MRI. 2010. Manuals for the Icelandic bottom trawl surveys in spring and autumn. Hafrannsóknastofnun Fjölrit 156.

Nickel, A.K. 2016. Trophic vulnerability of 0-group Atlantic cod (*Gadus morhua*) and saithe (*Pollachius virens*). A case study investigating the juveniles' feeding pattern and identifying valuable nursery habitats in the Icelandic Westfjords. Masters Thesis. University of Akureyri.

Ólafsdóttir, G.Á., Gunnarsson, G.S., and Karlsson, H. 2015. More rapid shift to a benthic niche in larger *Gadus morhua* juveniles. Journal of Fish Biology, 87: 480–486. doi: 10.1111/jfb.12719.

Ólafsdóttir, S.R., Danielsen, M., Ólafsdóttir, E., et al. 2020. Ástand sjávar 2017 og 2018. Haf- og vatnarannsóknir HV2020-40: 1–27.

Pálsson, Ó.K. 1974. Rannsóknir á fæðu fiskseiða við strendur Íslands. Náttúrufræðingurinn, 44: 1–21.

Pálsson, Ó.K. 1983. The feeding habits of demersal fish species in Icelandic waters. Rit Fiskideildar, Journal of the Marine Research Institute Reykjavik, 7: 1–60.

Pálsson, Ó.K., and Thorsteinsson, V. 2003. Migration patterns, ambient temperature, and growth of Icelandic cod (*Gadus morhua*): evidence from storage tag data. Canadian Journal of Fisheries and Aquatic Sciences, 60: 1409–1423.

Pálsson, Ó.K., and Björnsson, H. 2011. Long-term changes in tropic patterns of Iceland cod and linkages to main prey stock sizes. ICES Journal of Marine Science, 68: 1488–1499.

Pampoulie, C., Ruzzante, D.E., Chosson, V., et al. 2006. The genetic structure of Atlantic cod (*Gadus morhua*) around Iceland: insight from microsatellites, the *Pan I locus*, and tagging experiments. Canadian Journal of Fisheries and Aquatic Sciences, 63: 2660–2674.

Pampoulie, C., Jakobsdottir, K.B., Marteinsdottir, G., et al. 2008. Are vertical behaviour patterns related to the Pantophysin locus in the Atlantic Cod (*Gadus morhua* L.)? Behavior Genetics, 38: 76–81. doi: 10.1007/s10519-007-9175-y.

Pampoulie, C., Daníelsdóttir, A.K., Thorsteinsson, V., et al. 2012. The composition of adult overwintering and juvenile aggregations of Atlantic Cod (*Gadus morhua*) around Iceland using neutral and functional markers: A statistical challenge. Canadian Journal of Fisheries and Aquatic Sciences, 69: 307–320. doi: 10.1139/F2011-151.

Pampoulie, C., Skirnisdottir, S., Star, B., et al. 2015. Rhodopsin gene polymorphism associated with divergent light environments in Atlantic Cod. Behaviour Genetics, 45: 236–244.

Pardoe, H., Vainikka, A., Thórdarson, G., et al. 2009. Temporal trends in probabilistic maturation reaction norms and growth of Atlantic cod (*Gadus morhua*) on the Icelandic shelf. Canadian Bulletins of Fisheries and Aquatic Sciences, 66: 1719–1733.

Petursdottir, G., Begg, G.A., and Marteinsdottir, G. 2006. Discrimination between Icelandic cod (*Gadus morhua* L.) populations from adjacent spawning areas based on otolith growth and shape. Fisheries Research, 80: 182–189.

Righton, D.A., Andersen, K.H., Neat, F., et al. 2010. Thermal niche of Atlantic cod *Gadus morhua*: Limits, tolerance and optima. Marine Ecology Progress Series, 420: 1–13. doi: 10.3354/meps08889.

Sæmundsson, B. 1906. Lifnaðarhættir þorska. Úr fiskirannsóknum Bjarna Sæmundssonar 1905. Ægir, 2: 5–7.

Sæmundsson, B. 1913. Continued marking experiments on plaice and cod in Icelandic waters. Meddelelser fra Kommissionen for Havundersögelser Serie Fiskeri, 4: 3–35.

Sæmundsson, B. 1926. Fiskarnir (Pisces Islandiae). Bókaverslun Sigfúsar Eymundssonar, Reykjavik, Iceland.

Sæmundsson, B. 1933. Merkingar á fiskum. Ægir, 26: 277–282.

Sæmundsson, B. 1935. Þorskur merktur við Vestmannaeyjar, veiðist við Finnmörku. Ægir, 28: 32.

Saemundsson, K., Jonasson, J.P., Begg, G.A., et al. 2020. Dispersal of juvenile cod (*Gadus morhua* L.) in Icelandic waters. Fisheries Research, 232: 105721. doi: https://doi.org/10.1016/j.fishres.2020.105721.

Schmidt, J. 1907. Marking experiments on plaice and cod in Icelandic waters. Meddelelser fra Kommissionen for Havundersögelser Serie Fiskeri, 2: 1–25.

Schmidt, J. 1931. Den Atlantiske torsk (*Gaduc callarias* L.) og dens locale racer. Meddelelser fra Carlsberg Laboratoriet Bind 18.

Schopka, S. 1993. The Greenland cod (*Gadus morhua*) at Iceland 1941–90 and their impact on assessment. NAFO Science Council Studies, 18: 81–85.

Schopka, S.A. 1994. Fluctuations in the cod stock off Iceland during the twentieth century in relation to changes in the fisheries and environment. ICES Marine Science Symposia, 198: 175–193.

Sólmundsson, J., Jónsdóttir, I.G., Björnsson, B., et al. 2015. Home ranges and spatial segregation of cod *Gadus morhua* spawning components. Marine Ecology Progress Series, 520: 217–233.

Sólmundsson, J., Karlsson, H., Björnsson, H., et al. 2020. A manual for the Icelandic groundfish survey in spring 2020. Marine and Freshwater Research in Iceland HV 2020-08:61.

Sólmundsson, J., Jónsdóttir, I.G., and Hjörleifsson, E. 2021. Þorskur (*Gadus morhua*). *In*: Óskarsson, G.J. (ed.). Staða umhverfis og vistkerfa í hafinu við Ísland og horfur næstu áratuga. Haf- og vatnarannsóknir HV 2021-14: 63–72.

Stefánsson, U. 1962. North Icelandic waters. Rit Fiskideildar, 3: 1–269.

Stige, L.C., Yaragina, N.A., Langangen, Ø., et al. 2017. Effect of a fish stock's demographic structure on offspring survival & sensitivity to climate. PNAS 114: 1347–1352. doi: 10.1073/pnas.1621040114.

Tåning, A.V. 1934. Göngur þorsksins í Norður-Atlantshafi. Ægir, 27:31–35.

Thordardottir, T. 1986. Timing and duration of spring blooming south and southwest of Iceland. pp. 345–360. *In*: Skreslet, S. (ed.). The Role of Freshwater outflow in Coastal Marine Ecosystems. Springer, Berlin.

Thorisson, K. 1989. The food of larvae and pelagic juveniles of cod (*Gadus morhua* L.) in the coastal waters west of Iceland. Rapports et Procés-Verbaux des Réunions du Conseil International pour l'Exploration de la Mer., 191: 264–272.

Thorsteinsson, V., and Marteinsdottir, G. 1998. Size specific time and duration of spawning of cod (*Gadus morhua*) in Icelandic waters. ICES CM 1998/DD:5 1–18.

Thorsteinsson, V., Pálsson, Ó.K., Tómasson, G.G., et al. 2012. Consistency in the behaviour types of the Atlantic cod: repeatability, timing of migration and geo-location. Marine Ecology Progress Series 462: 251–260. doi: 10.3354/meps09852.

Våge, K., Pickart, R.S., Spall, M.A., et al. 2013. Revised circulation scheme north of the Denmark Strait. Deep-Sea Research Part I, 79: 20–39. doi: doi.org/10.1016/j.dsr.2013.05.007.

Van Geest, J.L., Burridge, L.E., Fife, F.J., et al. 2014. Feeding response in marine copepods as a measure of acute toxicity of four anti-sea lice pesticides. Marine Environmental Research, 101: 145–152. doi: 10.1016/j.marenvres.2014.09.011.

Willis, K.J., and Ling, N. 2003. The toxicity of emamectin benzoate, an aquaculture pesticide, to planktonic marine copepods. Aquaculture, 221: 289–297.

Chapter 6
Faroe Islands Cod Stocks

Petur Steingrund, Helga Bára Mohr Vang* and
Karin Margretha H. Larsen

Introduction

Historical Change in Abundance

Atlantic cod has played a major role in the Faroese society and other fishing nations for centuries. Numerous cod bones have been found in archaeological excavations at the village of Sandur dated to 800–1,200 AD, indicating that cod individuals were around 60 cm long (40–90 cm) (Church et al. 2005). Dried salted cod was exported to Denmark by the Danish Monopoly Trade in the 18th and 19th centuries (Degn 1929). Many nations fished for cod in Faroese waters, including sloops (smacks) from the Shetland Islands and fishing with handlines from 1809 to 1911 (Goodlad and Litt 2014). From 1872 to ca. 1939, Faroese sloops participated in the cod fishery, and they also fished with handlines (Hansen 1961; Patursson 1961). From 1898 to 1976, British steam- and later motor trawlers participated in the cod fishery (Jones 1966). After introducing the 200 nm EEZ in 1977, Faroese fishing vessels have taken the majority of the cod catch.

Landings in Faroese waters have been registered since 1903 (Figure 1) (ICES subdivision Vb) and have been split between Faroe Plateau (Vb1) and Faroe Bank (Vb2) since 1965 (Figure 2). The cod catch on Faroe Bank constituted approximately 10% of the total catch during those periods when the Faroe Bank was not closed to commercial fishing (for Faroese vessels), which happened in 1990–1992 and 2008–2021. Cod catches on the Faroe Plateau normally fluctuated between 20 and 40 thousand tons, except during the Second World War 1939–1945. However, in 1990–1994 and since 2004, the catches in most years have only fluctuated around 10 thousand tons (Figure 1).

Faroe Marine Research Institute, Nóatún 1, P.O. Box 3051, FO 110 Tórshavn, Faroe Islands.
* Corresponding author: Peturs@hav.fo

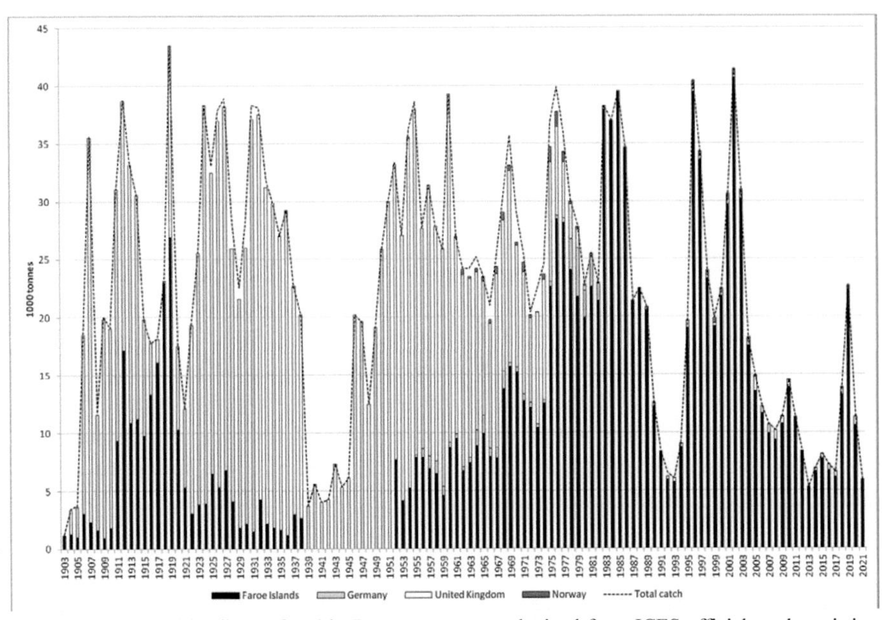

Figure 1. Historical landings of cod in Faroese waters as obtained from ICES official catch statistics. From 1965, the catch is for the Faroe Plateau only. Faroese catches from 1939 to 1951 were not registered. Source: ICES.

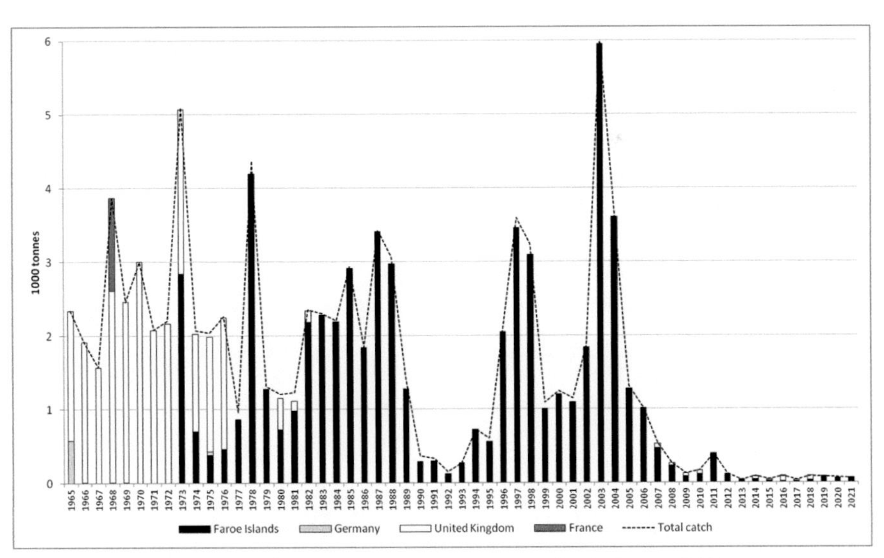

Figure 2. Historical landings of cod on Faroe Bank. Source: ICES.

Long-time series may reveal patterns that are not apparent in short-time series. In order to compile a time series of cod biomass back in time, catch per unit effort series were obtained from the following sources: catch per vessel for Shetland vessels 1859 to 1896, catch per vessel for Faroese vessels 1885 to 1914, catch per day absence from port for British trawlers 1906 to 1938, catch per million ton-hours for British trawlers 1924 to 1972 and finally absolute biomasses from 1959 to 2020 from the stock assessment. Working backward by correlating series that overlapped in time, the absolute biomass of cod was extended back to 1860 (ICES 2015, 2016). Hansen (1966, 1978) provided information about "good" and "bad" cod years in the second part of the 19th century that corresponded to biomasses of 220 and 81 thousand tons, respectively. A model combining the occurrence of good/bad years back in time with the export of dried cod provided rough estimates of cod biomass back to 1711 (ICES 2016) (Figure 3).

The biomass of cod in Faroese waters normally ranged between 100 and 200 thousand tons from 1861 to 1955 but declined to less than 50 thousand tons in 1991–94 and since 2006 (Figure 3). Although the biomass estimates are associated with uncertainty there are indications that the upper limit may lie somewhere between 200 and 250 thousand tons. There are also indications of periods when the biomass was below 50 thousand tons, e.g., around 1740 and 1813. Although this is comparable to the period since 2006, local people with their small boats may have experienced such periods as being devoid of cod. The exploitation rate was low prior to 1930 but has since been high, except in war periods and when the cod stock has been low (Figure 4).

The majority of the catch was taken by trawl before the extension of the 200 nm EEZ in 1977, when British vessels dominated the fishery. After 1977, Faroese vessels took the majority of the cod catch, and from 1985, they have been split by gear type (Figure 5). Trawl and longlines have taken around 90% of the catch. The share for each of them has ranged between 20% and 70%, where years with high exploitation rates correspond to a high share for longlines (Figure 4; Figure 5).

Distribution

There are three main populations or stocks of cod in Faroese waters: Faroe Plateau, Faroe Bank and Iceland-Faroe Ridge. Cod on Faroe Plateau and Faroe Bank are considered separate biological stocks. The evidence comes from morphometric and meristic characters as well as biochemical analyses (see Jákupsstovu and Reinert 1994; Joensen et al. 2000; Magnussen 1996). However, no significant genetic differences were found between cod on the Faroe Plateau and east of Iceland or

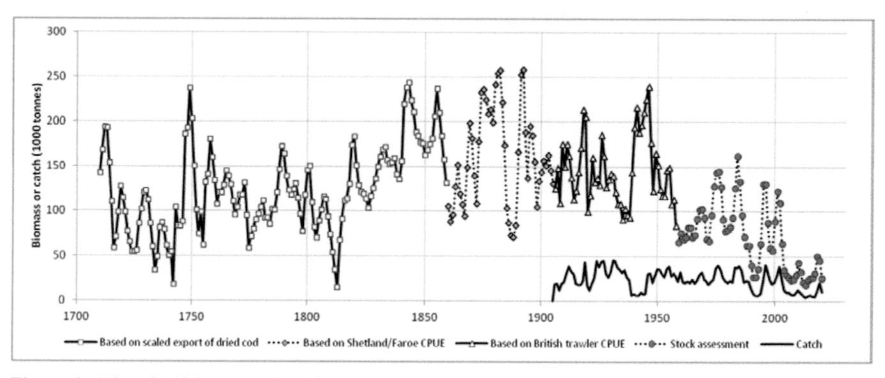

Figure 3. Historical biomass of cod in Faroese waters compiled from different sources. The catch from the Faroes or Faroe Plateau (since 1965) is also shown.

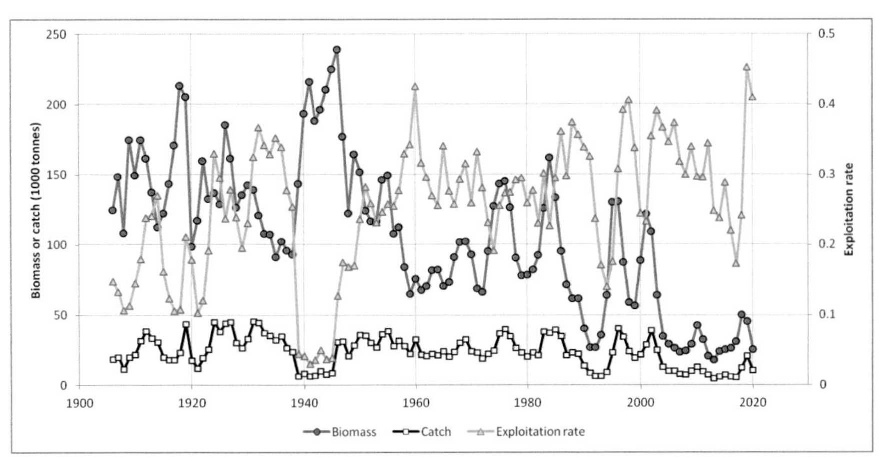

Figure 4. Biomass, catch and exploitation rate (catch divided by biomass) of cod in Faroese waters.

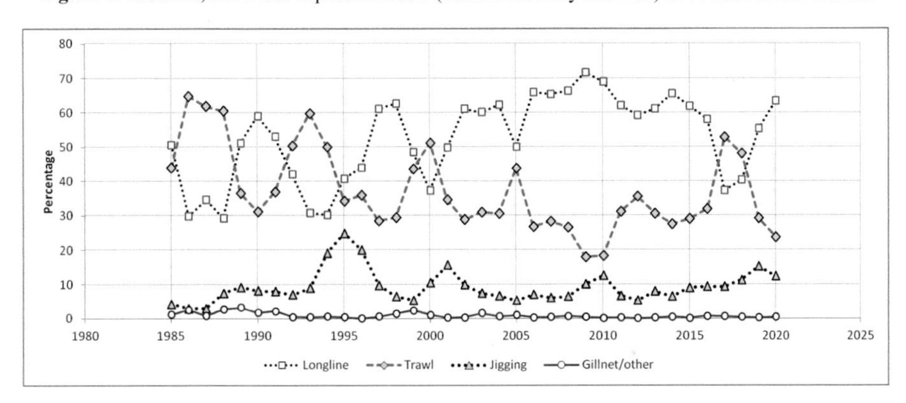

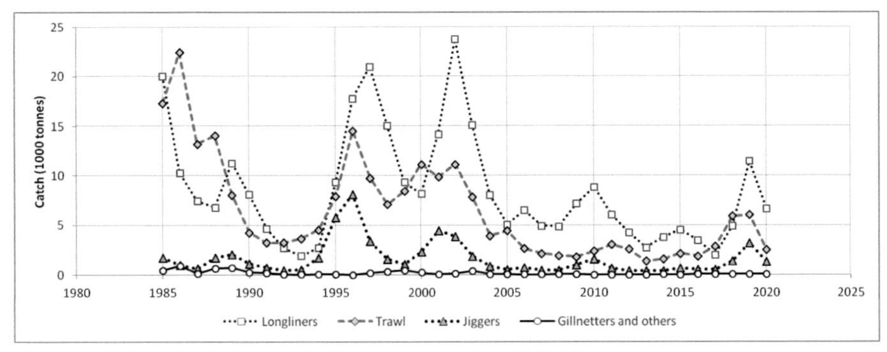

Figure 5. Cod catch split by gear for Faroese vessels since 1985. Upper panel: percentage, lower panel: tons. Source: ICES (2021).

between cod spawning north or west of the Faroe Islands (Pampoulie et al. 2008a). Cods on the Faroe-Iceland ridge (within the Faroese EEZ) are mainly considered part of the Icelandic cod stock and migrate to areas southwest of Iceland to spawn (ICES 2007).

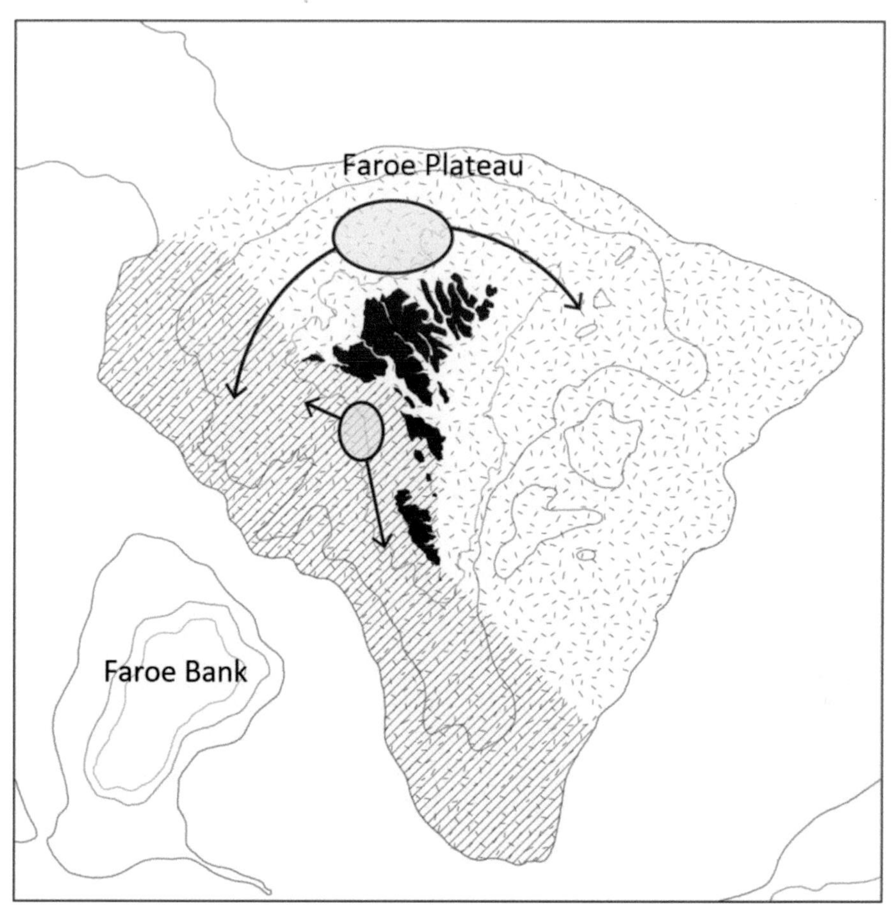

Figure 6. Spatial distribution of cod on Faroe Plateau. Ellipses show the major spawning areas. Cod spawning in the northern area originates from all parts of the Faroe Plateau shallower than 500 m, while cod spawning in the western area originate from the south western part of the Faroe Plateau. Arrows indicate directions of migration. The 100, 200 and 500 m depth contours are shown. Faroe Bank is located south west of the Faroe Plateau and is home to a separate cod stock that is not shown in the figure.

Cod on the Faroe Plateau, one order of magnitude larger than the other cod populations, spawn in March in two main locations north and west of the islands (Figure 6), where the depth is around 100 m. While cod spawning in the northern spawning area originates from the entire Faroe Plateau, cod spawning in the western spawning area comes from southwesterly areas (Figure 6). The depth distribution depends on size. Cod grow up in the nearshore areas for the first one to two years of life (Joensen and Tåning 1970; Joensen et al. 2005) and move gradually deeper with increasing size, occupying typical depths of 50 to 200 m although the largest cod (ages 8+, > 90 cm) are found slightly shallower (Steingrund and Ofstad 2010).

The spatial distribution varies between years, and medium-sized cod (four to six years old) may migrate to the nearshore areas when they are in poor condition, and this seems to hamper the survival of young cod (Steingrund et al. 2010). At the same time, more cod move to deeper waters (Steingrund and Ofstad 2010).

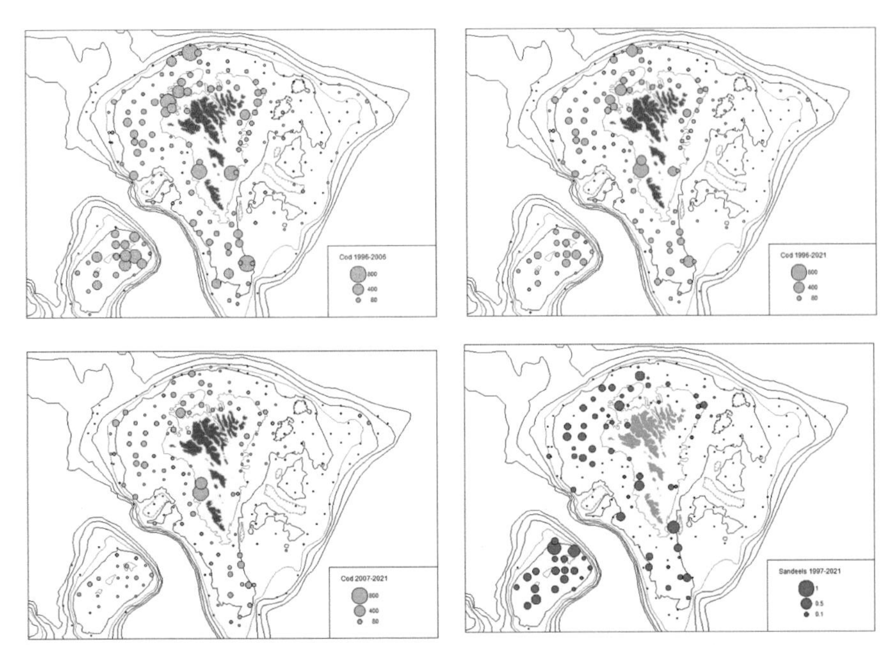

Figure 7. Spatial distribution of cod in kg per trawl hour in the August-September survey in periods with normal (1996–2006, upper left panel) and low (2007–2021, lower left) biomasses of cod. The distribution of sand eels as measured by the stomach content (percentage of predator weight) of cod (lower right) matches well the distribution of cod 1996–2021 (upper right). Also, 100–600 m depth contours are shown.

These migrations seem related to the general feature that cod stay where sand eels (Ammodytes) are abundant (Figure 7). In years when sand eels are abundant, cod stay in a narrower area around 70–130 m depth, but in years with few sand eels, cod spread out to both shallower and deeper areas. The largest cod are seemingly independent of these patterns.

Diet

The diet of larval and juvenile cod on the Faroe Plateau 1998–2005 is described by Jacobsen et al. (2020) [see also Gaard and Reinert (2002)]. Yolk-sac larvae consume phytoplankton, copepod eggs and nauplii. Early larval stages consume copepod eggs and calanoid nauplii, while late larval stages consume medium-sized copepod species, mainly *Pseudocalanus* sp., *Acartia* sp. and early-stage *Calanus finmarchicus*. Juvenile cod prey on *Temora longicornis* and late-stage *C. Finmarchicus*, as well as decapod larvae, in addition to *Oithona* sp. and barnacle larvae. Late larval and early juvenile cod apparently suffer from unfavorable feeding conditions since they feed on smaller prey than they normally prefer (Jacobsen et al. 2020). The food of juvenile cod in the littoral zone is mainly crustaceans and young of other fish species, e.g., saithe and sand eels (Joensen and Tåning 1970).

Stomach content investigations on adult cod were performed from 1949 to 1962 and showed that sand eels were most important, followed by Pandalid shrimps, but Norway pout was of little importance (Rae 1967). Norway pout was much more important in 1975–76 (Du Buit 1982). The stomach content of cod was low in 1991 compared with 1999 (Homrum 2007).

Stomach content investigations in spring and autumn 1997–2021, where the water depth is between 65 and 520 m, show that the diet depends much upon predator size. Small fish feed on equal proportions of crustaceans and fish, and the proportion of fish increases with fish size (Figure 8). The total stomach content (relative to predator weight) is smallest for medium-sized fish, which also seem to supplement their diet with low-caloric food such as brittle stars. Of fish, sand eels are most important for small fish, whereas Norway pout, blue whiting and flatfish increase in importance with fish size (Figure 9). The largest cod feed on a variety of large prey items, e.g., flatfish, that apparently can be overpowered by these large cod. There seems to be a critical size of cod, between 55 and 75 cm, where crustaceans may not be sufficient for such large cod while they still may be too small to catch sufficient amounts of fish prey. This is especially true when the abundance of sand eels and Norway pout is low (Sørensen 2021). In such years, e.g., in 2003 and 2019, some of these cod migrate close to land and hamper the recruitment by cannibalism since even small amounts of cod in cod stomachs may lead to significant reductions in year class strength (Steingrund et al. 2010).

The stomach content is also highly dependent upon water depth, as reflected by the natural occurrence of prey. In shallow waters, cod feed on equal proportions of crustaceans and fish. While the amount of crustaceans is constant with depth, the biomass of fish increases so that the total stomach content is highest in deep waters (Figure 10). Sand eels are the most important fish prey in shallow waters, and Norway pout at intermediate depths, especially in spring, while blue whiting dominates in the deepest areas (Figure 11). Of crustaceans, various species are preyed on in shallow waters (e.g., shrimps, swimming crabs, squat lobsters and hermit crabs), while the squat lobster *Munida sarsi* is the solely important crustacean prey below 150 m.

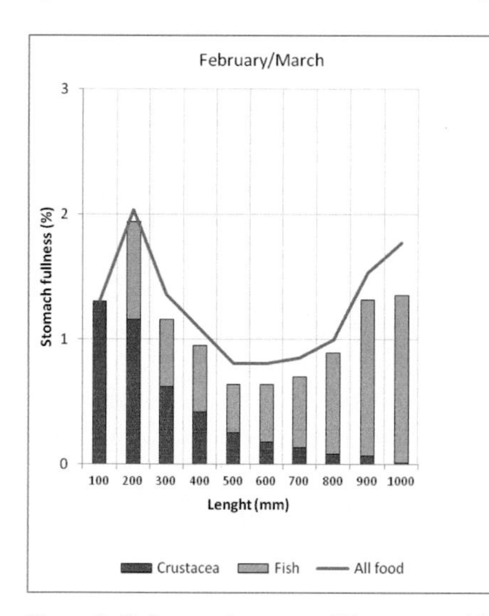

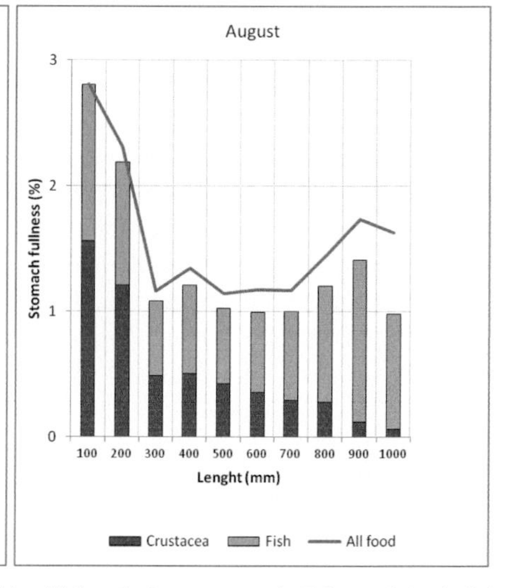

Figure 8. Cod stomach content of Crustacea and fish with length. From surveys in February/March (left panel) and August (right panel). Source: national database.

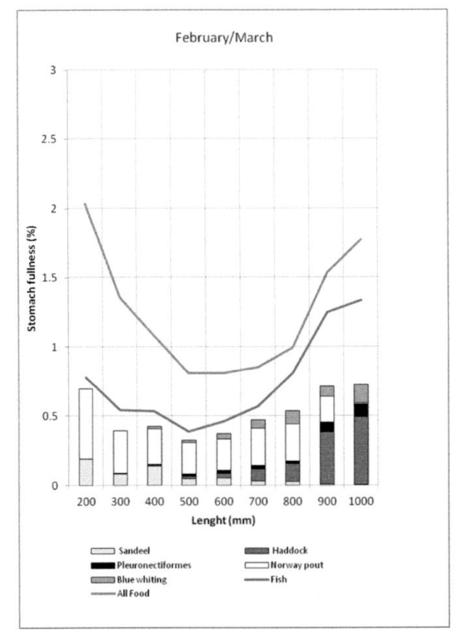

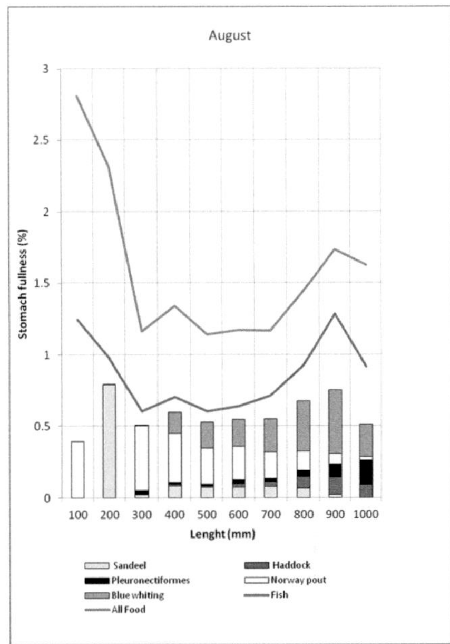

Figure 9. Cod stomach content of various species of fish with length. From surveys in February/March (left panel) and August (right panel). Source: national database.

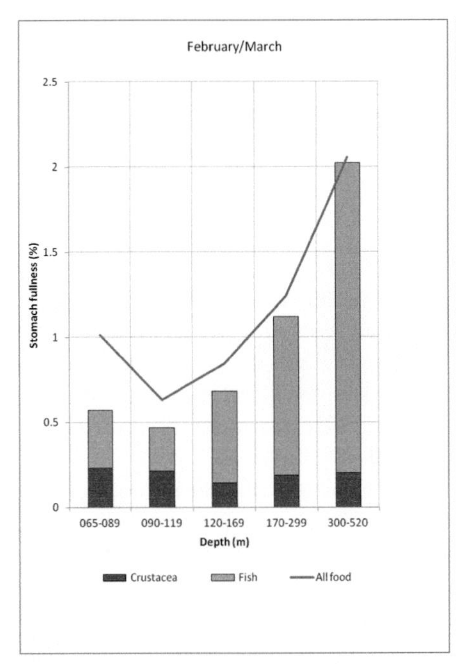

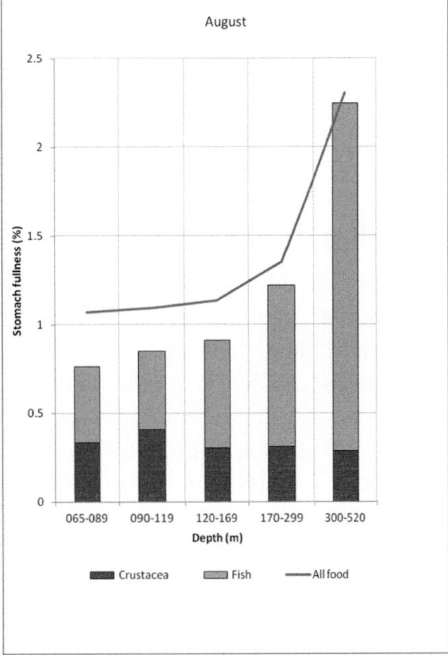

Figure 10. Cod stomach content of Crustacea and fish with depth. From surveys in February/March (left panel) and August (right panel).

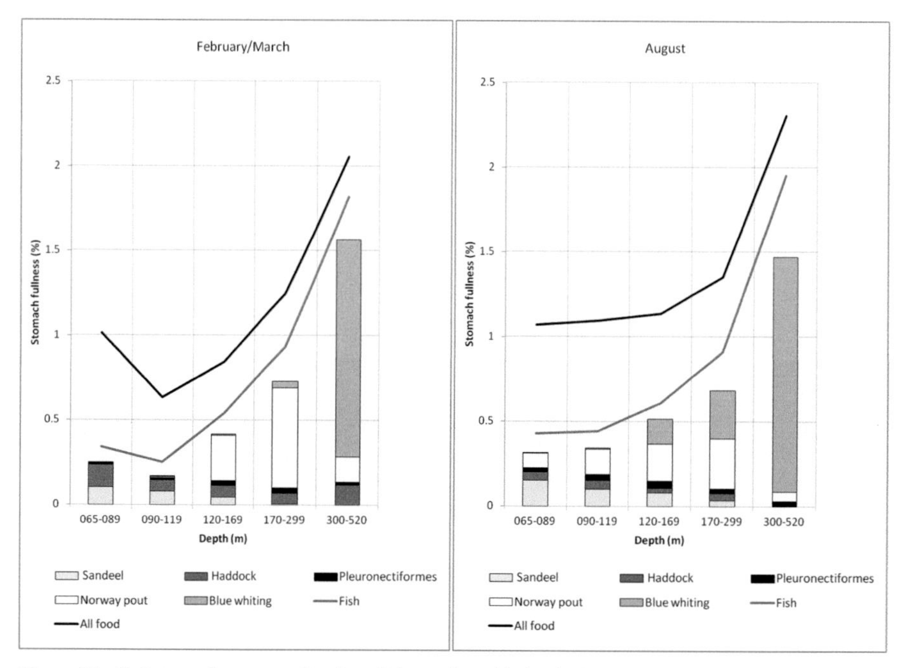

Figure 11. Cod stomach content of various fish species with depth. From surveys in February/March (left panel) and August (right panel). Source: national database.

The spatial distribution of cod seems to be linked to the cod diet, especially the individual size of the prey items. In shallow waters, all sizes of cod manage to prey on sand eels and congregate in these areas. At intermediate depths, a small part of the cod population is found and preys on Norway pout, while in the deepest areas, only large cod are found, and they feed on the relatively large blue whiting.

Individual Growth

Cod eggs on Faroe Plateau hatch in April, and the larvae have a weekly growth increment of 1–1.5 mm in the first month, 2–3 mm in the next month, and 4–5 mm in the third month. The juveniles of size 3.5–4.0 cm leave the pelagic phase in the second week of July. In late July, demersal stages about 4.5 cm long are found in the seaweed in the littoral zone. In mid-August, they are 5 cm, and in mid-September, they are about 7 cm (Jákupsstovu and Reinert 1994).

There is much more information about the growth of the later stages of cod. In Faroese waters, cod grow at different rates. While cod on Faroe Plateau has growth rate that is similar to other neighboring cod stocks, cod on Faroe Bank grow exceptionally fast and are among the fastest-growing cod stocks in the entire North Atlantic Ocean. For example, a typical three-year-old cod from Faroe Plateau is 53 cm long and weighs 1.6 kg, whereas a typical three-year-old cod from Faroe Bank is 77 cm long and 5.3 kg. The difference in growth occurs from age one to three, and Faroe Plateau cod catches up with Faroe Bank cod at age 10–11 years after rapidly growing 2 kg per year from age six to ten years (Figure 12). This corresponds to

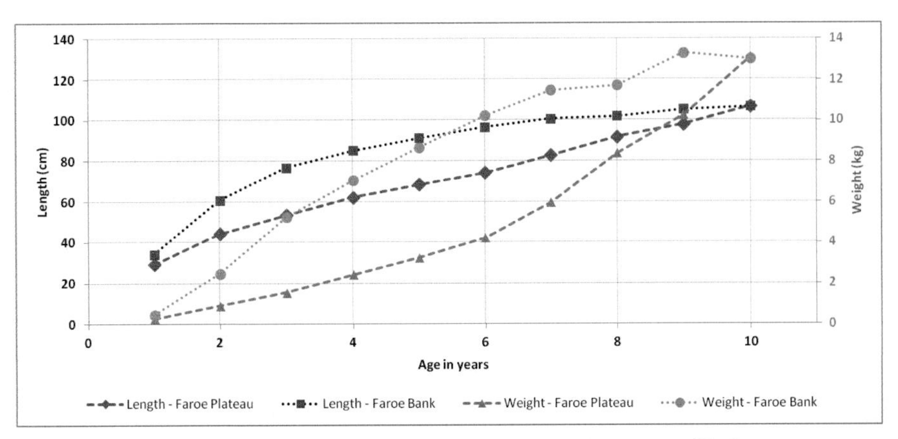

Figure 12. Individual growth of cod on Faroe Plateau and Faroe Bank as measured in the autumn surveys 1996–2021.

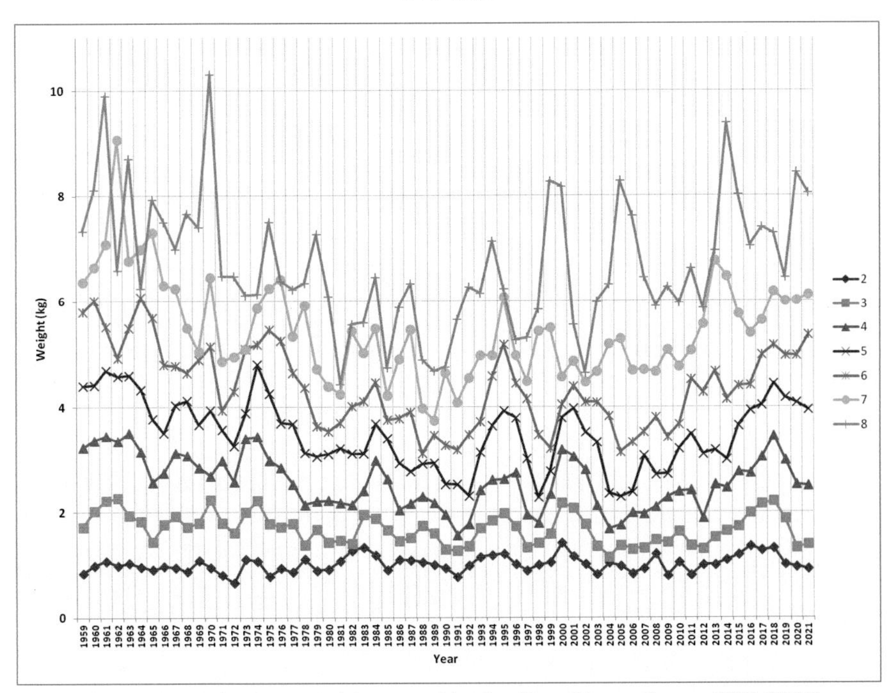

Figure 13. Weight of two to eight- year- old cod on Faroe Plateau. Source: ICES (2021).

when cod have passed the diet shift from a mixture of crustaceans and fish to a nearly fish-dominated diet (Figure 8). The individual weights of cod tend to be larger in deep waters (> 150 m) than in shallower waters. This is observed for ages two to seven years, while there is no difference for older cod (Steingrund and Ofstad 2010). The difference in growth between cod on Faroe Plateau and Faroe Bank could also partially be caused by different genetics since pen-reared cod originating from parents from Faroe Bank grew faster than cod that originated from Faroe Plateau, even though the rearing conditions were the same (Fjallstein and Magnussen 1996).

The individual weight-at-age fluctuates greatly between years on the Faroe Plateau (Figure 13) and is partially caused by variations in primary production and associated abundances of sand eels (Eliasen et al. 2011). There was a decrease in growth from the 1960s to the 1990s and a substantial increase after 2005 when the cod stock was exceptionally small, indicating effects of density-dependence as well as environmental effects.

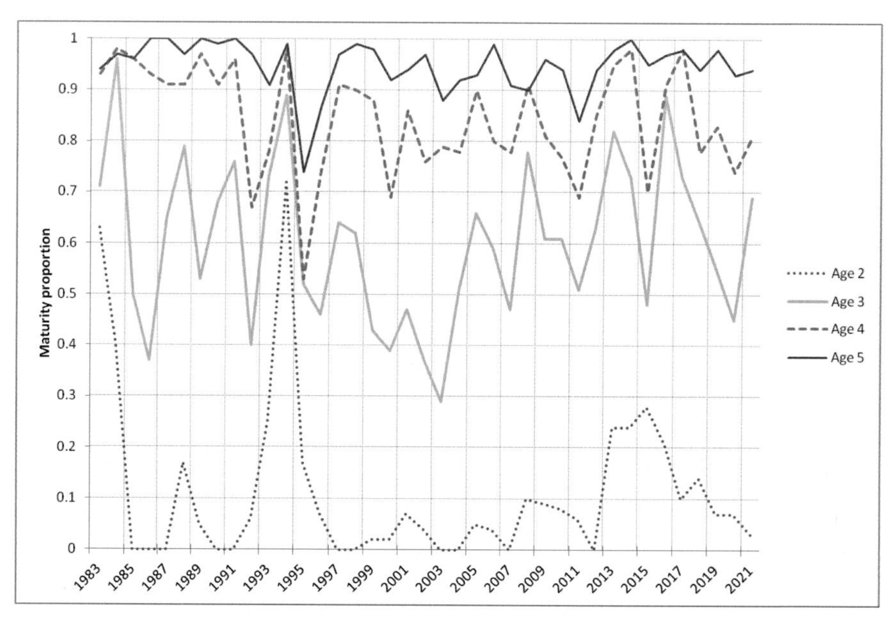

Figure 14. Maturity of cod on Faroe Plateau as measured in the spring groundfish survey in February-March.

Maturation

Cods on Faroe Plateau and Faroe Bank become sexually mature between two and four years old (Figure 14). While the maturity of age two seems to be positively correlated with the individual weight, there is no correlation for ages three or four.

Migration

Cods on the Faroe Plateau are local to the area since only around one in a thousand recaptured tagged cod migrate to other areas and only around one per cent the opposite way (Bedford 1966; Steingrund 2009). Also, cod on Faroe Bank are local to the area (Magnussen 1996). However, cod undertake migrations within these areas as demonstrated by many tagging experiments performed since 1909. Small (immature) cod usually stay within ten nautical miles from the tagging area (Strubberg 1916; Strubberg 1933; Joensen et al. 2005; Steingrund 2009), and this is typically close to land. As cod grow, they move to deeper waters (Steingrund and Gaard 2005; Steingrund and Ofstad 2010).

During the spawning season, sexually mature cod migrate to and from the spawning areas (Figure 6), and cod on the Faroe Plateau are grouped as 'accurate

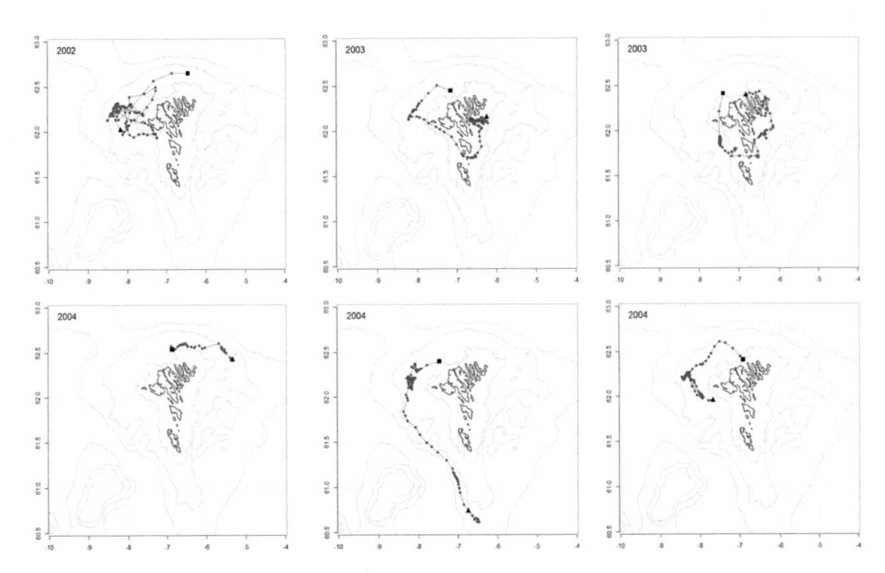

Figure 15. Modelled migration routes of three cod in a normal year (2004, lower panel) and three cod in years with food shortage (2002–2003, upper panel), data from Ottosen et al. (2017). These cod were marked and released on the northern spawning area in late March. The data is cut off on 30 June of the same year in order to facilitate comparisons between individuals.

homers' (Robichaud and Rose 2004). The migration routes are roughly parallel to land, where one branch goes westwards from the northern spawning area and one branch eastwards so that all areas on the Faroe Plateau are covered. Cods that spawn west of the islands tend to originate from the southwestern part of the Faroe Plateau.

Migration routes of individual cod, based on temperature/depth records from data storage tags (DST) combined with hydrographic model results (Ottosen et al. 2017), show that the horizontal migration pattern may be more complex than indicated in Figure 6. In total, 23 migration routes were reconstructed, showing that cod stayed within 10 m from the bottom over 90% of the time (Ottosen et al. 2017). During years with normal food conditions, cod moved along relatively straight migration routes. In years with food shortage, however, cod moved faster and in a more complex way that probably indicated that they searched for food and, on some occasions, were located close to land (Figure 15). The swimming velocity was fastest during the spawning season, decreased in the summer, and was lowest in the autumn (Ottosen et al. 2017).

The timing of the spawning migration can be exemplified by a cod that was at liberty for about two years after tagging (Figure 16). There was apparently limited movement during the feeding season that was located deeper than the typical spawning depth of around 100 m. At the beginning of February 2004, the cod started the spawning migration and was located on the spawning ground from March to mid-April, when it increased activity and apparently moved to the same feeding area as the year before. In mid-May, it apparently reached the feeding area and stayed there until the beginning of February 2005. In mid-February, it reached the spawning area and stayed there until late March, when it left the spawning area and probably headed

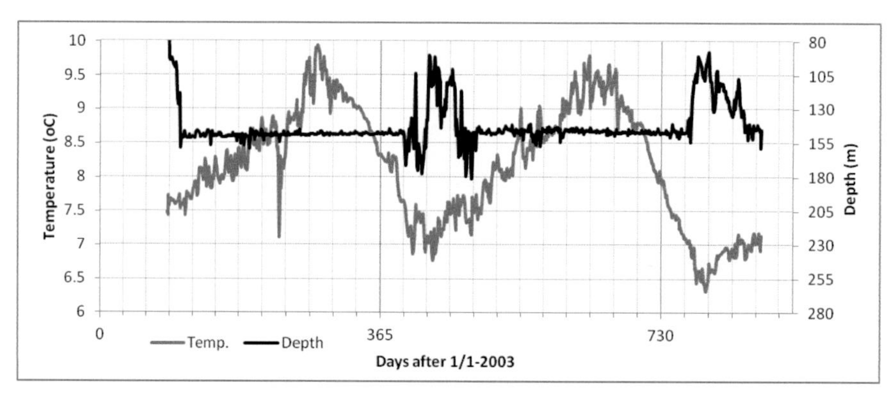

Figure 16. Daily temperature and depth information of a cod that was tagged 28. March 2003 on the spawning ground west of Faroe Islands and stayed at liberty for two years.

for the same feeding area again. The timing and probable location of the spawning migration apparently were not exactly the same in both years, indicating that spatial landmarks may not be the directional cues for the migration. This also applied to the individual in Figure 15 (mid-lower panel) that migrated back to the same spawning area where it was tagged along a slightly different route (see Figure 3 in Ottosen et al. 2017).

Along with most DST-tagged cod in Faroese waters, the individual referred to in Figure 16 showed a migration pattern during the feeding season that is characteristic for *Pan* I^{AA} genotypes, which is 10–30 times more common in Faroese waters than *Pan* I^{BB} individuals or heterozygotes (Case et al. 2005; Nielsen et al. 2007; Pampoulie et al. 2008a). In Icelandic waters where the frequency of the *Pan* I^{A} and *Pan* I^{B} loci is about equal, it has been demonstrated that *Pan* I^{AA} individuals grow fast, are typically found in shallower waters and exert much less variation in depth compared with slow-growing *Pan* I^{BB} individuals (Pampoulie et al. 2008b). The frequency of the *Pan* I allele is under natural selection, where the frequency of the *Pan* I^{B} allele increases with decreasing temperature, increasing depth and increased salinity (Case et al. 2005), although no changes have been observed over time in the southeastern part of the species distribution (Nielsen et al. 2007). Hence, the migration pattern of cod can be directly linked to genetics that, however, may be related to more genes than just the *Pan* I locus (see the chapter about Icelandic cod).

Spawning

Retention of eggs/larvae in combination with the occurrence of Calanus eggs/nauplii are regarded as the main factors that determine the location of the cod spawning areas on the Faroe Plateau (Gaard and Steingrund 2001). The main spawning area is located north of the islands, where the water depth is around 100 m, and a smaller spawning area is found west of the islands (Tåning 1940; Tåning 1943; Steingrund et al. 2004; Ottosen et al. 2018). The timing of the spawning is from February to April, with a peak in March (Jákupsstovu and Reinert 1994) and in former times, i.e., before 1960, from February to May, with a peak in April (Joensen and Tåning 1970).

The spawning migration to and from the spawning areas takes only a few days or weeks due to the short distances on the Faroe Plateau (Figure 15). The location of the northern spawning area varies in relation to temperature, being located to the west in cool years and to the east in warm years. In addition, the timing of the peak spawning occurs earlier in warm years. This is consistent with cod trying to find the coldest locations to spawn (Ottosen et al. 2018). No such relationships were, however, found for the western spawning area.

Juveniles and Nursery Grounds

After peak spawning time on the Faroe Plateau in late March, the eggs hatch in mid-April (Jákupsstovu and Reinert 1994; Jacobsen et al. 2020). The eggs/larvae are pelagic and drift with clockwise currents around the islands inside the tidal front (Jacobsen et al. 2019, see also the section on Ocean Currents). In July, the juveniles settle to the bottom in shallow areas, fjords, and sounds (Joensen and Tåning 1970; Jákupsstovu and Reinert 1994). They stay in the nearshore nursery areas until age one to two years (30–40 cm), sometimes longer, when they undertake a habitat shift to the feeding grounds of adult cod. At ages two to four, they become sexually mature and join adult cod in the spawning migration and spawning activities.

Cod on Faroe Bank

Faroe Bank, with its associated fish assemblage (Magnussen 2002), is located southwest of the Faroe Islands (Figure 6) and has a population of cod that is genetically distinct from other cod stocks (Magnussen 1996; Magnussen 2007; Jákupsstovu and Reinert 1994; Joensen et al. 2000) and therefore regarded as a distinct cod stock that belongs to ICES area 5b2. Cod on Faroe Bank is exceptional with regards to their extremely rapid growth rate in the first years of life (Figure 12), and reasons for this have been attributed to genetics (Fjallstein and Magnussen 1996), high temperatures (Brander 1995) and good feeding conditions (Magnussen 2007; Magnussen 2011). For coldwater fishes, environmental factors may account for 70–80% of the growth, while the remaining part can be ascribed to genetic properties (Gjedrem 2000). Cod from Faroe Bank has traditionally been in high demand due to its large size and quality of the flesh (Love 1974).

Catch statistics of cod from Faroe Bank have been available since 1965 (Figure 2). Annual catches fluctuated between 1,000 and 4,000 tons until 1990–1992 when the bank was closed to commercial fishing due to the poor state of the stock. After a recovery, the bank was again closed from 2008 to 2021 but reopened in 2022 to a limited commercial fishery. Reasons for the recent poor state of the stock are unknown, but high temperatures combined with low biomasses due to heavy fishing in the late 1980s and early 2000s may be contributing factors. The commercial fishery in 2022 was at a low level, while the surveys indicate that the cod stock on Faroe Bank has fully recovered (Figure 17).

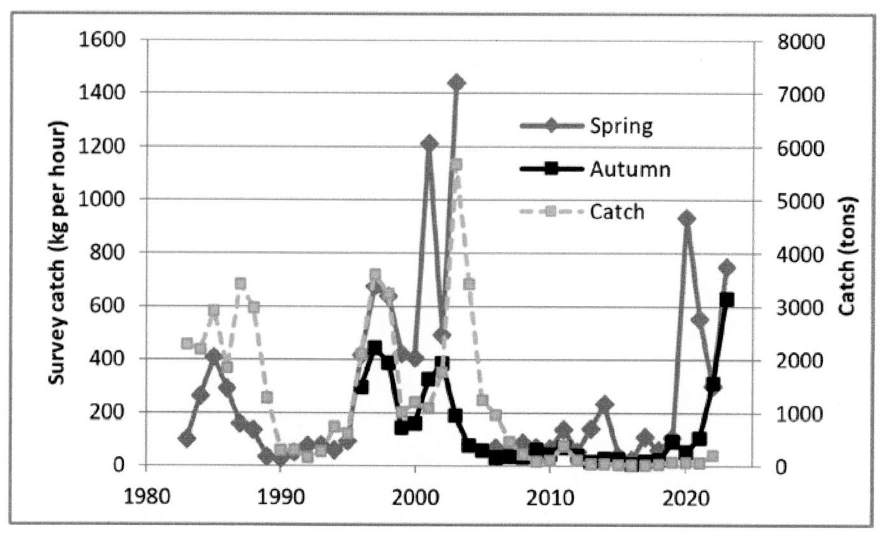

Figure 17. Survey catch rate and commercial catch of cod on Faroe Bank.

Stock Challenges

Ocean Currents

The upper layers of Faroese waters are dominated by Atlantic Water (AW). The North Atlantic Current transports relatively warm and saline AW northwards across the Iceland-Faroe Ridge (IFR), where it meets colder and fresher water from the East Icelandic Current and returned AW, which has been cooled and freshened in the Norwegian Sea (Figure 18). After crossing the IFR, the AW focuses on the Faroe Current just north of the Faroes. The Faroe Current runs eastwards toward Norway, and a part of it bifurcates into the Faroe Shetland Channel, where it recirculates and joins the AW transported with the Continental Slope Current along the Scottish slope. Both the Faroe Shelf and the Faroe Bank are, therefore, surrounded by AW (Hansen and Østerhus 2000).

On the Faroe Shelf, strong tidal currents mix the water column out to approximately 100 m bottom depth and create a clockwise residual circulation around the shelf. The well-mixed water, termed Faroe Shelf Water (FSW), is somewhat colder and fresher than the surrounding AW due to more effective winter cooling and higher precipitation over land (Larsen et al. 2008). This creates a front between the FSW and the AW, which partly isolates the FSW (Figure 18). The front is most pronounced around March when winter cooling levels off, and during spring and summer, it translates into a typical tidal front and moves closer to the shore, especially in the southwestern part of the shelf (Larsen et al. 2009; Eliasen et al. 2017). The spawning areas of the Faroe Plateau cod are located inside and close to the front (Figure 18).

The Faroe Bank has similar physical features as the Faroe Shelf but at a much smaller scale. The general circulation on the bank is clockwise (Hansen et al. 1998), and water on the shallow bank is known to become colder than off the bank (Hansen 2018). Also, the Faroe Bank cod spawns in the mixed water on the shallow part of the bank.

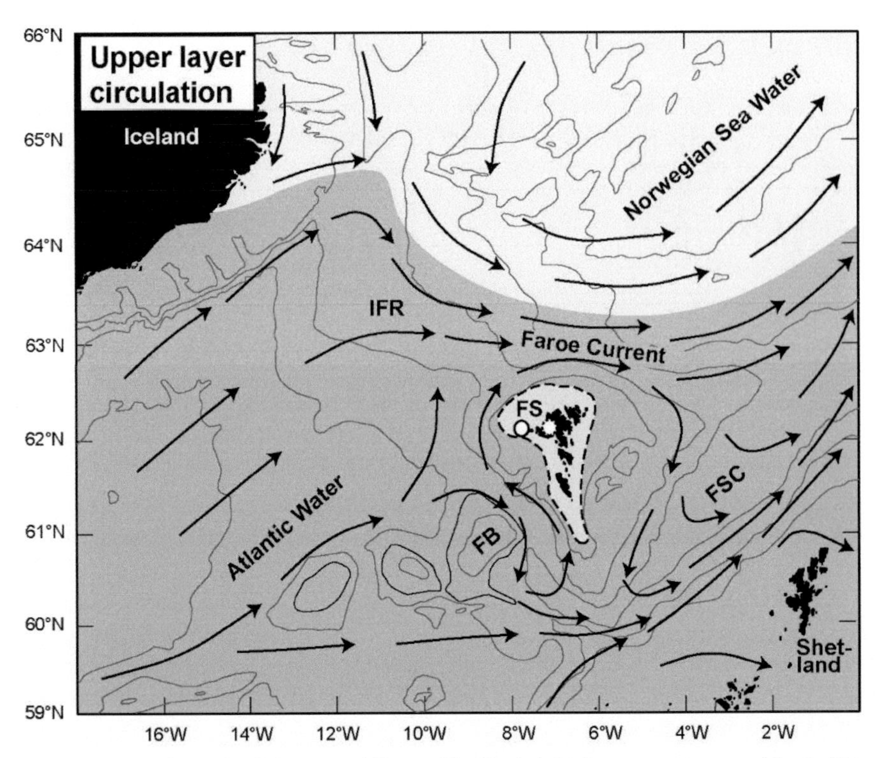

Figure 18. Upper layer circulation around Faroes. The North Atlantic current transports Atlantic Water towards the Faroes, but after crossing the Iceland-Faroe Ridge (IFR), it meets colder and fresher water in the Norwegian Sea. On the Faroe Shelf, the water is somewhat colder and fresher than the Atlantic Water, and this creates a front indicated by the dashed line. Black arrows indicate the general circulation in the upper layers. Atlantic Water is indicated with a dark grey area, and water of Arctic origin is indicated with a light grey area to the north. Abbreviations: Faroe Shelf (FS), Faroe Bank (FB), Iceland-Faroe Ridge (IFR) and Faroe-Shetland Channel (FSC). Coastal stations Mykines (white dot with full black line) and Oyrargjógv (white dot with dotted black line) are indicated on the map.

Temperature

The temperature of AW in Faroese waters is modulated by the strength of the subpolar gyre: when the gyre is weak, temperatures and salinities are high and vice versa (Hátún et al. 2005; Larsen et al. 2012). The long-term trend in AW properties is advected onto the shelf, such that long-term variability is similar to off-shelf. Figure 19 shows the long-term and seasonal temperature variation of FSW. Cold periods were observed in the late 1910s, 1960s and in the mid-1990s. In recent years, temperatures increased in the late 1990s and again around 2002–2003 and have remained relatively high since then, with an annual mean above 8°C. A recent abrupt freshening in the AW was observed in Faroese waters in 2016, but it was not accompanied by a similar reduction in temperatures (Holliday et al. 2020).

Seasonally, the temperature on the shelf varies between 6–7°C in January–March and 10–11°C in August–September (Figure 19b). There is also observed a

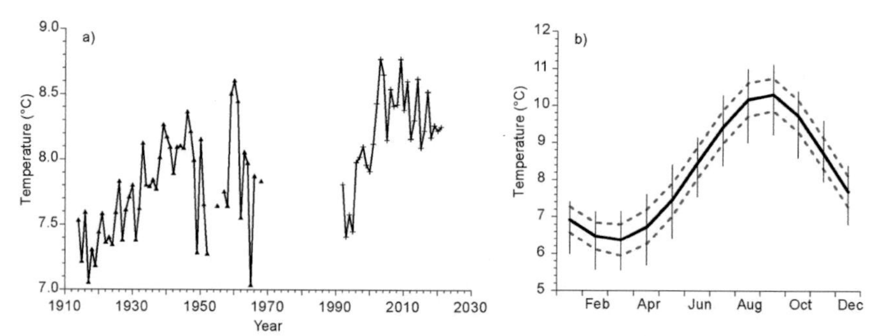

Figure 19. (a) Annual mean temperature observed at coastal stations Mykines (1914–1969) and Oyrargjógv (since 1991) representing FSW (see location on Figure 18) and (b) seasonal variation based on observations at coastal station Oyrargjógv. The thick black line is the monthly climatology for the period 1991–2020, the dotted lines are the climatology +/– 1 standard deviation, and the vertical bars are the minimum and maximum temperatures each month observed in the period. Updated from Larsen et al. (2008).

tendency for a phase shift in the seasonal variation, where the minima and maxima now occur about ½–1 month later in the year (Larsen et al. 2008).

Continuous measurements of temperature on the Faroe Bank do not exist, but observations from spring surveys (typically in March) indicate similar long-term variability as seen on the Faroe Shelf, with temperatures (in March) on the shallowest part of the bank ranging from 7.4°C in 1992 to 8.5°C in 2011 (Faroe Marine Research Institute, unpublished data).

Plankton

Phytoplankton forms the basis of the food chain on the Faroe Shelf, and annual variations in the magnitude of the spring bloom can be followed up the food chain to zooplankton, fish larvae/juveniles, sand eels, demersal fish and seabirds (Eliasen et al. 2011, Gaard et al. 2002, Steingrund and Gaard 2005, Jacobsen et al. 2019). Specifically, both recruitment and weight-at-age of cod and haddock have shown a prominent positive relationship with the magnitude of the spring bloom (Gaard et al. 2002). Phytoplankton seems, via zooplankton production (eggs/nauplii), to regulate the survival of cod larvae/juveniles that may determine the year class strength of cod later in life (Jacobsen et al. 2019; Jacobsen et al. 2020). The mechanism could also be more indirect where phytoplankton stimulates the abundance of fish larvae/juveniles, including sand eels, that act as food or reduce predation and/or cannibalism on juvenile cod (Jacobsen et al. 2019; Steingrund et al. 2010; Figure 20). Somewhat surprisingly, fish larvae/juveniles have a top-down effect on zooplankton biomasses during summer, causing a strong negative relationship with cod recruitment (ICES 2021; Jacobsen et al. 2019). The potential influx of zooplankton from the Norwegian Sea, e.g., via the East Icelandic Current, might also play a role in recruiting cod on the Faroe Plateau (Faroe Marine Research Institute, unpublished data).

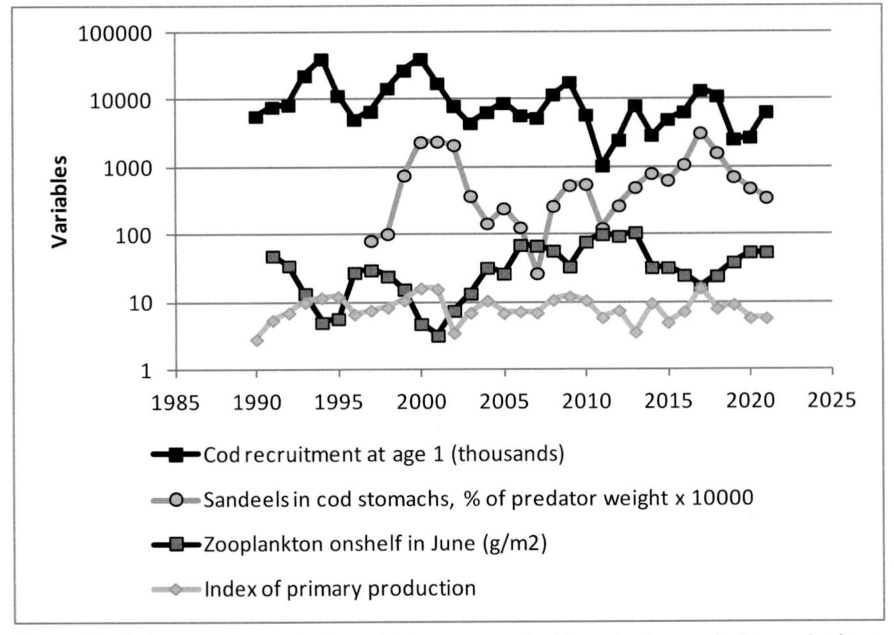

Figure 20. Cod recruitment on the Faroe Plateau compared with sand eels, zooplankton and primary production on the Faroe Shelf. Source: ICES (2021) and for sand eels: national database. Note the logarithmic scale.

Nursery Grounds

Cods grow up close to land, including high-production areas like estuaries and/or the littoral zone that may be covered with dense forests of macroalgae. The construction of harbors, industrial places or roads in such areas may reduce the extent of these areas and, over time, negatively affect cod recruitment.

Trophic Interaction

Cod in Faroese waters are preyed on by many organisms, including anglerfish (*Lophius piscatorius*) that 2001–2005 consumed 7,543 tons of cod annually, corresponding to 11.2% of the total cod stock biomass (Ofstad et al. 2013). On the other hand, cod prey on juvenile anglerfish, and a large cod stock could, therefore, reduce anglerfish recruitment and thereby its own natural mortality. This could be regarded as an example of a "cultivation effect" (Walthers and Kitchell 2001). Haddocks prey on many of the same food organisms (crustaceans) as small cod (Sørensen 2021), but large cod also consume many haddock (Figure 9; Figure 11). Therefore, a large cod stock could reduce the food competition between small cod and haddock. High fishing mortality leading to few large predatory cod could consequently increase the natural mortality of small cod from anglerfish and increase the food competition from haddock.

Climate Change

A negative relationship has been observed between temperature and the biomass of several cod stocks. This could be coincidental where the temperature has increased due to climate change whereas cod biomass has decreased due to, e.g., too high fishing pressure. However, increased temperatures could negatively affect the survival of cod larvae. It has been found that cod in the North Atlantic usually spawn at temperatures between 3°C and 7°C (Righton et al. 2010; Rose 2005). The temperature in the spawning places in Faroese waters is close to or higher than this upper limit (Figure 16; Figure 19). Ottosen et al. (2018) found that cod in the northern spawning area seem to spawn earlier and in more easterly locations when the temperature is higher than normal, seemingly in order to avoid too high temperatures during spawning. This did not happen in the western spawning area, and other factors could, therefore, not be ruled out. Temperature might be an even greater problem on Faroe Bank, which is 1°C higher than on Faroe Plateau from February to April (Hansen 2018). Another way high temperatures might negatively affect cod is through their sand eel prey. Sand eels might lose much fat when overwintering in too high temperatures (Eliasen et al. 2011; Eliasen 2013) and, therefore, be less profitable food for cod. *Calanus finmarchicus* may encounter higher overwintering temperatures in the Norwegian Sea or elsewhere (Conzález-Pola et al. 2020), causing them to be less profitable food (Jónasdóttir et al. 2019) for a variety of organisms on the adjacent shelves. Increasing temperatures could also affect the genetics of cod by, e.g., decreasing the frequency of the dominating "coldwater" hemoglobine allele Hb-I(2) in the cod population at the Faroes (Sick 1965; Jamieson and Birley 1989; Imsland and Jónsdóttir 2003). In contrast to cod populations in Iceland or in Norwegian waters that, in principle, can move to areas with suitable temperatures when temperatures are rising, cod in Faroese waters (on either the Faroe Plateau or Faroe Bank) are left with no places to go under increasing temperatures and are therefore more vulnerable to climate change. The only option to mitigate the effects of climate change could, therefore, be to keep the biomass of cod in Faroese waters high (Brander 2005) by a low fishing mortality.

Stock Management

Stock Assessment

An age-based stock assessment for Faroe Plateau cod is performed annually in ICES (2021). The input data to the assessment is a time series of catch in numbers at age and tuning abundance indices from the groundfish surveys in March and August. In addition, information about individual growth (weight-at-age) and maturity is used. The assessment provides absolute estimates of the population numbers of all ages (2–10+) in the stock back to 1959. Multiplying these numbers with weight-at-age data and maturity data gives estimates of the biomass of the total stock as well as the spawning stock. Fishing mortalities are calculated as the decrease in cohort population numbers.

Stock biomasses were reconstructed back in time in ICES (2015) and ICES (2016) (see Figure 1). As already mentioned, the biomass of cod on the Faroe Plateau has decreased markedly over time. The ratio between the catch and the fishable biomass (age three and older)—another measure of fishing pressure—was low at the beginning of the 20th century, was high around 1960 and decreased in the 1970s. It again increased and reached high levels in the late 1980s and in the 2000s, followed by near-collapses in the stock (Steingrund 2009). Although fishing undoubtedly has affected the stock dynamics, environmental variables, such as primary production (Steingrund and Gaard 2005), have likely also contributed to the low current biomass of cod on the Faroe Plateau.

Effort Management System

The cod fishery at the Faroes was regulated by technical regulations and area restrictions in the 1980s. The area restrictions secured that trawl and longline fisheries were separated. A quota system was introduced in 1994 in the aftermath of the collapse of the national economy of the Faroe Islands in 1992 and the poor state of the cod/haddock stocks. In 1995–96, the cod stock recovered, and the catch rates were much higher than expected, especially in relation to the low quotas. Partly as a result of this, an effort management system was introduced in June 1996 (Jákupsstovu et al. 2007), leaving the Faroe Islands the only country in the region that does not rely on quotas as the main tool to regulate demersal fisheries. The fishing fleet was separated into five groups: (1) large single trawlers fishing saithe and redfish; (2) large pair trawlers fishing saithe and redfish; (3) large longliners fishing cod, haddock, ling and tusk; (4) small longliners fishing cod, haddock, ling and tusk; (5) small boats fishing cod and haddock with longlines and jigging reels. These groups were allocated a number of fishing days per fishing year (1st September to 31st August the year after) that allowed them to land the whole catch without considering species composition. In 2005, a group of small single trawlers fishing for cod, haddock and flatfish within the 12 nm fishery limit during summer was separated from Group 4 to form Group 4T.

In 1995 work was done to set the initial number of fishing days. The fishing mortality in the coming years proved to be much higher than expected (ICES 2021). From 1996 to 2020, the Faroese Parliament decided the final number of fishing days per fishing year. Normally, around 40% of the allocated fishing days were not used (ICES 2021). Over the years, the number of allocated days has been reduced by around half since its introduction in 1996 (ICES 2021).

Harvest Control Rule

Work was done in 2011 to construct a formal management plan for the cod/haddock and saithe fisheries based on the effort management system, but the plan was never implemented. In 2018, the work continued, partially in an attempt to get the Faroese fisheries MSC certified. A management plan for cod, haddock and saithe on the Faroe Plateau—based on the existing effort management system—was constructed in 2019 (Anon. 2019) and implemented in 2021 (ICES 2021). Faroe Bank is not a part of the management plan.

The management plan operates with two groups of fleets, those fishing saithe (Group 2) and those fishing cod and haddock (Group 3, 4, 4T and 5). Group 2 is regulated by the stock status of Saithe, while Groups 3–5 are regulated by the stock status of cod and haddock. If the spawning stock is below a certain limit or the harvest rate (defined as catch over fishable biomass) is too high, the number of fishing days is reduced by 5% (harvest rate is used instead of fishing mortality because it is easier to understand). If the spawning stock is above a certain limit and the harvest rate is below a certain limit, the number of fishing days is increased by 5%. Otherwise, the number of fishing days is unchanged. For the cod/haddock fisheries, the harvest control rule is complicated by having to consider the stock status of both cod and haddock, i.e., reducing days if either of the spawning stocks is too low or if either of the harvest rates is too high or increasing days if both spawning stocks are high enough and both harvest rates are low enough. The management plan was evaluated by Faroese authorities in autumn 2023, and the revised management plan is supposed to be sent to ICES for evaluation in 2024.

Management Challenges

The effort management system has the important beneficial property that there is no reason to discard fish, and this is believed to lead to accurate catch statistics. Also, the mixed fish problem is partially solved because there are no quotas. Another requirement for a successful management system is that the fishing industry accepts the system. As with all management systems, this system also has limitations and challenges. Regulating fishing mortalities (harvest rates) with a number of fishing days may be a challenge because the number of hooks or trawl hours is not specified, and it might be difficult to reach the desired fishing mortality. In addition, natural conditions affect the efficiency of longlines. For example, the catchability with longlines may vary interannually depending on food conditions since cod are likelier to take longline baits when hungry (ICES 2021; Steingrund et al. 2009) (Figure 21).

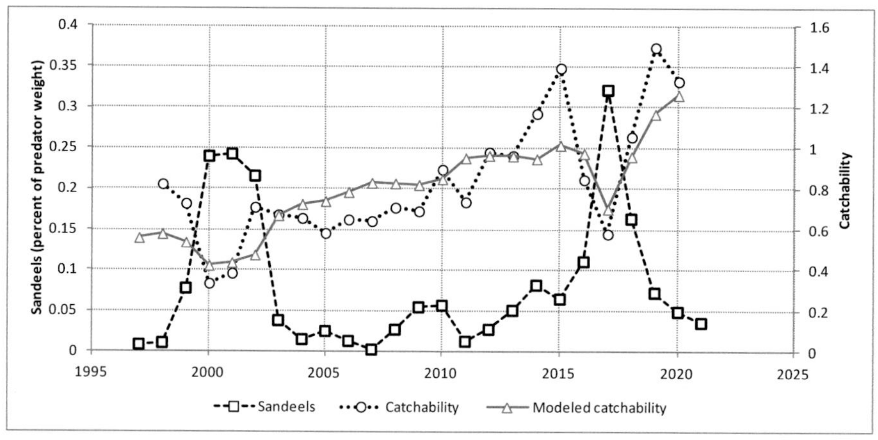

Figure 21. Abundance of sand eels and catchability with longlines (Group 3). The catchability is calculated as partial fishing mortality per 10,000 fishing days. The catchability (log-transformed) is modelled by year and sand eels and indicates an increase in gear efficiency of 1–2% per year.

Since longlines take between 30% and 70% of the cod catch, this affects the overall fishing mortality on cod. Another feature of the effort management system concerns increases in gear efficiency that may be 1–2% per year (Figure 21), although the harvest control rule will take this into account by reducing the number of fishing days when harvest rates become too high or spawning stock biomasses too low. However, the value of the catch may not necessarily be optimized because the highest catches per fishing day will be obtained when cod are in poor condition in connection with the spawning season when the unit price may be lower than, e.g., during the autumn when cod are in better condition.

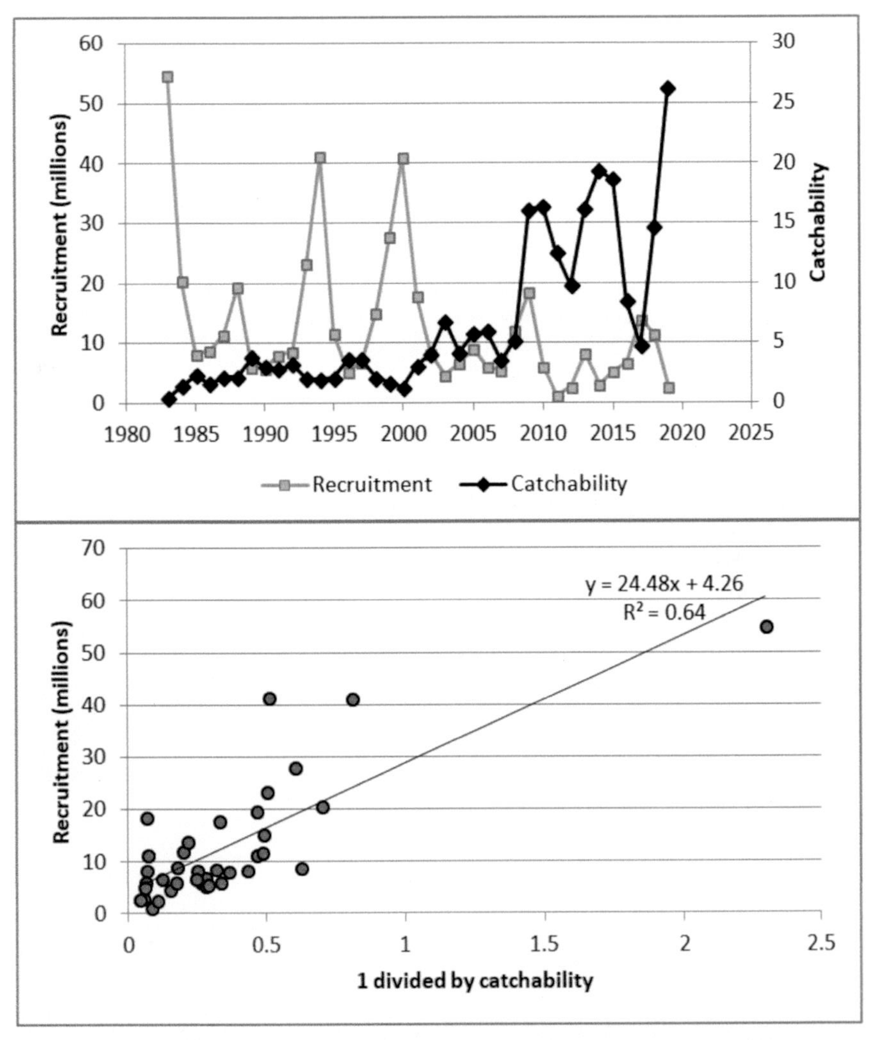

Figure 22. Relationship between recruitment at age one and catchability with longlines for small longliners operating close to land. Catchability is calculated as catch per unit effort in June-October (kg/day) for age 3+ cod divided by the assessment biomass of age 3+ cod at the beginning of the year (thousand tons) and is a measure of food shortage and/or cannibalism. Modified from ICES (2021) and Steingrund et al. (2010).

The current harvest control rule has benefits and drawbacks. While a constraint to change the number of fishing days by only 5% from year to year is desirable for the fishing industry, it may be a drawback if rapid adjustments are required, e.g., if the cod stock declines due to poor environmental conditions. In such situations, though, the management plan opens up the possibility to apply other measures. Relaxing the constraint from 5% to 10% or even 20% may solve this problem, but it could, during normal years, lead to unwanted oscillations in the number of fishing days.

Another challenge is the observed decreasing recruitment of cod over time, which seems to result from food shortage for adult cod or increased spatial overlap between adult cod and juvenile cod (Steingrund et al. 2010) (Figure 22). The food shortage could result from lower productivity of the Faroe Shelf ecosystem (Steingrund and Gaard 2005) or other processes, including the effects of rising temperatures. These questions are crucial to resolve in the future in order to obtain a sustainable management of cod in Faroese waters.

References

Anon. 2019. Fiskidagaskipan og umsitingarætlan. Frágreiðing og tilmæli frá arbeiðsbólkinum at gera uppskot til umsitingarætlan og at eftirmeta skipanina í fiskiskapinum eftir botnfiski undir Føroyum (Fisheries management plan and harvest control rule for cod, haddock and saithe in Faroese waters) (in Faroese). https://lms.cdn.fo/media/12444/fiskidagaskipan-og-umsitingar%C3%A6tlan-fr%C3%A1grei%C3%B0ing-fr%C3%A1-arbei%C3%B0sb%C3%B3lki-mai-2019.pdf?s=k4aSmLc1W4eXb_RidnITC-WYwDA.

Bedford, B.C. 1966. English cod tagging experiments in the North Sea. ICES CM 1966/G:9.

Brander, K.M. 1995. The effect of temperature on the growth of Atlantic cod (*Gadus morhua* L.). ICES Journal of Marine Science, 52: 1–10.

Brander, K.M. 2005. Cod recruitment is strongly affected by climate when stock biomass is low. ICES Journal of Marine Science, 62: 339–343.

Case, R.A.J., Hutchinson, W.F., Hauser, L., et al. 2005. Macro and micro-geographic variation in pantophysin (PanI) allele frequencies in NE Atlantic cod *Gadus morhua*. Marine Ecology Progress Series, 301: 267–278.

Church, M.J., Arge, S.V., Brewington, S., et al. 2005. Puffins, Pigs, Cod and Barley: Palaeoeconomy at Undir Junkarinsfløtti, Sandoy, Faroe Islands. Environmental Archaeology, 10: 179–197. DOI: 10.1179/env.2005.10.2.179.

Degn, A., 1929. Oversigt over Fiskeries og Monopolhandelen paa Færøerne 1709–1856. Felagið Varðin. Thorshavn, 1929. (Overview of the fisheries and the monopoly trade on the Faroes 1709–1856, in Danish).

Du Buit, M.-H. 1982. Essai sur la prédation de la morue (*Gadus morhua*, L.), l'eglefin (*Melanogrammus aeglefinus* (L.)) et du lieu noir (*Pollachius virens* (L.)) aux Faeroes. Cybium, 8: 13–19.

Eliasen, K., Reinert, J., Gaard, E., et al. 2011. Sandeel as a link between primary production and higher trophic levels on the Faroe shelf. Marine Ecology Progress Series, 438: 185–194.

Eliasen, K. 2013. Sandeel, Ammodytes spp., as a Link between Climate and Higher Trophic Levels on the Faroes Shelf (Doctoral dissertation, PhD Thesis, Faroe Marine Research Institute).

Eliasen, S.K., Hatun, H., Larsen, K.M.H., et al. 2017. Faroe shelf bloom phenology - the importance of ocean-to-shelf silicate fluxes. Cont. Shelf Res., 143: 43–53. https://doi.org/10.1016/j.csr.2017.06.004.

Fjallstein, I., and Magnussen, E. 1996. Growth of Atlantic cod (*Gadus morhua* L.) of Faroe Bank strain and Faroe Plateau strain in captivity. ICES CM F12: 1–16. Available at www.nvd.fo/uploads/tx_userpubrep/Growth_of_Atlantic_cod_ICESPaperCM96F_12.pdf.

Gaard, E., Hansen, B., Olsen, B., et al. 2002. Ecological features and recent trends in physical environment, plankton, fish and sea birds in the Faroe plateau ecosystem. pp. 245–265. *In*: Sherman,K., and H.-R. Skjoldal (eds.). Large Marine Ecosystem of the North Atlantic. Elsevier. 449 pp.

Gaard, E., and Reinert, J. 2002. Pelagic cod and haddock juveniles on the Faroe Plateau: distribution, diets and feeding habitats, 1994–1996. Sarsia, 87: 193–206.

Gaard, E., and Steingrund, P. 2001. Reproduction of Faroe Plateau cod. Spawning grounds, egg advection and larval feeding. Fróðskaparrit (Annales Societatis Scientarum Færoensis), 48 : 87–103.

Gjedrem, T. 2000. Genetic improvement of cold-water fish species. Aquaculture Research, 31: 25–33.

González-Pola, C., Larsen, K.M.H., Fratantoni, P., et al. 2020. ICES Report on Ocean Climate 2019. ICES Cooperative Research Reports No. 350. 136 pp. https://doi.org/10.17895/ices.pub.7537.

Goodlad, J.H., and Litt, M. 2014. The Shetland cod fishery from 1811 to 1909. A study in historical geography. A thesis presented for the degree of Doctor of Philosophy at the University of Aberdeen, 2014.

Hansen, B. 2018. Bottom temperature and echo-sounder corrections around the Faroe Plateau. Faroe Marine Research Institute, technical report no. 18–02. http://www.hav.fo/PDF/Ritgerdir/2018/TecRep1802.pdf.

Hansen, J.S., 1961. Havið og vit. 1. partur. Nordoyggja skipa- og handilssøga. Egið forlag, Klaksvík, 1961. (The Sea and Us, first part. The maritime and trade history in the northern islands in the Faroes, in Faroese).

Hansen, J.S. 1966. Havið og vit. 2. partur. Minniligir dagar. Egið forlag, Klaksvík, 1966 (The Sea and Us, second part. Memorable days, in Faroese).

Hansen, J.S. 1978. Tey byggja land, 5. partur. Húsa og Mikladals sóknir, Klaksvík. (Building land, fifth part. Villages Húsar and Mikladalur in Kalsoy, Faroe Islands, in Faroese).

Hansen, B., Meldrum, D., and Ellett, D. 1998. Satellite-tracked drogue paths over Faroe Bank and the Iceland–Faroe Ridge. ICES Cooperative Research Report, 225: 150–161.

Hansen, B., and Østerhus, S. 2000. North Atlantic–Nordic Seas exchanges. Prog. Oceanogr. 45: 109–208. doi: 10.1016/S0079-6611(99)00052-X.

Hátún, H., Sandø, A.B., Drange, H., et al. 2005. Influence of the Atlantic Subpolar Gyre on the Thermohaline Circulation. Science, 309: 1841–1844. doi: 10.1126/science.1114777.

Homrum, E., í. 2007. Condition and diet of cod (*Gadus morhua* (L.)) on the Faroe Plateau in January 1991 compared to March 1999. Master Thesis, University of Gothenburg. 44 pp.

Holliday, N.P., Bersch, M., Berx, B., et al. 2020. Ocean Circulation Causes the Largest Freshening Event for 120 Years in Eastern Subpolar North Atlantic. Nat. Com., 11: 585. doi: 10.1038/s41467-020-14474-y.

ICES, 2007. Report of the North-Western Working Group (NWWG), 24 April-3 May, 2007, ICES Headquarters, Copenhagen, Denmark. ICES CM 2007/ACFM:17.

ICES, 2015. Report of the North-Western Working Group (NWWG), 28 April–5 May, 2015, ICES Headquarters, Copenhagen, Denmark. ICES CM 2015/ACOM:07.

ICES, 2016. Report of the North-Western Working Group (NWWG), 27 April–4 May, 2016, ICES Headquarters, Copenhagen. ICES CM 2016/ACOM:08.

ICES, 2021. Northwestern Working Group (NWWG). ICES Scientific Reports. 3:52. 556 pp. https://doi.org/10.17895/ices.pub.8186.

Imsland, A.K., and Jónsdóttir, Ó.D.B. 2003. Linking population genetics and growth properties of Atlantic cod. Reviews in Fish Biology and Fisheries 13: 1–26.

Jacobsen, S., Gaard, E., Hátún, H., et al. 2019. Environmentally Driven Ecological Fluctuations on the Faroe Shelf Revealed by Fish Juvenile Surveys. Front. Mar. Sci., 6: 559. doi: 10.3389/fmars.2019.00559.

Jacobsen, S., Klitgaard Nielsen, K., Kristiansen, R., et al. 2020. Diet and prey preferences of larval and pelagic juvenile Faroe Plateau cod (*Gadus morhua*). Marine Biology, 167: 122. https://doi.org/10.1007/s00227-020-03727-5.

Jamieson, A., and Birley, A.J. 1989. The demography of a haemoglobin polymorphism in the Atlantic cod, *Gadus morhua* L. Journal of Fish Biology, 35 (supplement A): 193–204.

Jákupsstovu, S.H., Cruz, L.R., Maguire, J.-J., et al. 2007. Effort regulation of the demersal fisheries at the Faroe Islands: a 10-year appraisal. ICES Journal of Marine Science, 64: 730–737.

Jákupsstovu, S.H.í, and Reinert, J. 1994. Fluctuations in the Faroe Plateau cod stock. ICES Marine Science Symposia, 198: 194–211.

Joensen, H., Steingrund, P., Fjallstein, I., et al. 2000. Discrimination between two reared stocks of cod (*Gadus morhua*) from the Faroe Islands by chemometry of the fatty acid composition in the heart tissue. Marine Biology, 136: 573–580.

Joensen, J.S., Steingrund, P., Henriksen, A., et al. 2005. Migration of cod (*Gadus morhua*): tagging experiments at the Faroes 1952–1965. Fróðskaparrit, 53: 100–135.

Joensen, J.S., and Tåning, Å.V. 1970. Marine and Freshwater Fisheries. Vald Pedersen Bogtrykkeri, Copenhagen.

Jónasdóttir, S.H., Wilson, R.J., Gislason, A., et al. 2019. Lipid content in overwintering *Calanus finmarchicus* across the subpolar eastern North Atlantic ocean. Limnol. Oceanogr., 64: 2029–2043. https://doi.org/10.1002/lno.11167.

Jones, B.W. 1966. The cod and the cod fishery at Faroe. Ministry of Agriculture, Fisheries and Food. Fishery Investigations. Series II, 24(5): 32.

Larsen, K.M.H., Hansen, B., and Svendsen, H. 2008. Faroe shelf water. Cont. Shelf Res. 28: 1754–1768. doi: 10.1016/j.csr.2008.04.006.

Larsen, K.M.H., Hansen, B., and Svendsen, H. 2009. The Faroe shelf front: properties and exchange. J. Mar. Sys., 78: 9–17. doi: 10.1016/j.jmarsys.2009.02.003.

Larsen, K.M.H., Hátún, H., Hansen, B., et al. 2012. Atlantic Water in the Faroe Area: Sources and Variability. ICES J. Marine Sci., 69: 802–808. doi: 10.1093/icesjms/fss028.

Love, R.M., Robertson, I., Lavéty, J., et al. 1974. Some biochemical characteristics of cod (*Gadus morhua* L.) from the Faroe Bank compared with those from other fishing grounds. Comparative Biochemistry and Physiology, 47B: 149–161.

Magnussen, E. 1996. Electrophoretic studies of cod (*Gadus morhua*) from Faroe Bank and Faroe Plateau compared with results found in other distribution areas. ICES C.M. 1996/G:10. 18pp.

Magnussen, E. 2011. Food and feeding habits of cod (*Gadus morhua*) on the Faroe Bank. ICES Journal of Marine Science, 68(9): 1909–1917. doi:10.1093/icesjms/fsr10.

Magnussen, E. 2002. Demersal fish assemblages of the Faroe Bank: species composition, distribution, biomass spectrum and diversity. Marine Ecology Progress Series, 238: 211–225.

Magnussen, E. 2007. Interpopulation comparison of growth patterns of 14 fish species on Faroe Bank: are all fish on the bank fast-growing? Journal of Fish Biology, 71: 453–475.

Nielsen, E.E., MacKenzie, B.R., Magnussen, E., et al. 2007. Historical analysis of *Pan* I in Atlantic cod (*Gadus morhua*): temporal stability of allele frequencies in the southeastern part of the species distribution. Canadian Journal of Fisheries and Aquatic Sciences, 64: 1448–1455. Doi:10.1139/F07-104.

Ofstad, L.H., Steingrund, P., and Pedersen, T. 2013. Feeding ecology of anglerfish *Lophius piscatorius* in Faroese waters. In: Anglerfish *Lophius piscatorius* L. in Faroese waters: Life history, ecological importance and stock status. PdD dissertation, University of Tromsø, Norway, January 2013.

Ottosen, K.M., Pedersen, M.W., Eliasen, S.K., et al. 2017. Migration patterns of the Faroe Plateau cod (*Gadus morhua*, L.) revealed by data storage tags. Fisheries Research, 195: 37–45.

Ottosen, K.M., Steingrund, P., Magnussen, E., et al. 2018. Distribution and timing of spawning Faroe Plateau cod in relation to warming spring temperatures. Fisheries Research, 198: 14–23.

Pampoulie, C., Steingrund, P., Stefánsson, M.Ö., et al. 2008a. Genetic divergence among East Icelandic and Faroese populations of Atlantic cod provides evidence for historical imprints at neutral and non-neutral markers. ICES Journal of Marine Science, 65: 65–71.

Pampoulie, C., Jakobsdóttir, K.B., Marteinsdóttir, G., et al. 2008b. Are Vertical Behaviour Patterns Related to the Pantophysin Locus in the Atlantic Cod (*Gadus morhua* L.)? Behavioral Genetics (2008) 38: 76–81. DOI 10.1007/s10519-007-9175-y.

Patursson, E., 1961. Fiskiveiði – fiskimenn 1850–1939. Minningarrit Føroya Fiskimannafelags 1911–14. november 1961. (Catches-fishermen 1859–1939. Memorandum for Føroya Fiskimannafelag 1911 to 14. November 1961, in Faroese).

Rae, B.B. 1967. The food of cod on Faroese grounds. Marine Research, 6. 23 pp.

Righton, D.A., Andersen, K.H., Neat, F., et al. 2010. Thermal niche of Atlantic cod *Gadus morhua*: Limits, tolerance and optima. Mar. Ecol. Prog. Ser., 420: 1–13. doi: 10.3354/meps08889.

Robichaud, D., and Rose, G.A. 2004. Migration behaviour and range in Atlantic cod: inference from a century of tagging. Fish and Fisheries, 5: 185–214.

Rose, G.A. 2005. On distributional responses of North Atlantic fish to climate change. ICES Journal of Marine Science, 62: 1360–1374.doi:10.1016/j.icesjms.2005.05.007.

Sick, K. 1965. Haemoglobin polymorphism of cod in the North Sea and the North Atlantic Ocean. Hereditas, 54(3): 49–73.

Steingrund, P., and Gaard, E. 2005. Relationship between phytoplankton production and cod production on the Faroe shelf. ICES Journal of Marine Science, 62: 163–176.

Steingrund, P., Hansen, B., and Gaard, E. 2004. Cod in Faroese waters. ICES Cooperative Research Report, 274: 50–55.

Steingrund, P. 2009. The near-collapse of the Faroe Plateau cod (*Gadus morhua* L.) stock in the 1990s: The effect of food availability on spatial distribution, recruitment, natural production and fishery. Dr. Philos. thesis at the University of Bergen, 320 pp. http://hdl.handle.net/1956/3697.

Steingrund, P., Clementsen, D.H., and Mouritsen, R. 2009. Higher food abundance reduces the catchability of cod (*Gadus morhua*) to longlines on the Faroe Plateau. Fisheries Research, 100: 230–239.

Steingrund, P., Mouritsen, R., Reinert, J., et al. 2010. Total stock size and cannibalism regulate recruitment in cod (*Gadus morhua*) on the Faroe Plateau. ICES Journal of Marine Science, 67: 111–124.

Steingrund, P., and Ofstad, L.H. 2010. Density-dependent distribution of Atlantic cod (*Gadus morhua*) into deep waters on the Faroe Plateau. ICES Journal of Marine Science, 67: 102–110.

Strubberg, A.C. 1916. Marking experiments with cod at the Færoes. Meddelelser fra Kommissionen for Danmarks Fiskeri- og Havundersøgelser, serie: Fiskeri, 5(2): 1–125.

Strubberg, A.C. 1933. Marking experiments with cod at the Faroes. Second report. Experiments in 1923-1927. Meddelelser fra Kommissionen for Danmarks Fiskeri- og Havundersøgelser, serie: Fiskeri, 9(7): 1–36.

Sørensen, B. 2021. Growth and spatial distribution of cod and haddock on the Faroe Plateau and relationship with the amount of forage fish. Biology thesis, University of Copenhagen, Faculty of Science, Department of Biology, November 2021. 73 pp.

Tåning, Å.V. 1940. Migration of cod marked on the spawning places off the Faroes. Meddelelser fra Kommissionen for Danmarks Fiskeri- og Havundersøgelser, serie: Fiskeri, 10(7): 1–52.

Tåning, Å.V. 1943. Fiskeri- og havundersøgelser ved Færøerne (Fisheries- and oceanographic investigations at the Faroes, in Danish). Fiskeri- og Havundersøgelser, 12: 1–127.

Walters, C., and Kitchell, J.F. 2001. Cultivation/depensation effects on juvenile survival and recruitment: implications for the theory of fishing. Canadian Journal of Fisheries and Aquatic Sciences, 58: 39–50. DOI: 10.1139/cjfas-58-1-39.

CHAPTER 7

Northwest European Shelf Cod Stocks; North Sea, West of Scotland, Irish Sea and Celtic Sea

Peter J. Wright,[1,*] *Helen Dobby*[2] and *Clive Fox*[3]

Stock Descriptions

Cod on the northwest European shelf were managed as four stocks: 7e-k, which includes the Celtic Sea and Western Channel, Irish Sea (Division 7a), west of Scotland (Division 6a), and the North Sea (sub-area 4), together with the Skagerrak (Division 3a) and eastern English Channel (Division 7d; Figure 1a). For brevity, the stocks composed of more than one ICES sub-area are referred to by the largest of these: the Celtic Sea for 7e-k and the North Sea for 3a, 4, and 7d. The latter was originally considered to be three separate areas for stock assessment purposes but were combined in 1996 based on tag-recapture evidence of mixing (ICES 1997). These cod stocks have been heavily exploited for decades, with catches and spawning stock biomass (SSB) now at low levels (Pope and Macer 1996; Cook et al. 1997; ICES 2022a,b). In the early 2000s, the European Commission implemented a series of recovery plans that led to a reduction in fishing mortality, although with the recent exception of Irish Sea cod, northwest European shelf stocks have generally continued to be overexploited. In 2023, the west of Scotland and North Sea were combined in a multi-stock assessment model in an attempt to better reflect population structure (ICES 2022c,d). In the following discussion, we address the challenges facing the recovery and management of these stocks; however, first, we begin by describing the region and the biology and ecology of cod inhabiting these waters.

[1] Marine Ecology and Conservation Consultancy, Ellon, AB41 8YH.
[2] Marine Directorate, Scottish Government, Marine Laboratory, 375 Victoria Road, Aberdeen AB11 9DB.
[3] Scottish Association for Marine Science, Oban, Argyll, PA37 1QA.
* Corresponding author: p.j.wright_mecc@outlook.com

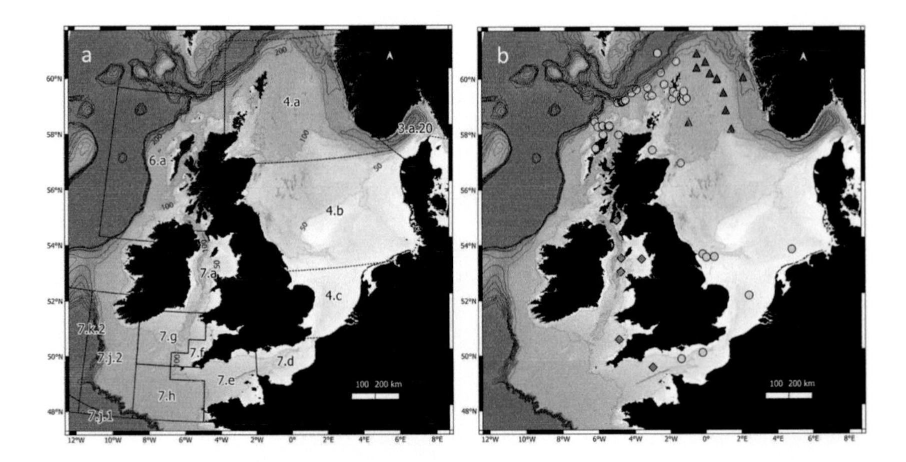

Figure 1. (a) Northwest European Stock areas: North Sea, Skagerrak and Eastern Channel (4a–c, 3a20, 7d), West of Scotland (6a), Irish Sea (7a), Celtic Sea and Western Channel (7e–k). (b) Locations of genetically assigned cod to Viking (▲), Dogger (●) and Celtic demes (◊) based on Heath et al. (2014) and Wright et al. (2021).

Population Structure

The genetic structuring of cod in the northeast Atlantic shelf seas appears to have developed due to recolonisation following the last glacial maximum (Pampoulie et al. 2008), although other analyses suggest pre-glacial genetic divergence (Bigg et al. 2007). Based on microsatellite and single nucleotide polymorphisms (SNPs), evidence for the present population structure within the northwest European shelf indicates three demes or populations. However, these biological units did not align with ICES stock boundaries (Nielsen et al. 2009; Poulsen et al. 2011; Heath et al. 2014; Sodeland et al. 2016; Fairweather et al. 2018; Wright et al. 2021) (see Figure 1b). The genetic differences are associated with large genomic regions of divergence where SNPs are in persistently high linkage disequilibrium among several Atlantic cod populations and migratory types (Bradbury et al. 2013; Hemmer-Hansen et al. 2013; Berg et al. 2015). These markers provide evidence of barriers to gene flow between the northern 100–200 m deeper region and the shallower shelf region of the North Sea and west of Scotland, except for the Clyde Sea (Heath et al. 2014; Wright et al. 2021). The northeast North Sea population centred around the Viking Bank has been termed the Viking deme, while the remaining North Sea and northern part of the west of Scotland have been termed the Dogger deme (Heath et al. 2014). No genetic differentiation is associated with the 4°W border between the west of Scotland and North Sea cod stocks (Heath et al. 2014; Wright et al. 2021) (see Figure 1b triangles). The third deme encompasses the Celtic and Irish Sea, as well as those from the Clyde, within the west of Scotland stock (Heath et al. 2014; Fairweather et al. 2018).

Further structuring that suggests demographic independence over the time scales relevant to management has been indicated based on the transport of early-life stages (Heath et al. 2008) and juvenile and adult fidelity (Wright et al. 2006a, 2006b; Righton et al. 2007; Neat et al. 2014; Wright et al. 2018). This led ICES (2020a, 2022c) to conclude that the North Sea stock consists of reproductively isolated populations of Viking and Dogger cod, with the Dogger deme exhibiting spatial heterogeneity, particularly in the west North Sea between Flamborough and further north extending to the northern part of Division 6a. The implications of this evidence to the current management units are discussed in the final section of this chapter.

Spawning and Early Life-Stage Distribution

Spawning areas have been identified from the distribution of spawning adults in trawls and the presence of early-stage eggs in plankton surveys. The spawning grounds tend to occur over coarse sand and mixed seabed at depths less than 150 m and a current velocity of < 1.1 ms^{1}. Comparing historical maps of cod spawning locations with more recent data suggests that spawning areas have not moved substantially over at least 100 years, so their locations may be anchored to local geographic features, facilitating the successful drift of early-life stages towards frontal zones. This has been hypothesised as a mechanism to increase the chances of larvae encountering good feeding conditions as primary and secondary production is often elevated or concentrated in oceanic fronts (Dickey-Collas et al. 1996; Munk et al. 2002; Lelievre et al. 2014).

Spawning areas are found in the eastern English Channel (Fox et al. 2008; Lelievre et al. 2014), German and Southern Bights (Brander 1994; Fox et al. 2008; Höffle et al. 2017), off Flamborough, the Moray Firth, Shetland Isles, and in the northeastern North Sea (Fox et al. 2008; González-Irusta and Wright 2016). Historically, large spawning areas were found in the Forties grounds and off the east UK coast (Graham 1934; Raitt 1967; West 1970), but surveys during this century found little evidence of spawning in the western and central North Sea (Fox et al. 2008; González-Irusta and Wright 2016; Höffle et al. 2017). Egg densities and mature cod distribution do not suggest large spawning grounds in the Skagerrak, although many fjordic populations have been identified (Knutsen et al. 2003).

Spawning areas may be continuous between the northern North Sea (4a) and west of Scotland (6a) stocks as the distribution of late-stage cod eggs (stages 3–6) extends across the 4°W boundary (Raitt 1967; Heath et al. 1994). Major spawning areas are found off the Outer Hebrides between the Butt of Lewis and North Rona, in the Minch and the Firth of Clyde (Raitt 1967; Wright et al. 2006a), including large aggregations on the sill at the edge of the North Channel (Armstrong et al. 2005; Wright et al. 2006b). Based on the incidence of spawning cod in research trawl surveys, the importance of spawning areas in much of the Minch and Inner Hebrides has declined in recent years (Holmes et al. 2008).

In the Irish Sea, spawning tends to occur in coastal waters less than 50 m deep. The most concentrated aggregations of eggs have been found in the western Irish Sea and between the Isle of Man and Cumbria and the Isle of Man and Wales (Heffernan et al. 2004; Fox et al. 2005), and these spawning areas appear stable over time (Brander 1994). Most cod spawning in the Celtic Sea occurs off northern Cornwall, southeast Ireland, and a little in the western English Channel (ICES 2020b).

In terms of timing, cod spawn around the coldest part of the year across the four stock areas (Brander 1994; González-Irusta and Wright 2016) at temperatures from 5.5–9°C (Neat et al. 2014; Höffle et al. 2017). Based on embryonic development rates, these temperatures would lead to hatching between 11 to 17 days post-fertilisation (Thompson and Riley 1981; Geffen et al. 2006). While eggs are passively transported, larvae become active swimmers and undertake diel movements, generally moving deeper at night (Nielsen and Munk 2004). The larvae can initially survive using energy reserves from the yolk sac, but this is depleted in four to six days, after which they must begin to feed (Thompson and Riley 1981), initially on phytoplankton and small zooplankton. Later, copepods dominate the larval diet (Heath and Lough 2007; Rowlands et al. 2008).

The southern and eastern central North Sea is shallow, and despite potentially extensive egg and larval transport from the eastern English Channel and Southern Bight across to the Jutland current, there appears to be little exchange with the deeper northern North Sea (Heath et al. 2008; Romagnoni et al. 2020). Salinity fronts arising from the mixing zone between the outflows of major European river basins and more saline offshore water also appear to confine the dispersal of the early-life stages (Munk et al. 2002).

Most Atlantic inflow into the northern North Sea is via the Faroe–Shetland channel and the Norwegian trench, although circulation is largely wind-driven during the winter spawning period. Southern flows along the Norwegian trench lead to the transport of cod's early life stages from the northeast North Sea spawning grounds into the Skagerrak (Huserbråten et al. 2018). The northern region of the west of Scotland stock and the northwest North Sea are well connected hydrodynamically so that according to Heath et al. (2008), eggs spawned in the Clyde tended to be retained in the stock area while those spawned in the northern west of Scotland region tended to be transported into the northwest North Sea.

The patterns of circulation within the Irish Sea are mainly influenced by local seasonal dynamics such as tides, waves and wind, but in the western Irish Sea, a seasonal density-driven gyre leads to anticlockwise circulation (Horsburgh et al. 2000). Typically, the gyre circulation begins at the end of April while cod and haddock larvae are hatching. Cod eggs released off the Irish east coast are thus thought to be entrained into this density-driven gyre (Dickey-Collas et al. 1997; Hill et al. 1997). The situation in the eastern Irish Sea is less well studied, but this area has weaker residual tidal flow with less transport potential (Howarth 1984). A biophysical model of cod eggs and larval transport for the Irish Sea indicated possible dispersal of 60–100 km from their spawning locations (van der Molen et al. 2007), with the extent being influenced by spawning time, local environmental conditions, and inferred swimming behaviours.

In the eastern Celtic Sea, cod eggs and larvae have been recorded close to the Saltees and the Smalls fishing grounds and in the southern entrance to the Bristol Channel in plankton surveys in 1938–39, 1953, 1990 and 2000 (Brander 1994; Dransfeld et al. 2004), again suggesting that cod spawning locations are also relatively persistent in this region.

Metamorphosis to the pelagic juvenile life stage occurs around 2–3 months after hatching (Fahay 1983). In the North Sea, the International 0-group Gadoid Survey monitored the pelagic 0-groups in June–July from 1974 to 1983 and found most cod off the coast of Denmark and in depths > 100 m in the northern North Sea, with smaller numbers off the northeast UK coast

(Holden 1981; Sparholt et al. 2015). Although few observations were made in the shallower, southern North Sea, the available data indicate that pelagic 0-group were scarce in the German and Southern Bight, consistent with a net northward transport from those spawning areas and consistent with later biophysical models of predicted distribution at the end of the larval phase (Kvile et al. 2018).

While settling 0-group cod were once common in offshore waters, the highest densities reported usually occur in shallow and coastal areas (Riley and Parnell 1984; Heessen 1993; Gibb et al. 2007). As pelagic juveniles are active swimmers capable of making large vertical migrations (Bailey 1975), the difference between the pelagic and demersal 0-group distribution might reflect movement inshore between these juvenile stages. In the 1960s–80s, substantial concentrations of settled 0-group cod tended to occur close to the Danish-German coast and extended over the Dogger Bank and other shallow areas (Daan 1978; Heessen 1993). However, since the 1990s, the importance of this region has declined, and the highest densities occur in the Skagerrak (ICES 2022a). In the western Irish Sea, 0-group pelagic cod distribution also reflected the close coupling between recurrent hydrographic features and spawning locations (Dickey-Collas et al. 1996).

Most empirical evidence on the connectivity of early-life stages has come from otolith micro-chemistry. Differences in trace element composition within the otolith can be related to the physico-chemical properties of the waters that the fish inhabit (Walther and Thorrold 2006), and so, by comparing elemental composition at different life stages within and between individuals, it is possible to infer connections among life stages. In the North Sea, clustering of early larval chemical signals indicated more than one contributing larval source, with differences between southern and northern sites (Wright et al. 2018), consistent with the hydrographic segregation predicted from biophysical models (Heath et al. 2008; Kvile et al. 2018; Romagnoni et al. 2020). Comparison of the otolith chemistry of settled juveniles with that of age 2–3 adults from the same year-class indicated that the Eastern Channel and southern North Sea spawning cod originate mostly from the southeast North Sea Dutch and German coastal nurseries, with a small contribution from the Skagerrak (Wright et al. 2018). In contrast, adults in the northeast North Sea mostly came from the Skagerrak, with some being of more local origin. As many settled juveniles in the Skagerrak had a larval signature consistent with the northeast North Sea, the apparent connectivity with the northeast North Sea may reflect a return migration. This would explain reports of high numbers of juveniles in the Skagerrak having genetic signatures of a North Sea origin (Knutsen et al. 2003; Knutsen et al. 2011). Similar otolith analyses for the west of Scotland and Shetland have indicated that almost all adults originated from local nurseries (Wright et al. 2006a).

Late Juvenile and Adult Movements

As seen elsewhere (Robichaud and Rose 2004), cod in the northwest European shelf generally exhibit seasonal migration or resident behaviour, the latter being most common in coastal areas (Neuenfeldt et al. 2013). Tag-recapture experiments since the 1950s have shown that North Sea cod do not disperse throughout the area but mostly remain within five regions (Anonymous 1971). Tagging with archival

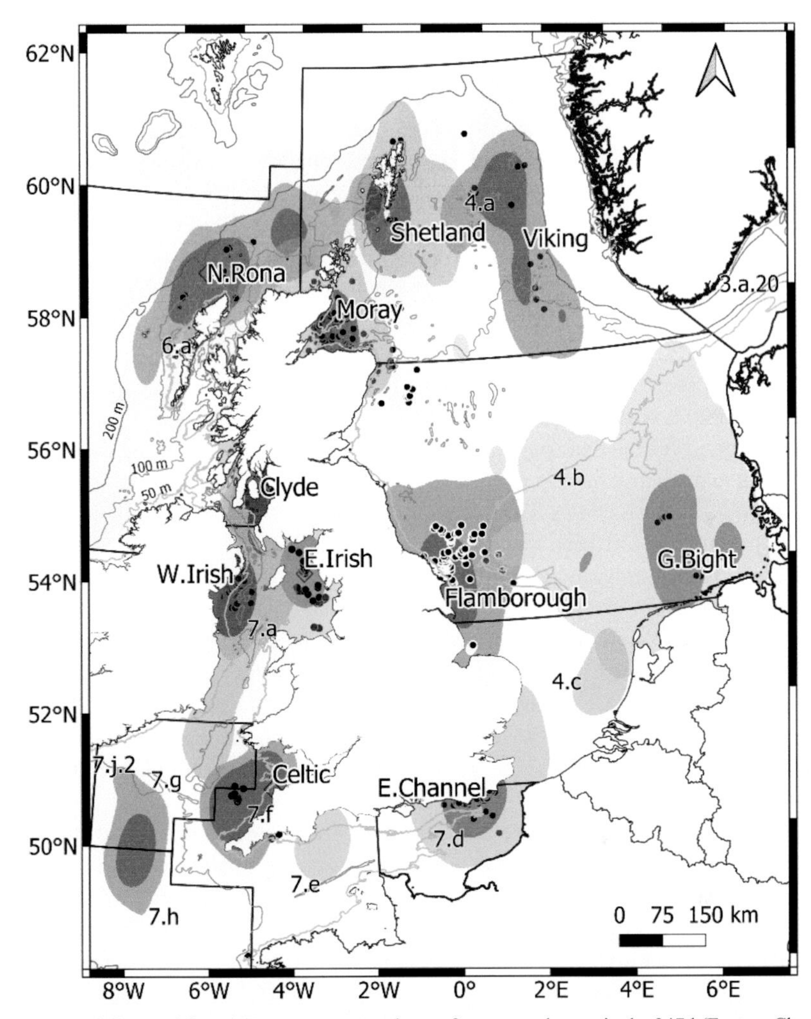

Figure 2. Cod dispersal from UK tag-recapture releases from named areas in the 347d (Eastern Channel, German Bight, Flamborough, Moray, Shetland and Viking), 6a (North Rona and Clyde), 7a (East and West Irish Sea) and 7e-k stocks (Celtic Sea-Western Channel). The darker and lighter shaded regions per release area reflect the 50% and 90% recapture distribution kernels after 90+ days and the points indicate release locations.

tags since the early 2000s has also provided estimates of home ranges derived from tidal and temperature geolocation methods (Righton et al. 2010; Neat et al. 2014). During the second and third quarters of the year, many tagged cod from the Eastern Channel and Southern Bight spawning areas migrated to feeding grounds in the central North Sea, where they would intermix with mostly resident cod (Righton et al. 2007; Neat et al. 2014). High levels of resident behaviour have been indicated in the Moray Firth (Wright et al. 2006a, 2006b) and coastal waters of Shetland (Neat et al. 2006). Cod from the Viking deme generally remains all year in the deep cool northeast North Sea (Wright et al. 2006a; Neat et al. 2014), and this is reflected in the seasonally persistent genetic structure (Wright et al. 2021).

Cod tagged off the northern Scottish coast at the boundary between the west of Scotland 6a and the North Sea 4a have been recaptured between the Outer Hebrides and the Moray Firth (Wright et al. 2006b). Similarly, cod tagged off the Outer Hebrides moved to sites off the northern mainland coast of Scotland and Papa Bank in the North Sea (Easey 1987). Hence, the interchange of cod between the north of 6a and 4a appears consistent with the lack of genetic differences (Wright et al. 2021). Year-round fidelity of cod in the coastal regions of Clyde has been indicated from tag-recapture experiments, with the average displacement distance from one spawning season to the next being < 90 km (Wright et al. 2006a). Tag-recapture experiments have also found strong fidelity at the Cape grounds, off the Irish coast (Ó Cuaig and Officer 2007). However, some cod tagged in both these southern areas of the 6a stock were found to move into the Irish Sea via the North Channel (Ó Cuaig and Officer 2007; Neat et al. 2014).

Cod tagged at Irish Sea spawning areas had restricted ranges during the spawning season, but many subsequently moved up into the North Channel or to the south of the stock area (Neat et al. 2014). There also appears to be some separation of adult cod tagged on east and west Irish Sea spawning areas, although some mixing may occur at the early-life stages (van der Molen et al. 2007). An Irish and Celtic Seas tagging programme run from 1997–2000, using external and data storage tags, revealed that although there was some movement between stock areas, movement from the Irish Sea into the Celtic Sea was limited, although a later study found some movement from the Irish into the Celtic Sea (ICES 2022b). No cod tagged in the Celtic Sea in the first tagging programme was recovered from the Irish Sea (Connolly and Officer 2001). Examples of the movement within the stock areas and population demes are shown in Figure 2.

The three genetic demes explain the scale of movements of all life stages and suggest that population structuring is related to a combination of early retention and regional site fidelity, with potential natal philopatry. In addition, a further level of population structuring is evident from the scale of movements from otolith and tagging studies that suggest little demographic connectivity between the southern and northwest components of the Dogger deme. Indeed, there may be finer scales of demographic isolation within a metapopulation type of structure (Wright et al. 2006a, 2006b). This was best indicated by a temporal genetic analysis of cod caught off Flamborough Head between 1954 and 1998 that found evidence for a change in the population composition, commensurate with a gradual loss of the original population and a partial recovery due to immigration (Hutchinson et al. 2003).

Development Rate and Productivity

The cod stocks inhabiting the northwest European shelf have the highest growth rates (Brander 1995; Righton et al. 2010) and condition factors (Rätz and Lloret 2003) seen for any cod stock. For example, a four-year-old cod in the Celtic Sea can be 12x heavier than the same-age cod off Labrador (Brander 1995). Comparing the four northwest European shelf stocks, mean-catch weights at age 4 for the period 2004–2019 were 3.9, 3.8, 5.8 and 7.2 kg for the North Sea (3a, 4, 7d), west of Scotland (6a), Irish Sea (7a) and Celtic Sea (7e-k), respectively (Figure 3). Hence, growth is generally higher

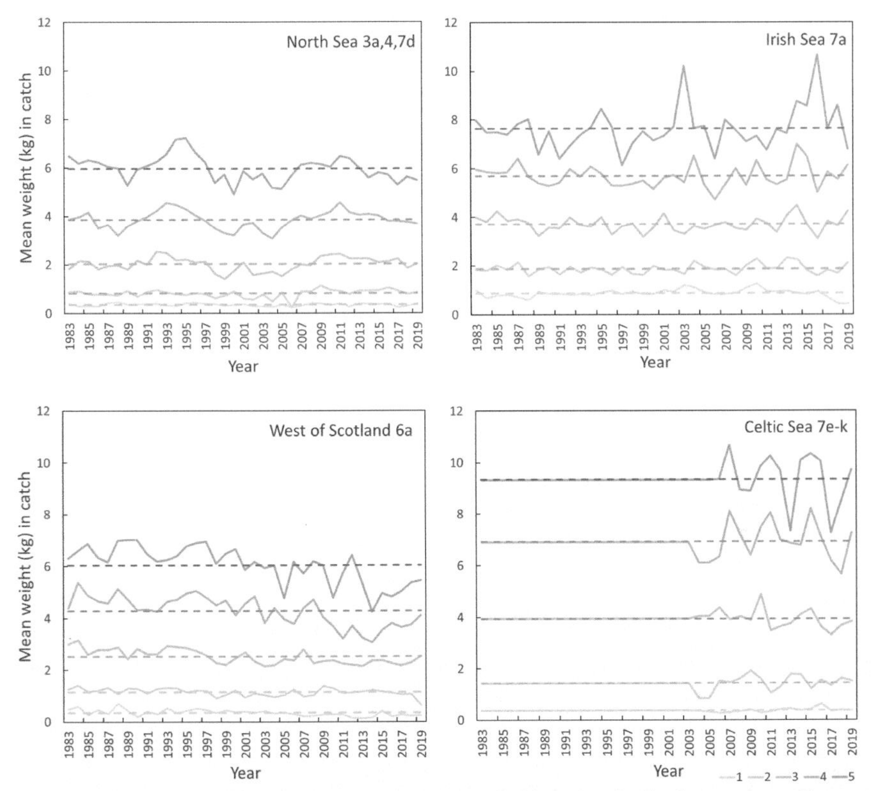

Figure 3. Inter-annual variation in mean catch weights (kg) at ages 1–5 for the four northwest European shelf stocks. Dashed lines indicate the average weight for the period. Note that constant weights were assumed for 7e-k stocks until 2003.

in the south of the region. The high growth rates of northwest European shelf cod stocks have been related to the productive and food-rich ecosystem and temperatures close to the optimum for growth (Dutil and Brander 2003). While there is variation in growth rates, extrapolation of annual length increments from otoliths has suggested little change in the North Sea since medieval times, although those < 50 cm from the southern North Sea grew slightly faster in recent times (Bolle et al. 2004). Similarly, Daan (1974) did not find evidence of much change in the growth rate of North Sea cod between pre-war years and 1970. Sexual size dimorphism has been found in cod in the northwest European shelf stocks, with growth rate and maximum length in males being generally lower and linked to the smaller size at maturity commitment (Armstrong et al. 2004; Yoneda and Wright 2004; Keyl et al. 2015).

Inter-annual growth variation within stocks has been related to temperature, prey availability and intra-specific competition. Temperature effects during the first year appear to have a strong influence on the subsequent weight-at-age of cod, as seen in the North Sea and Irish Sea, with positive effects on the first year of life (Brander 2000; Armstrong et al. 2004; Rindorf et al. 2008; Ikpewe et al. 2021). Prey limitation effects have often been inferred from the sub-optimal growth rates in the field relative to cod-fed *ad libitum* in experiments over a range of temperatures (Dutil and Brander 2003). In the North Sea, evidence for direct prey effects on growth has come from a positive effect of sand eel (*Ammodytes*) biomass during the first year of cod growth and similarly for biomass of demersal fish prey in cod older than two years (Rindorf et al. 2008).

Evidence of competition through density-dependent effects on cod growth has often been proposed, but the evidence for northwest European shelf stocks is equivocal. Macer (1983) and Van Alphen and Heessen (1984) found a correlation between total biomass and mean length-at-age, although only in fish younger than three years. Results from Macer (1983) imply that juvenile cod compete strongly with adults, while the data from Van Alphen and Heessen (1984) suggest strong within-year-class competition during the first three years of life. Some studies have reported an inverse correlation between growth and stock abundance for North Sea cod (Houghton and Flatman 1981; Cook et al. 1999). However, inconsistency in the proposed mechanisms of density-dependent growth and a lack of a relationship with stock density led several to question the importance of density-dependent growth (Daan 1974; Rijnsdorp et al. 1991; Armstrong et al. 2004). For example, large changes in the stock biomass of Irish Sea cod in the 1970s and 1990s were not associated with changes in average growth rates (Armstrong et al. 2004). In reality, density-dependent processes may differ with age, as Rindorf et al. (2008) reported that growth in the North Sea between ages 1 and 2 declined with decreasing cod density, whereas growth of cod older than two years increased.

Despite the focus on explaining stock level growth, intra-stock differences in annual growth rates have long been reported, including differences between the northern and southern North Sea, with cod in the latter area starting off growing faster but slowing down at a younger age and reaching a smaller maximum length compared with those in the northern North Sea (Daan 1974; Rijnsdorp et al. 1991). Direct comparisons between growth rates in late juvenile and adult cod and the temperature they experienced derived from archival tags indicate that cod growth is positive up to 10–12°C, after which it declines, leading to varying growth within stock areas (Righton et al. 2010).

Maturity

Across the northwest European shelf, cod mature between one and four years old. The males tend to mature at slightly smaller sizes and younger ages than the females (Nash et al. 2010). Except for the Viking deme, maturation and spawning of males may begin as early as the first year (Armstrong et al. 2004; Yoneda and Wright 2004; Nash et al. 2010). Females in the Irish Sea, west of Scotland, the Celtic Sea and much of the North Sea may begin maturing as early as age 1 and spawning at age 2, but in the Viking deme, it is at least a year later (Armstrong et al. 2004; Yoneda and Wright 2004; Wright et al. 2021). The ages- and lengths-at-50% maturity of the North Sea and Irish Sea cod are similar to the southern stocks in the northwest Atlantic (Brander 1995) (stocks described in Chapter 3 of this book). The fast growth of northwest European shelf cod stocks can partly explain the early age at maturity in these stocks as weight-at-age 2 in cod from the Celtic, Irish Sea and west of Scotland stocks is comparable to that of 4–5 year-olds from some boreal stocks, although the latter generally also mature at larger sizes. Ultimately, maturity schedules also need to be considered in the context of selection pressure (Stearns and Crandall 1984), as due to the relatively high juvenile mortality, early maturation of the northwest European stocks is required to achieve population growth (Wright 2014).

There have been substantial changes in length and age at maturity in all the northwest European stocks, with a general tendency towards smaller size and younger age at maturity, although inter-annual variation is often high (Armstrong et al. 2004; Yoneda and Wright 2004; Nash et al. 2010). Over a century ago, 50% of cod throughout the North Sea matured at > 64 cm total length (Holt 1893), while between 1968 and 2002, the length-at-50% maturity of North Sea cod females ranged between 37 and 80 cm, and the age-at-50% maturity ranged from about 2 to 4.5 years (Nash et al. 2010). Differences in maturity at length have been found between cod from the Viking deme and northwest North Sea (Yoneda and Wright 2004), as well as the northwest and southern North Sea in both the field and when raised under a common environment (Harrald et al. 2010).

Some of the inter-annual differences in maturation reflect changes in growth rate, as faster-growing cod tend to mature earlier (Godø and Moksness 1987). For example, in the Irish Sea, the increase in the proportion of mature at age 2 cod coincided with rising temperature and a decline in recruitment and stock biomass, suggesting better growth conditions favoured earlier commitment to maturity (Armstrong et al. 2004). Data from cod aquaculture provides some support for maturation being a physiological response to improved growth conditions as Barents Sea cod raised at a similar temperature to that experienced by Irish Sea cod, together with high feeding, were reported to mature at a similar age and length (Svåsand et al. 1996). While faster-growing fish will reach the critical size and condition needed to mature at an earlier age, this threshold is also under genetic control. To distinguish between growth effects and changes in maturation thresholds, Heino et al. (2002) developed the probabilistic maturation reaction norm that describes the probability of immature fish maturing at a certain age and size. While this approach does not fully account for direct environmental effects on gonad development, such as temperature (Tobin and Wright 2011), it has been widely used to infer changes in the size threshold at which fish commit to maturation. A highly significant decline in length at a 50% probability of maturing (Lp50) was found for North Sea cod since the 1980s (Marty et al. 2014). When viewed at the population level, this stock level change reflected large declines in Lp50s in the northwest and southern North Sea Dogger deme but only a minor change in the Viking deme (Wright et al. 2011). At the same time, there was a large reduction in the spawning biomass in the northwest and southern North Sea but not in the Viking deme (Holmes et al. 2014). The magnitude of the Lp50 change in the northwest North Sea resulted in cod maturing close to the minimum landing size of 35 cm. Similar declines in the Lp50 were reported for west of Scotland cod, with those from the Clyde maturing at consistently smaller sizes than those from the shelf and as young as age 1 (Hunter et al. 2015). As with similar studies of other cod stocks, the change in Lp50 found in the North Sea could be explained as an evolutionary response to the prolonged high fishing mortality (Devine et al. 2012). Theoretical studies suggest that the reversal of fisheries-induced maturation selection after fishing pressure is relaxed will occur considerably more slowly than the changes associated with its development (Heino et al. 2015).

Compared with other cod stocks, those in northwest Europe have a very high potential fecundity relative to their size, with the highest reported in the Irish Sea (Thorsen et al. 2010). By ages 4 or 5, cod in the Irish and North Sea have already attained the lifetime fecundity equivalent to that of many later maturing and long-lived cod stocks (Thorsen et al. 2010; Wright and Rowe 2019). Adult age composition also influences population fecundity since female size has a positive effect on fecundity per body weight (Yoneda and Wright 2004).

Stock Dynamics and Ecological Processes

Recruitment

Numbers at age 1 at the start of the year are used to estimate recruitment in the North Sea and West of Scotland, while in the Celtic Sea, age 1 or 0 have been used and in the Irish Sea, just the latter. The increase in North Sea cod biomass during the 1960s and 1970s 'gadoid outburst' period was due to several large year-classes (Hislop 1996), but since the 1980s, there has been a long-term trend towards low recruitment (ICES 2022a, 2022b). Inter-annual variation in recruitment for all four stocks up to 2022 is shown in Figure 4. The highest levels of recruitment occurred in the early decades of assessment, and except for the Celtic Sea, recruitment has been particularly low since 2000.

Estimates of the productivity in terms of survivors per spawner biomass (Minto et al. 2014) or survivors per egg (Wright et al. 2014) (Figure 5) show little long-term change in the west of Scotland, Irish and Celtic Sea stocks, but in the North Sea productivity has generally declined since the mid-1980s. Large-scale physical forcing, including the North Atlantic Oscillation and Atlantic Multidecadal Oscillation, has been implicated in year-class variation (Stige et al. 2006; Bentley

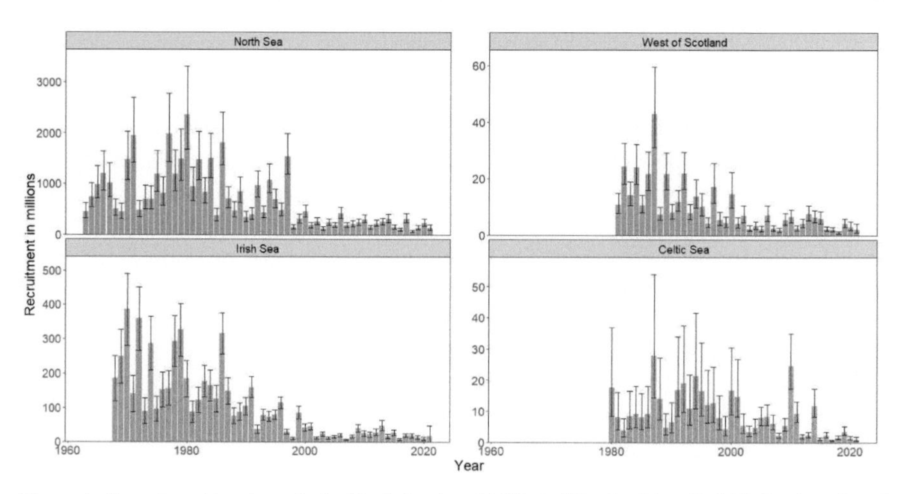

Figure 4. Recruitment trends are in the North Sea (age 1), West of Scotland (age 1), Irish Sea (age 0) and Celtic Sea (age 1).

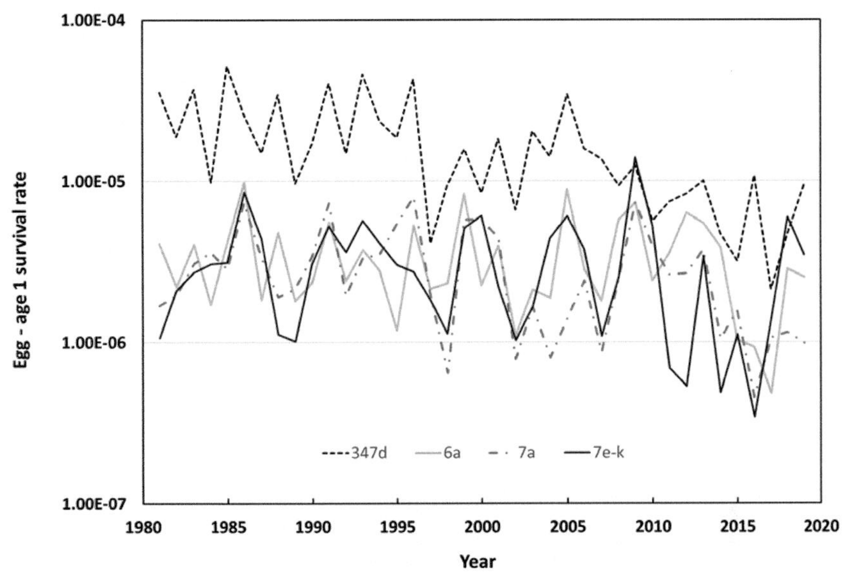

Figure 5. Inter-annual estimates of survival from egg to age 1 (quarter 1) for the four cod stocks : North Sea (347d), West of Scotland (6a), Irish Sea (7a) and Celtic Sea (7ek) (updated from Wright, 2014).

et al. 2020) with correlated recruitment among the Celtic, Irish Sea and to some extent west of Scotland stocks (Myers et al. 1995; Minto et al. 2014).

The reproductive potential of a stock is generally assumed to be of major importance to recruitment. Assessments and most studies of year-class variability have used spawning stock biomass (SSB), the total mass of mature fish, as the index of egg production. Ricker, Beverton-Holt and smooth hockey-stick stock-recruitment relationships (SRR) have all been found to provide satisfactory fits and used to determine the target and limit fishing mortality reference points Fmsy and Fcrash. However, there remains large variability around the SRR fits, suggesting a strong influence of other factors in determining recruitment success. SRRs also assume stationarity and do not effectively deal with shifts in the productivity of the ecosystem or the stock itself (Nash et al. 2009), which are evident from changes in the population growth rate (Wright 2014).

The classical SRRs are often expressed in terms of parameters related to stock productivity and resilience, making use of a 'steepness' parameter defined as the proportion of virgin recruits produced by 20% of the virgin spawning stock (Mace and Doonan 1988). The value of this parameter is critical to estimating fishing mortality limit reference points, with high steepness implying a resilient stock that can withstand higher fishing pressure. Based on the 2022 assessment, fishing mortality limit reference points estimates, Celtic Sea cod has the greatest resilience of the four northwest European stocks (Flim=1.13). In comparison, estimates for Irish Sea cod and North Sea cod are relatively low (0.43 and 0.58, respectively), and west of Scotland cod has an intermediate value (0.73).

Until recent years, a fixed maturity at age key was assumed in all four stocks which, due to a general trend towards earlier maturity at age since the 1980s

(Armstrong et al. 2004; Yoneda and Wright 2004; Nash et al. 2010; Wright et al. 2011), had led to an underestimate of SSB until changes were incorporated into North Sea and Irish Sea assessments within the last decade. High fishing mortality has led to a reduction in the mean age and age diversity of spawners (Morgan et al. 2013; Wright 2014), which reduces the number of eggs produced per spawning biomass, as larger females have a higher fecundity per body weight (Oosthuizen and Daan 1974; Yoneda and Wright 2004). Positive correlations between survival to age 1 and mean spawner age found in the North Sea and Irish Sea cod (Wright 2014) suggest some form of parental effect, possibly reflecting the differences in age-specific fecundity, egg viability (Trippel et al. 1997) or the positive effect of age composition on spawning time and duration (Morgan et al. 2013; González-Irusta and Wright 2016). Estimates of egg production for the northwest European cod stocks have been derived from annual estimates of maturity at age from surveys and fecundity at weight relationships (Wright et al. 2014; Kell et al. 2016). However, while these more realistic estimates of egg production change the perception of productivity, they do not account for more inter-annual variation in recruitment than an SSB calculated using varying maturity at age (Kell et al. 2016).

Several studies have noted a negative effect of temperature on inter-annual recruitment (O'Brien et al. 2000), consistent with the widely reported detrimental effect of increased temperature on recruitment in cod stocks near their southern limit (Planque and Frédou 1999). In the North Sea, weak year-classes have also occurred during cold years, but only when the spawning stock biomass was low, while strong year-classes have been associated with lower-than-average temperatures during the first half of the year, coincident with the timing of the early-life stages (O'Brien et al. 2000). This negative relation between temperature and survival rate is still evident based on updated assessments (ICES 2022a, 2022b) and the stock egg estimation method used by Wright (2014), with a similar trend as indicated by a significant negative log-normal regression and stock related differences in intercept (Figure 6).

There have been substantial changes in the abundance and distribution of zooplankton, particularly the larger *Calanus finmarchicus*, associated with warming, and this has been hypothesised to provide a mechanistic link with reduced cod recruitment (Beaugrand et al. 2003). Olsen et al. (2011) found evidence for density dependence in a Ricker-type SRR for North Sea cod at low zooplankton abundance, while at intermediate to high levels, the model indicated a Beverton-Holt type relationship, where the recruitment curve levelled out more slowly as zooplankton abundance improved. In addition to this zooplankton effect, a negative effect of increasing sea temperature was still evident. This combination of a negative temperature effect during spring, together with positive effects of zooplankton and spawning biomass, has also been found in other statistical analyses of North Sea cod recruitment (Nicolas et al. 2014). Predation on settled cod may also be important as an analysis of the interplay between temperature and food web effects on cod recruitment at low stock sizes concluded that while temperature may be the major driver, above a certain threshold, predation became more important (Kempf et al. 2009). Mechanistic explanations have remained elusive despite such correlations being identified and even incorporated into stock-recruit relationships.

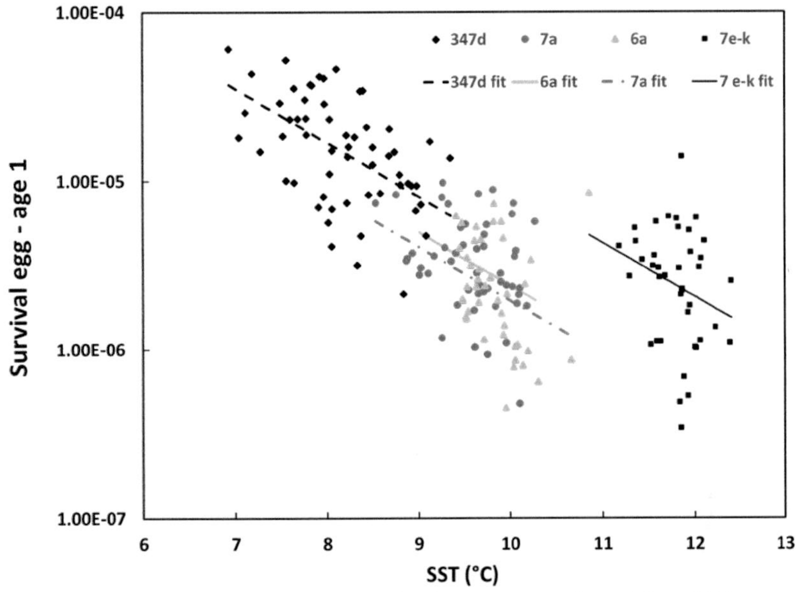

Figure 6. Relationship between proportion surviving between egg and age 1 from 2022 assessments (ICES, 2022a; 2022b) and January–June sea surface temperature (SST) from Hadley SST. Data for the four stocks are shown together with a log-normal regression, with different stock intercepts.

Several mechanistic models of the physical and/or prey conditions that cod larvae may experience have been developed in attempts to better understand the processes driving cod recruitment at the stock and population levels. Biophysical models of egg and larval transport highlight the influence of large-scale oceanographic processes on drift. Retention of cod larvae appears to be high in the shallow south of the North Sea (Heath et al. 2008) but is lower in years with a positive anomaly of the North Atlantic Oscillation (NAO) winter index (Huserbråten et al. 2018). During positive NAO, when westerly winds dominate, there is higher transport of cod spawned in the northeast North Sea into the Skagerrak (Jonsson et al. 2016) and to the western Norwegian coast and continental shelf (Huserbråten et al. 2018; Kvile et al. 2018).

Given that the timing of spawning occurs close to the onset of zooplankton production (Brander 1994; Blanchard et al. 2005; Fox et al. 2005; Wright 2005) and the link between growth and mortality (Anderson 1988), prey limitation after first feeding might be expected to have a major effect on early cod survival (Cushing 1990). Pitois and Fox (2008) predicted annual trends in early larval growth across the four northwest European stocks, using prey fields derived from Continuous Plankton Recorder (CPR) data and sea surface temperature. Their predicted larval growth did not decline and was not correlated with year-class strength in any of the stocks, suggesting that early growth conditions were not a key driver of recruitment. However, the authors acknowledged that later growth and survival might be impacted by changes in prey availability, as proposed by Beaugrand et al. (2003). Using coupled size spectrum and hydrodynamic models to simulate the prey field and transport of cod larvae in the North Sea, Daewel et al. (2015) and Huebert et al.

(2018) did find evidence for food-limited growth around the peak hatch time, which was partly correlated with subsequent recruitment strength. All these studies used proxies for larval prey because of the scarcity of information on the early-life stages of copepods that cod prey upon (Heath and Lough 2007). However, a trophodynamic individual-based model study of a single-year class of Irish Sea cod that used direct measures of the larval prey found that larval growth was mostly prey-limited in that area (Pitois and Armstrong 2014).

Predation

While bottom-up effects of prey on year-class strength are likely to be important, it is also essential to consider the relevance of top-down influences on early survival (Pitois and Armstrong 2014; Akimova et al. 2019). Kristiansen et al. (2011) modelled 30 days of larval growth and survival starting from the first feeding and included size-dependent, temperature-independent predation mortality. Their main finding was that bottom-up effects associated with the cumulative temporal overlap between larvae and prey strongly influenced survival. However, this focus on just 30 days after hatching may ignore the important effects of predation mortality before recruitment is measured. Moreover, predation pressure itself is probably not independent of temperature as the activity rates and even the abundance of predators will be affected by this environmental factor. For this reason, Akimova et al. (2016) modelled the cumulative predation mortality of cod from spawning through the first year of life, with a particular focus on the interplay between growth potential and predation, both of which increased with rising temperature. By comparing modelled size distributions of juveniles to those obtained from the surveys, they inferred that there was selection for fast growth, with mortality during the larval phase accounting for most of the inter-annual variation. Unfortunately, the models of both Kristiansen et al. (2011) and Akimova et al. (2016) suggested better cod survival in warmer years, which appears contrary to evidence for a negative correlation of year-class strength with temperature. Atlantic herring (*Clupea harengus*) predation on cod early-life stages is widely seen as a potentially important influence on cod recruitment (Fauchald 2010; Minto et al. 2014), and this species was found to have the largest predation effect on North Sea cod in a partial ecosystem model (Speirs et al. 2010). Akimova et al. (2019) combined an individual-based model of the early-life stages of cod (eggs, larvae and age-0 juveniles) with the distribution of predators in the North Sea from 1991 to 1997. They found predator distribution was important to the inter-annual variability in predicted cod survival, while variability in hydrography had a minor role. Atlantic herring and grey gurnard (*Eutrigla gurnadus*) emerged as the most important predators, with an 8–10 fold difference in survival predicted for cod originating from spawning grounds in the central and southeastern North Sea compared to spawning areas in the north and the Southern Bight. These spatial differences in predation may help explain the relatively large decline in 0-group cod off the Danish and German coasts and Dogger Bank since the 1990s (ICES 2022a).

Predation is generally thought to be the dominant source of age 0 and 1 cod mortality. Cod, being cannibalistic, is also one of the main predators of its own juveniles. In 1981 and 1991, ICES conducted a North Sea-wide sampling of fish stomach contents, and these data have formed the basis of estimates of natural mortality by the ICES Working Group on Multi-Species Stock Assessment Methods (ICES 2021a). These multispecies analyses are updated every three years in so-called 'key runs', with cod, whiting, and haddock as both predators and prey, herring, sprat and sand eel as prey only, saithe and mackerel as only predators, sole and plaice with no predator-prey interactions and 'external predators' (eight species of seabirds, starry ray, grey gurnard, North Sea horse-mackerel, western horse-mackerel, hake, grey seals and harbour porpoise). Except for the external predators, the dynamics of all species within the stock area are estimated within the model. The MSVPA results found an increase in 0-group predation from 1990–2018 (Figure 7), with grey gurnard contributing to 60% of the total predation mortality. Predation on age 2 and 3 cod has also increased since around 2000 before flattening off around 2005, with that in the latter age largely due to grey seals.

Despite the thorough treatment of a range of important predator and prey species, the MSVPA approach implicitly assumes constant overlap between predators and prey, which can be a source of bias given known differences in distribution. For example, grey gurnard is rarely found below 14°C, and their range has not yet expanded into important cod nursery areas, such as the Skagerrak (Kempf et al. 2013). Therefore,

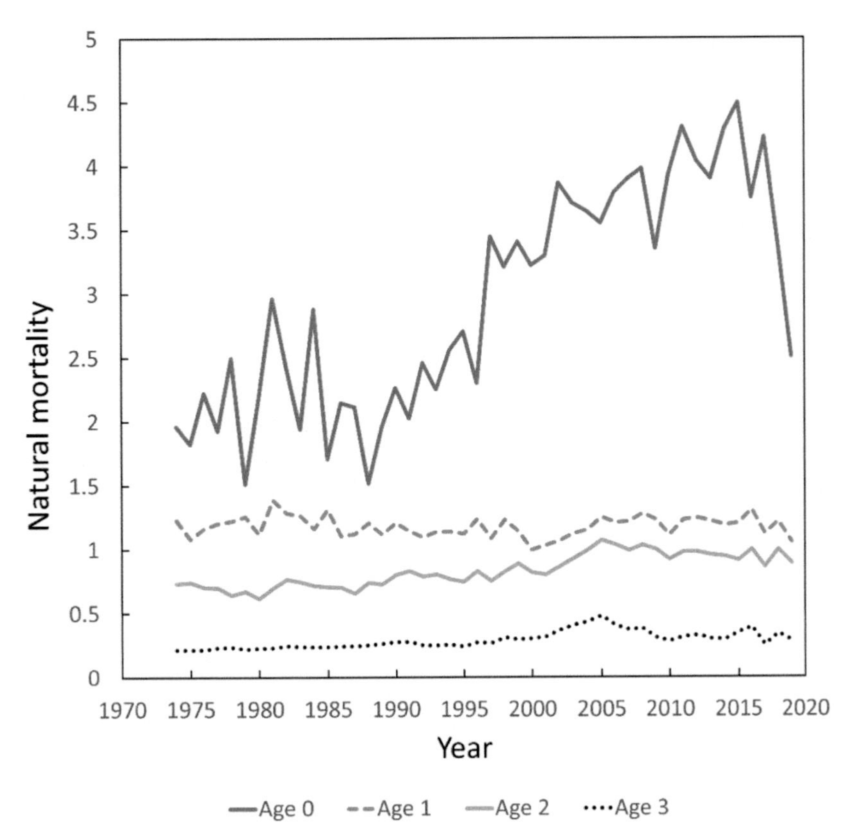

Figure 7. Annual natural mortality estimates for North Sea cod from ICES MSVPA 2020 key run for ages 0–3.

estimates of grey gurnard predation may be overstated. Furthermore, despite extensive stomach content data collection, difficulty in detecting patch-scale events could have important implications for natural mortality. For example, repeated fine-scale sampling was needed to follow the local extirpation of a dense 0-group cod aggregation by whiting (*Merlangius merlangus*) (Temming et al. 2007).

Direct estimates of predation are not used in the stock assessments for other northwest European cod stocks. However, given that predation is size dependent, the relationship of Lorenzen (1996) was used to vary natural mortality as a function of stock weight-at-age for the Celtic Sea and west of Scotland cod, the former using time invariant values and the latter time-varying. Due to the general decline in stock mean weights at age in the west of Scotland, natural mortality shows an increasing trend, consistent with the trend in the North Sea values. In the Irish Sea cod assessment, natural mortalities are estimated using a combination of approaches the Lorenzen (1996) relationship for younger ages, alongside an estimate derived from tag-recapture methods (Brownie and Pollock 1985; Hoenig et al. 1998) for all ages greater than age two. Two periods of tagging data were considered (1970–72 and 2017–18). Independent analyses of these data provided relatively consistent estimates of natural mortality, although the estimates at older ages are substantially greater than those used in the other cod stocks (ICES 2022b).

Cod are also consumed by marine mammals, especially the two species of seal found in the northwest European shelf. Concern about seal predation on cod and other commercial fish species has long been a politically sensitive issue, with several UK culls in the 1960s and 1970s until they became too unpopular to continue (Lambert 2002). The abundance of grey seals (*Halichoerus grypus*) has increased since the 1960s, and the total population was estimated at around 141,000 in 2016 (SCOS 2017). Numbers to the west of Scotland appear to have been stable since the 1990s and in Orkney since the 2000s, although they are still increasing in the North Sea (Thomas et al. 2019). Harbour seals (*Phoca vitulina*) are less numerous (approximately 43,500 in 2016), and they have been declining at colonies around the northern North Sea since 2000 (SCOS 2017).

While sand eels dominate the diet of seals in the North Sea and many parts of the west of Scotland stock, cod are frequent prey, with ages 1 to 3 being primarily taken (Hammond and Harris 2006; Wilson and Hammond 2019). In the west of Scotland stock, grey seal predation on cod is significant (Hammond and Harris 2006), and this may be impairing the ability of the cod stock to recover (Cook et al. 2015). Using a stock assessment model that included seal predation as a functional response under varying levels of fishing and seal population size, Cook and Trijoulet (2016) predicted that cod stock recovery appeared possible, although sensitive to relatively small increases in either fishing mortality or seal predation, a conclusion further supported by a Bayesian state-space model (Trijoulet et al. 2018). However, these models assumed that the west of Scotland cod stock is a well-mixed population, which is inconsistent with the evidence that it is actually comprised of a number of different resident and migratory groups (Wright et al. 2006a, 2006b). Furthermore, seal foraging occurs mostly in coastal areas (Russell et al. 2017), including rocky areas unsuitable for trawling, while the main fishery exploiting cod occurs along the shelf edge in the north, which appears to be a component of a wider northwest North Sea cod sub-population (Wright et al. 2021).

Climate Change

Since cod are essentially a cooler water species, climate change is likely to have an adverse effect on most aspects of the recruitment process, including earlier onset of spawning (McQueen and Marshall 2017), decreased spawning extent (González-Irusta and Wright 2016), trophic mismatch between early-life stages and their prey (Beaugrand and Kirby 2010) and predation linked to the northward expansion of Lusitanian species (Kempf et al. 2009). Due to the narrow thermal range of ≤ 8°C of cod during the spawning season (Righton et al. 2010) and the fact that temperatures > 9°C are beyond the optimum for larval cod growth (Steinarsson and Björnsson 1999), future winter–spring warming may be expected to impact the reproductive success of populations, particularly in the southern North Sea, Celtic and Irish Seas.

Some have suggested that climate effects could already explain the lack of cod recovery in the early 2000s. For example, on the basis that cod were not observed much above annual mean bottom temperatures of 12°C (Dutil and Brander 2003), Drinkwater (2005) predicted that the Celtic Sea and the English Channel cod would disappear with an additional 1°C temperature increase above that from a 1961–90 baseline, while the Irish Sea stock and cod in the southern North Sea would decline. Similarly, Clark et al. (2003) and Kirby et al. (2009) proposed that rebuilding North Sea cod to historical levels might be impossible due to ecosystem changes related to climate-induced effects. Despite these warnings, a reduction in fishing mortality in line with single-stock management advice during the 2000s was associated with a partial recovery of the North Sea stock despite an average warming of just over 1°C above the baseline period (Brander 2018). However, although this recovery did support the leading role of fishing mortality in maintaining the low stock size, the improvement was not seen throughout the North Sea. The southern population, in particular, does not appear to have responded positively.

A northward shift in the distribution of cod in the North Sea since the 1990s has been widely reported (Hedger et al. 2004; Perry et al. 2005; Engelhard et al. 2014; Núñez-Riboni et al. 2019). This has often been interpreted as arising from an active movement of individual fish away from the warm southern North Sea, but this is incorrect. Electronic tagging has shown that even in the warmest temperatures (19°C), cod in the southern North Sea do not migrate to cooler northern waters but rather reduce their activity (Turner et al. 2002; Neat and Righton 2007). Moreover, geographic attachment of sub-populations (Wright et al. 2006a, 2006b; Righton et al. 2007) within the North Sea was found to lead to very different thermal exposures of cod (Neat et al. 2014; Wright et al. 2021) (Figure 8). An alternative suggestion for the northward shift based on the passive transport of larvae followed by fidelity of later life stages (Rindorf and Lewy 2006) was also not supported by subsequent evidence for larval retention in the south (Heath et al. 2008; Kvile et al. 2018), natal migrations (Wright et al. 2018) and genetic structuring (Heath et al. 2014).

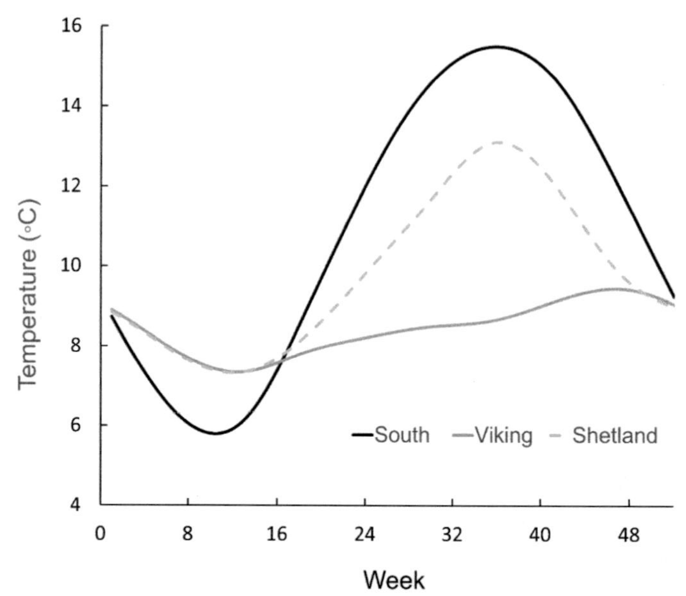

Figure 8. Comparison of temperature experienced by data storage tagged cod that were released and recaptured in three different regions of the North Sea. South refers to the southern region of the North Sea, corresponding to the shelf <50 m depth. Viking and Shetland release areas are shown in Figure 2. Regional smooths are derived from data presented in Neat et al. (2014).

The behavioural and geographic attachment evidence highlights the importance of spatial differences in fishing pressure and temperature in interpreting these distributional shifts. The relative stability of spawner biomass in the northeast Viking deme up until the early 2000s, compared to the marked decline in other population areas of the North Sea (Holmes et al. 2014), was associated with both lower fishing mortality (Engelhard et al. 2014) and cooler waters. Warming does appear to be reducing the extent of the preferred habitat for cod, especially in the southern and central North Sea, where cod have experienced the warmest temperatures in the northwest European shelf (Neat et al. 2014; Núñez-Riboni et al. 2019). Cod avoid foraging habitat when temperatures exceed 15°C if cooler, deeper areas are within a daily movement range (Freitas et al. 2015, 2016), but most historic nursery areas in the southeast North Sea now regularly exceed this temperature.

Growth in the southern North Sea will have been affected by the warming temperatures as cod regularly experience some months of the year when temperatures far exceed those associated with optimal growth (Righton et al. 2010) and even foraging (Turner et al. 2002). Such periods have been linked to changes in otolith zone formation (Millner et al. 2011). Nevertheless, cod still grow more rapidly in the southern North Sea than in the north (Righton et al. 2010).

Changes in marine communities associated with climate change are likely to affect cod predation, such as the expansion of grey gurnard distribution in the North Sea (Kempf et al. 2013). An ecosystem model for the west of Scotland also predicted that warming would have a positive effect on existing competitors of cod that are more tolerant of warm waters, such as whiting, *Merlangius merlangus*, in parallel with declines of cod and grey seals (Serpetti et al. 2017). Hence, the widely reported northward distribution shift is likely to result from a combination of regional fishing mortality, decreasing habitat suitability, changes in predation pressure and reduced recruitment success in cod inhabiting the southern North Sea.

While cod are certainly at risk due to ocean warming, it is important to recognise that they have persisted during warmer conditions in the past. Based on otoliths and bones found in Danish middens from 7000–3900 BC, cod was an important part of the diet of Mesolithic Stone Age man when sea temperatures were estimated to be 2–2.5°C above current levels (Enghoff et al., 2007), but this was a time when fishing mortality would have been minimal. In addition to warming, ocean acidification is an increasing threat and is already well above levels in the Mesolithic Stone Age (Ruddiman 2003; Bouttes et al. 2012). Using an end-of-21st century CO_2 concentration, predicted under the extreme IPCC RCP 8.5 scenario (~ 1,100 μatm), Stiasny et al. (2016) found a doubling of daily mortality rates in cod in the first 25 days post-hatching compared to the present day CO_2 concentrations, which was predicted to reduce survival for western Baltic and Arcto-Norwegian cod larvae to 8% and 24% of current levels respectively. Epigenetic effects may also become important as the offspring of cod subjected to high CO_2 water were found to do better in high CO_2 conditions, at least when prey density was high (Stiasny et al. 2018).

Stock Management and its Challenges

Historically, fishing mortality has been the most important factor driving changes in the northwest European shelf cod stocks. In the North Sea (sub-area 4), between 1903 and 1965, landings were mostly around 100,000 tonnes, except during World War II when they were substantially lower. However, following that period, landings increased 2–3 times until the mid-1980s before subsequently declining (Pope and Macer 1996). Estimates of fishing mortality from the 1920s to the time when annual assessments by ICES began in the 1960s suggest that fishing mortality was around 0.5 for ages 2–5, except during the war, followed by an upward trend to around 1.0 by 1990 (Daan et al. 1994; Pope and Macer 1996). Above-average recruitment in the 1960s-70s and especially large year-classes in 1969, 1970, 1976 and 1979 helped sustain high catches during the so-called 'gadoid outburst'. However, because of the increasing fishing mortality, these strong year-classes had comparatively little impact on spawning stock biomass, as most were caught before they reached maturity (Daan et al. 1994; Hislop 1996).

Expansion of European fisheries during the 1970s and 1980s resulted in high fishing pressure on stocks of cod, haddock, whiting and saithe caught in the mixed demersal fisheries of the North Sea. This overcapacity became the key problem for sustainable management of these stocks (Gulland 1987), an issue that already had been raised by Graham (1934) following the increase in demersal fish abundance during the First World War. Following on from the development of scientifically informed fisheries management after the Second World War (Beverton and Holt 1957), up to the 1980s, most North Sea fisheries policy focussed on short-term tactical measures to decide on a Total Allowable Catch (TAC) rather than promoting long-term sustainability. There was little consideration of the stock-recruitment relationship in demersal fish management and the risks of recruitment failure were generally underestimated (Gulland 1987). Annual catch quotas for many demersal stocks attempted to maintain fishing at the level of the previous year, which resulted in above-optimal exploitation levels for these stocks for most of the period from the 1970s until the 2000s. According to Beverton (1998), this over-exploitation reflected a reliance on quotas that were not suited for mixed species fisheries due to the difficulty in forecasting incoming recruitment, fishers' disregard of landing limits and underreporting of catches. Similar declines associated with overcapacity were seen in the Irish Sea and West of Scotland cod stocks during the 1970s and 1980s.

In 1983, the European Commission Common Fisheries Policy (CFP) was established and became the main framework for managing European fish stocks. TACs were the pivotal tool of catch regulation in EU fisheries up until 2002. TACs were favoured by the EC because they allowed stability in the allocation to member states. TACs were agreed upon at the annual Council of Ministers, supplemented by various technical measures, including closed areas/seasons, effort regulations and gear regulations. However, as discards were permitted in EU waters and most fisheries targeted several demersal species, there was a strong incentive to keep fishing while the quota for other species remained, leading to discarding and, in some cases, unreported and illegal landings.

From 1990, ICES advised cuts in fishing mortality on cod, and there were repeated calls from scientists for stringent management action to reduce fishing mortality (Hislop 1996; Cook et al. 1997). ICES consistently advised that TACs alone could not protect cod stocks and that cuts in effort were also needed (Bannister 2004). Following the UN Rio declaration in 1992, the first precautionary reference points for SSB (B_{pa}) and fishing mortality (F_{pa}) were proposed by ICES in 1998 (González-Laxe 2005). These were intended to keep stock size above the level of spawning biomass below which a stock is considered to have reduced reproductive capacity (B_{lim}) and fishing mortality below the associated fishing mortality (F_{lim}) while accounting for assessment uncertainty. For northwest European cod stocks, F_{pa} ranged around ~ 0.6–0.7, but these precautionary limits were never intended to be management targets. In late 2002, with the aim of restoring the stock to above such a safe level in the fastest possible time, ICES advised the closure of the North Sea fisheries because the cod stock was so far below B_{lim}. Political considerations during annual negotiations often led to cod TACs being set above ICES scientific advice (O'Leary et al. 2011), although the actual catch was often less and more in line with scientific advice (Cook et al. 2012). However, inaccurate catch forecasts (Cook et al. 2012) and sometimes biased catch information (Patterson 1998) tended to lead to catch advice from ICES that was too high.

The historical trends in fishing mortality relative to precautionary and MSY reference limits highlight that in the North Sea, after fishing mortality increased in the 1960s and 70s, it remained very high until the early 2000s. Following the introduction of recovery measures, it then declined until 2013, and following an increase in 2018–19, it has since declined again (Figure 9). The SSB declined for much of the period since the early 1970s, except for a small increase following improved recruitment coupled with a slight dip in fishing mortality in the mid-1990s. Nevertheless, with low recruitment since 1998 and continued high mortality rates, SSB continued to decline to its lowest level in 2005. The reduction in fishing mortality up to 2013 seemed to lead to an increase in the SSB to close to Bpa but subsequently declined in recent years.

The assessment period for the west of Scotland stock was shorter but showed high fishing mortality from the mid-1980s for most years and a corresponding decline in SSB. While there was some decline in fishing mortality since the mid-2000s, this is not to the same extent as estimated for North Sea cod, and fishing mortality remained at around F_{lim} (ICES 2022b). In the Irish Sea, there was a period in the 1960s and 1970s when fishing mortality was below $F_{lim,}$ and SSB was above B_{pa} before a period of overfishing associated with the decline in SSB. Fishing mortality reduced substantially in the early 2000s and has been below Fmsy since 2012, but the SSB has failed to recover due to continued low recruitment. In the Celtic Sea, the F_{lim} is high, and the stock sustained three decades of fishing mortality, removing over 50% of the stock aged 2 to 5 per year. However, a decline in SSB during the early 2000s and recent poor recruitment coupled with increasing fishing mortality has resulted in a further decline in SSB to below B_{lim}. It is also important to realise that in the stock assessments, total mortality is partitioned between assumed natural mortality and fishing mortality. Natural mortality has varied over time for both the North Sea and West of Scotland stock assessments, but this is not the case for the other stocks. If natural mortality has increased more than assumed in the assessment models, this can lead to an erroneous perception that fishing mortality has increased or reduced

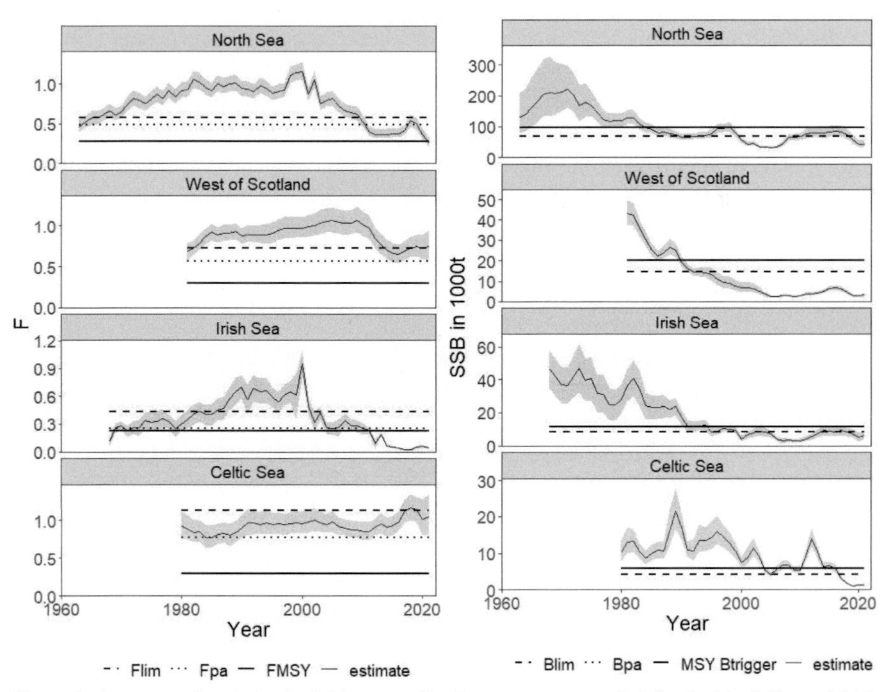

Figure 9. Inter-annual variation in fishing mortality (average over ages 2–4 for the North Sea and Irish Sea and ages 2–5 for West of Scotland and Celtic Sea) and spawning stock biomass in the North Sea, West of Scotland, Irish Sea and Celtic Sea and the western English Channel.

less than it has. For example, in the west of Scotland, stock natural mortality may not account sufficiently for increased predation from seals and hence could have had this effect (Cook and Trijoulet 2016).

In the early 2000s, the EU brought in emergency measures and stock recovery plans, which included restrictions on effort in terms of days at sea and the setting of multiannual catch limits with more strategic biomass targets. Due to the serious depletion of the North Sea, west of Scotland and Irish Sea cod, the EU agreed that stringent action was required to rebuild these stocks, and so the first cod recovery plan was put in place, which aimed at applying a common approach across stocks (European Council Regulation (EC) No 423/2004). The cod recovery plan aimed to move the stocks to above B_{pa} by setting TACs consistent with 30% annual increases in SSB. Importantly, it also allowed measures that restricted fishing efforts. The minimum cod end mesh size in the whitefish trawl fleet was also increased from 100 mm to 120 mm. However, ICES (2005) stated that the recovery plans for cod were not precautionary because they would not restore the stock to above B_{lim} in the fastest possible time and continued to advise against any fishing. By 2005, the new measures and plan had led to around a 37% reduction in fishing mortality on cod in the North Sea and involved a substantial reduction in the UK demersal trawling fleets (Horwood et al. 2006). In the North Sea, the planned reduction in F to about 0.65 y^1 proposed by the plan might have allowed recovery under the high early-life history survival rates seen in the 1960s–1970s but not under the levels seen since the mid-1990s (Kell et al. 2005; Horwood et al. 2006; Wright 2014).

In 2008, the European Commission amended the cod recovery plan (Council Regulation (EC) No 1342/2008) by replacing biomass targets with those based on optimum fishing rates. This was intended to provide a high sustainable yield under a new system of effort management with effort ceilings (kilowatt-days) for groups of vessels or fleet segments to be managed by the Member States. The latter allowed for incentives on efforts that promoted cod avoidance and gave more local control. Member States, in discussion with the local fishing industry, established several cod avoidance schemes, including temporary spatial measures and gear modifications. To protect the industry, a restriction on year-to-year variation in the TACs of $\leq 20\%$ was also employed (Anonymous 2009).

An ICES/STECF review group convened in 2011 concluded that the objectives of the EU Cod Management Plan were not met as F did not decline to the levels envisaged, with no reductions evident in the Irish Sea and West of Scotland stocks, and only a 1.5% average annual reduction for the North Sea (Kraak et al. 2013). Part of the problem may have been discarding rates that were uncertain but estimated to be 30–50% of the total catch. In the west of Scotland, catches were sometimes several times larger than the TAC. Area misreporting of cod landings caught in this stock area but declared in the northern North Sea and elsewhere by the Scottish fleet appears to have been a problem (ICES 2020c). High levels of discards of marketable North Sea and West of Scotland cod occurred post-2007 because of more effective enforcement to stop illegal landings, the relatively strong 2005 year-class in the northern North Sea, and the low TAC (Fernandes et al. 2011).

Continuation of the cod recovery measures did lead to a substantial reduction in fishing mortality in the North Sea and an increase in SSB from its lowest point in 2006.

Based on the 2016 assessment, North Sea cod appeared to have recovered almost to the biomass reference point, B_{pa} and well above B_{lim} (ICES 2017), while fishing mortality was estimated to be close to F_{msy}. On this basis, the Marine Stewardship Council awarded their certificate of sustainability in 2017. Acting on ICES' advice, the European Commission dropped the cod recovery plan in 2017, abandoning days-at-sea controls and raising limits on allowed catches. Unfortunately, subsequent ICES assessments revealed a significant retrospective bias in that SSB had not actually risen so substantially since 2006 (remaining only just above B_{lim}) while fishing mortality had remained well above F_{msy} and then increased. Subsequently, ICES advice for heavy restrictions on TAC was not fully implemented by the EC or the subsequently devolved UK, and the MSC withdrew their certificate of sustainability in 2019.

Both seasonal and year-round closed areas have been used as additional measures to reduce fishing mortality on northwest European cod stocks. For closed areas to be successful in reducing fishing mortality, they need to target aggregations where there are much higher catch rates than elsewhere and be coupled to an overall restriction on effort such that redistribution of fishing activity does not negate the reduced mortality caused by the closure (Horwood et al. 1998). Fishing on cod spawning aggregations is often associated with much higher catch rates, which was seen in the Clyde spawning area before a seasonal closure was implemented in 2001 (Clarke et al. 2015). Due to the rich population structure of cod, spawning closures might also be useful in reducing the risk of over-exploitation of specific spawning components of the stock (van Overzee and Rijnsdorp 2015), such as the many resident coastal groups evident from tagging and otolith chemistry (Neat et al. 2006; Wright et al. 2006a, 2006b; Righton et al. 2007).

In the northwest European stocks, seasonal spawning closures were introduced in the Irish Sea in 2000, west of Scotland in 2001 and Celtic Sea in 2005. The Irish Sea measures initially included two closed areas in the eastern and western Irish Sea to protect spawning cod during the spawning season and applied to all fishing activities, except derogations for *Nephrops* trawl and beam trawlers in defined parts of the closed areas. From 2001 onwards, the closure encompassed the majority of the western Irish Sea spawning aggregation but excluded the eastern Irish Sea. Reviews of the Irish Sea closure concluded that it probably played a role, in conjunction with other measures, in reducing fishing opportunities for cod (Kelly et al. 2006; STECF 2007), but there was no sustained recovery of the stock. The reasons for the failure of the stock to rebuild are uncertain, but continued discarding in derogated *Nephrops* trawl fisheries may have played a role as pelagic stage juvenile cod are known to move into the stratified waters overlying the western Irish Sea muddy grounds (Dickey-Collas et al. 1996; Kelly et al. 2006). The Celtic Sea cod closure was proposed by French, Irish and UK fishermen as an alternative to days-at-sea limits to reduce fishing mortality. By displacing fishing activity away from spawning aggregations off north Cornwall and hence making vessels less efficient at catching cod, this closure is thought to have played a role in the reduction in the fishing effort of French trawlers in the Celtic Sea (STECF 2007). However, once again, there is little indication that these measures have had the desired effect on the Western Channel and Celtic Sea cod stock (Figure 9).

In the west of Scotland, another closed area, known as the 'windsock', was established as an emergency measure in 2001. The location of the area was based on high pre-closure landings, and discussions involved scientists, the industry and the EC. The initial closure period was for the spawning period, but from 2003, it became a year-round closure and remained in force until the cessation of most EU-closed areas in 2019. While surveys did find that cod tended to be larger and more numerous inside the closed area (Jaworski and Penny 2009) there was no recovery in the stock SSB. However (ICES 2022b), did note that in the recent decade, most landings from this stock were concentrated around the 'windsock' closed area. It is very difficult to evaluate the effectiveness of cod-closed areas as they are typically emergency measures put in place without sufficiently clear goals and adequate monitoring. Through a comparison with changes at other spawning areas west of Scotland, Clarke et al. (2015) concluded that the Clyde seasonal closure did not lead to a recovery of the local sub-population. However, the closure stopped vessels displaced from the Irish Sea closure targeting spawning cod, and it remains the most important spawning area locally.

Other cod avoidance schemes have included the use of temporary closed areas. The Scottish Conservation Credits (SCC) scheme, introduced in 2008, used real-time closures (RTCs) to discourage vessels from areas of high cod abundance. These were applied to areas for 21 days when detected catch rates of cod exceeded a trigger level. The scheme was incentivised by rewarding participation with additional days at sea. Compliance with the closures was good, but the assessed catch savings of cod from RTCs appeared to be less than predicted (Holmes et al. 2011; Needle and Catarino 2011). A similar real-time closure scheme in English waters in the North Sea and Eastern Channel was also established, and both UK schemes have continued, latterly as part of the UK National North Sea cod avoidance plan. The initial size of RTCs did not consider the extent of cod movement throughout the closure, and so following subsequent estimates from geolocated archive tagging (Neat et al. 2006; Neat et al. 2014), the area of Scottish RTCs was increased by fourfold in 2010, and this is the size adopted by the UK National North Sea cod avoidance plan since 2020.

In 2009, the European Community and Norway agreed to implement the RTCs scheme in the North Sea and Skagerrak to protect juvenile cod, haddock, whiting and saithe. These closures were intended to reduce catches of undersized fish and avoid any incentive to discard. Many gear developments have also been undertaken to reduce the capture of undersized cod. In particular, square mesh panels (SMP) allow for the improved escape of small cod and are now mandatory in *Nephrops* and other trawl fisheries around the British Isles. However, the efficiency of release is strongly affected by the location of the SMP in the trawl net, so the maximum benefit may not be obtained in all fisheries (Melli et al. 2019). Sorting grids and separator veils are also mandatory in some fisheries, particularly those using finer mesh sizes to catch shrimp and *Nephrops*. They have been extensively trialled in the Kattegat and are legally required as part of cod recovery measures (Madsen and Valentinsson 2010; Madsen et al. 2017). However, their uptake has been resisted in other areas as they are perceived to interfere with fishing operations and can be prone to clogging (Polet 2002). Most gear modifications have focussed on reducing the retention of small cod, although reducing the capture of larger cod could deliver faster conservation benefits because of the lower natural mortality rates on these larger fish and the fact that they are already mature. Several high-tech innovations are being trialled, including in-trawl cameras and trials of artificial intelligence-driven selection gates, which might allow the selective release of larger cod (O'Neill et al. 2019). These systems are very much at the cutting edge of technological innovation, and whether they will prove robust and reliable enough for long-term use on commercial fishing vessels remains to be seen.

While fishing mortality has the greatest impact of any human activity on cod stocks, settled 0-group cod are mostly concentrated in nearshore waters (Gibb et al. 2007) with emergent structures, including maerl and seagrass, providing important shelter (Kamenos et al. 2004; Lilley and Unsworth 2014). During the last century, most of these areas of habitat were degraded, and it is unclear whether this has also affected early cod survival. Measures are now starting to conserve and, in some places, restore such habitat through the northeast Atlantic network of marine protected areas (OSPAR-Commission 2015).

It has been recognised that the science underpinning management advice needs to be improved if assessments and management actions are to reflect the relevant scales of population dynamics. As previously described, the North Sea and West of Scotland cod stock management boundaries do not reflect the population structure, and this disparity is relevant due to differing dynamics among the sub-populations (Heath et al. 2014; Holmes et al. 2014; ICES 2020a, 2022d). Population dynamic simulations have indicated large differences in fishing mortality among adults in these sub-populations during the period of cod recovery plans (Jardim et al. 2018). Lower adult mortality in the northwest North Sea population may suggest that local management measures had been more effective in that region. The slower development rates in the Viking cod may also mean that they are more susceptible to overfishing, so North Sea-wide reference points may not be appropriate for that population. Genetic and tagging evidence also indicates that the northern offshore component of the west of Scotland stock is part of this northwest North Sea population (Heath et al. 2014; Wright et al. 2021), and current high local biomass at the border of these two stocks (ICES 2022a, 2022d) is likely to lead to area misreporting of landings. The northern offshore component of the west of Scotland cod has accounted for most of the landings from Division 6a since the mid-2000s, and there is an almost complete absence of cod in the rest of the western shelf. The only other significant west of Scotland component remaining is around the Clyde and northern Irish coast, which are well connected to cod in the Irish Sea (Heath et al. 2014; Neat et al. 2014).

Following workshops to explore ways to take better account of population structure in assessments (ICES 2020a, 2022d), ICES has merged the west of Scotland (Division 6a) and the North Sea stocks (3a, 4, 7d) into a northern shelf stock and divided this region into three sub-divisions that more closely reflect population structure (see Figure 10a). Assessments of sub-stock trends have been made using an age-based multi-stock state-space assessment model that uses catches and surveys and a period when sufficiently spatially resolved data were available (ICES 2023). The SSB estimates indicate some differences in trends among sub-stocks (Figure 10b). The northwestern sub-stock was the largest component with a more stable SSB between 1983 and 1997, followed by a large decline to 2005 and then a generally

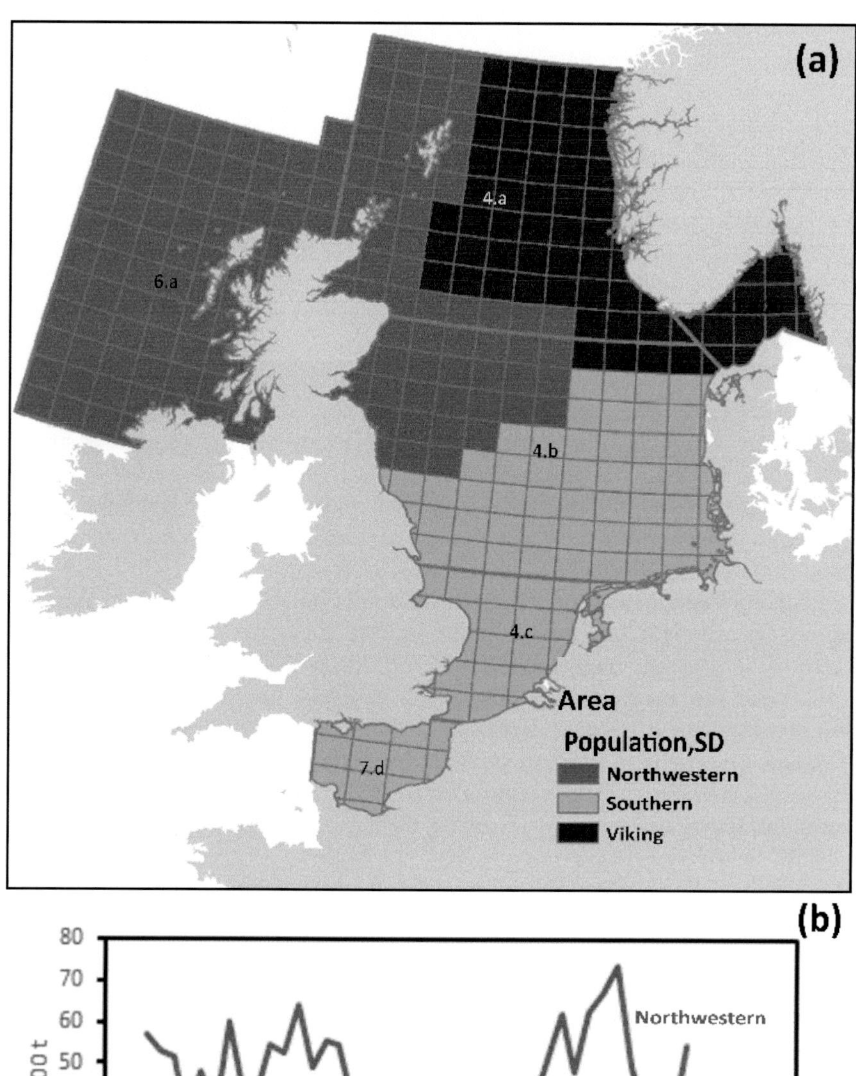

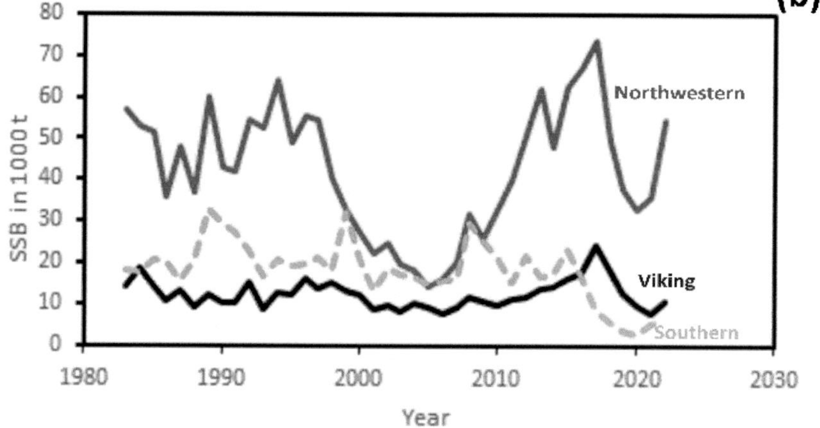

Figure 10. **(a)** Northern shelf cod stock, showing three substock components: Northwestern, Viking and Southern and (b) estimated spawning stock biomass (tonnes) in the 2023 assessment (ICES 2023).

increasing trend, except for 2017–2020. Throughout the 1983–2016 period, Viking was the smallest sub-stock with little trend in SSB, while there appears to have been a sharp decline in SSB in the southern stock since 2016.

Cod occur throughout the northern shelf stock area, and there is uncertainty about the most appropriate sub-stock boundaries and the extent of movements outside the spawning season. Further, these sub-divisions do not fully reflect the many groups of resident and migratory ecotypes evident from tagging studies (Wright et al. 2006a, b; Neat et al. 2014). ICES recognises that the latest assessment's ability to resolve the dynamics of the different sub-stocks is currently limited. As such, ICES views this as an intermediate step between a single-stock assessment and a full mixed-stock model approach. Nevertheless, the progress is encouraging and begins to explain spatial differences in cod abundance.

The northwest European cod stocks are extremely productive and well adapted to the very high levels of mortality that all life stages naturally experience. However, climate change adversely affects activity, growth and early survival conditions. Together with ecological changes to fish communities, climate change will affect the potential for stock recovery and the magnitude of achievable stock sizes. Several studies have highlighted the need to consider environmental conditions in estimating future productivity and fishing mortality reference points (Clark et al. 2003; Cook and Heath 2005; Kell et al. 2005). In order to account for such productivity changes, reference points for North Sea cod (and subsequently northern shelf cod sub-stocks) have been derived using the SRR based on a truncated series of SSB and recruitment data (ICES 2021b, ICES 2022d). Attempts to incorporate the environment in forecasting stock dynamics suggest that relatively simple indices of temperature may be a sufficient proxy to account for changes in productivity in the near future (Clark et al. 2003; Koul et al. 2021) while Bentley et al. (2021) proposed a new fishing mortality reference point, F_{eco}, that scales F_{msy} based on sea surface temperature and applied this to the Irish Sea as a case study.

Whatever improvements there are in future scientific advice, more effective measures to reduce fishing mortality will be needed if depleted cod stocks are to have a chance to recover. Some of the northwest European stocks and new northern shelf cod sub-stocks are now largely within the UK EEZ, and following Brexit, the UK has increased freedom to implement management measures outside of the Common Fisheries Policy. Other stocks are shared, and collaboration with the EU and other coastal states in their management will be required as obligated under international law (Fernández et al. 2022). The main management problems relate to the mixed nature of the fisheries catching cod, making it challenging to reduce fishing mortality on a single species. While recent cod avoidance measures appear to have reduced northern shelf cod fishing mortality, further spatial and technical measures to reduce cod catches from other depleted populations are probably the only way to rebuild these stocks sustainably. If this can be achieved, it might allow commercial cod fishing to have a future around the northwest European shelf.

Acknowledgements

We gratefully acknowledge the ICES community and the Scottish Government project SU02A0.

References

Akimova, A., Hufnagl, M., Kreus, M., et al. 2016. Modeling the effects of temperature on the survival and growth of North Sea cod (*Gadus morhua*) through the first year of life. Fisheries Oceanography, 25: 193–209.

Akimova, A., Hufnagl, M., and Peck, M.A. 2019. Spatiotemporal dynamics of predators and survival of marine fish early life stages: Atlantic cod (*Gadus morhua*) in the North Sea. Progress in Oceanography, 176.

Anderson, J.T. 1988. A review of size dependent survival during pre-recruit stages of fishes in relation to recruitment. Journal of Northwest Atlantic Fishery Science, 8 : 55–66.

Anonymous. 1971. ICES North Sea Roundfish Working Group.

Anonymous. 2009. Reform of the Common Fisheries Policy: Green Paper. European Commission. 24 pp.

Armstrong, M., Beveridge, D., Cotter, J., et al. 2005. Fisheries Science Partnership programmes off the NE and SW coasts of England and in the Irish Sea: experiences gained and future directions. https://www.ices.dk/sites/pub/CM%20Doccuments/2005/Y/Y0605.pdf.

Armstrong, M.J., Gerritsen, H.D., Allen, M., et al. 2004. Variability in maturity and growth in a heavily exploited stock: cod (*Gadus morhua* L.) in the Irish Sea. ICES Journal of Marine Science, 61: 98–112.

Bailey, R.S. 1975. Observations on diel behaviour patterns of North Sea gadoids in the pelagic phase. Journal of the Marine Biological Association of the United Kingdom, 55: 133–142.

Bannister, R.C.A. 2004. The rise and fall of cod (*Gadus morhua*, L.) in the North Sea. Management of Shared Fish Stocks: 316–338.

Beaugrand, G., Brander, K.M., Lindley, A.J., et al. 2003. Plankton effect on cod recruitment in the North Sea. Nature, 426: 661–664.

Beaugrand, G., and Kirby, R.R. 2010. Spatial changes in the sensitivity of Atlantic cod to climate-driven effects in the plankton. Climate Research, 41: 15–19.

Bentley, J.W., Lundy, M.G., Howell, D., et al. 2021. Refining fisheries advice with stock-specific ecosystem information. Frontiers in Marine Science, 8.10.3389/fmars.2021.602072.

Bentley, J.W., Serpetti, N., Fox, C.J., et al. 2020. Retrospective analysis of the influence of environmental drivers on commercial stocks and fishing opportunities in the Irish Sea. Fisheries Oceanography, 29: 415–435.

Berg, P.R., Jentoft, S., Star, B., et al. 2015. Adaptation to low salinity promotes genomic divergence in atlantic cod (*Gadus morhua* L.). Genome Biology and Evolution, 7: 1644–1663.

Beverton, R. 1998. Fish, Fact and Fantasy: a Long View. Reviews in Fish Biology and Fisheries, 8: 229–249.

Beverton, R.J.H., and Holt, S.J. 1957. On the Dynamics of Exploited Fish Populations, U.K. Ministry of Agriculture and Fisheries, London. 533 pp.

Bigg, G.R., Cunningham, C.W., Ottersen, G., et al. 2007. Ice-age survival of Atlantic cod: agreement between palaeoecology models and genetics. Proceedings of the Royal Society London Series B, 275: 163–172.

Blanchard, J.L., Heffernan, O.A., and Fox, C.J. 2005. North Sea cod. pp. 76–88. International Council for the Exploration of the Seas.

Bolle, L.J., Rijnsdorp, A.D., van Neer, W., et al. 2004. Growth changes in plaice, cod, haddock and saithe in the North Sea: a comparison of (post-) medieval and present-day growth rates based on otolith measurements. Journal of Sea Research, 51: 313–328.

Bouttes, N., Paillard, D., Roche, D.M., et al. 2012. Impact of oceanic processes on the carbon cycle during the last termination. Climate of the Past, 8: 149–170.

Bradbury, I.R., Hubert, S., Higgins, B., et al. 2013. Genomic islands of divergence and their consequences for the resolution of spatial structure in an exploited marine fish. Evolutionary Applications, 6: 450–461.

Brander, K.M. 1994. The location and timing of cod spawning around the British Isles. ICES Journal of Marine Science, 51: 71–89.

Brander, K.M. 1995. The effect of temperature on growth of Atlantic cod (*Gadus morhua* L.). ICES Journal of Marine Science, 52: 1–10.

Brander, K.M. 2000. Effects of environmental variability on growth and recruitment in cod (*Gadus morhua*) using a comparative approach. Oceanologica Acta, 23: 485–496.

Brander, K.M. 2018. Climate change not to blame for cod population decline. Nature Sustainability, 1: 262–264.

Brownie, C., and Pollock, K.H. 1985. Analysis of multiple capture-recapture data using band-recovery methods. Biometrics: 411–420.

Clark, R.A., Fox, C.J., Viner, D., et al. 2003. North Sea cod and climate change—modelling the effects of temperature on population dynamics. Global Change Biology, 9: 1669–1680.

Clarke, J., Bailey, D.M., and Wright, P.J. 2015. Evaluating the effectiveness of a seasonal spawning area closure. ICES Journal of Marine Science, 72: 2627–2637.

Connolly, P., and Officer, R. 2001. The use of tagging data in the formulation of the Irish Sea Cod Recovery Plan. ICES CM.

Cook, R., and Heath, M. 2005. The implications of warming climate for the management of North Sea demersal fisheries. ICES Journal of Marine Science, 62: 1322–1326.

Cook, R.M., Holmes, S.J., and Fryer, R.J. 2015. Grey seal predation impairs recovery of an over-exploited fish stock. Journal of Applied Ecology, 52: 969–979.

Cook, R.M., Kunzlik, P.A., Hislop, J.R.G., et al. 1999. Models of growth and maturity for north sea cod. Journal of Northwest Atlantic Fishery Science, 25: 91–99.

Cook, R.M., Needle, C.L., and Fernandes, P.G. 2012. Comment on "Fisheries mismanagement". Marine Pollution Bulletin, 64: 2265–2266.

Cook, R.M., Sinclair, A., and Stefánsson, G. 1997. Potential collapse of North Sea cod stocks. Nature, 385: 521–522.

Cook, R.M., and Trijoulet, V. 2016. The effects of grey seal predation and commercial fishing on the recovery of a depleted cod stock. Canadian Journal of Fisheries and Aquatic Sciences, 73: 1319–1329.

Cushing, D. 1990. Plankton production and year-class strength in fish populations: an update of the match/mismatch hypothesis. pp. 249–293. *In:* Advances in Marine Biology. Elsevier.

Daan, N. 1974. Growth of North Sea cod, *Gadus morhua*. Netherlands Journal of Sea Research, 8: 27–48.

Daan, N. 1978. Changes in cod stocks and cod fisheries in the North Sea. Rapports et Proces-Verbaux des Reunions (Denmark).

Daan, N., Heessen, H.J.L., and Pope, J.G. 1994. Changes in the North Sea cod stock during the twentieth century. ICES Marine Science Symposium, 198: 229–243.

Daewel, U., Schrum, C., and Gupta, A.K. 2015. The predictive potential of early life stage individual-based models (IBMs): an example for Atlantic cod *Gadus morhua* in the North Sea. Marine Ecology Progress Series, 534: 199–219.

Devine, J.A., Wright, P.J., Pardoe, H.E., et al. 2012. Comparing rates of contemporary evolution in life-history traits for exploited fish stocks. Canadian Journal of Fisheries and Aquatic Sciences, 69: 1105–1120.

Dickey-Collas, M., Brown, J., Fernand, L., et al. 1997. Does the western Irish Sea gyre influence the distribution of pelagic juvenile fish? Journal of Fish Biology, 51: 206–229.

Dickey-Collas, M., Gowen, R.J., and Fox, C.J. 1996. Distribution of larval and juvenile fish in the Western Irish Sea: Relationship to phytoplankton, zooplankton biomass and recurrent physical features. Marine and Freshwater Research, 47: 169–181.

Dransfeld, L., Dwane, O., McCarney, C., et al. 2004. Larval distribution of commercial fish species in waters around Ireland.

Drinkwater, K.F. 2005. The response of Atlantic cod (*Gadus morhua*) to future climate change. ICES Journal of Marine Science, 62: 1327–1337.

Dutil, J.-D., and Brander, K. 2003. Comparing productivity of North Atlantic cod (*Gadus morhua*) stocks and limits to growth production. Fisheries Oceanography, 12: 502–512.

Easey, M.W. 1987. English cod tagging experiments to the north of Scotland 1977–1979. ICES C.M. 1987/G:48.

Engelhard, G.H., Righton, D.A., and Pinnegar, J.K. 2014. Climate change and fishing: a century of shifting distribution in North Sea cod. Global Change Biology, 20: 2473–2483.

Enghoff, I.B., MacKenzie, B.R., and Nielsen, E.E. 2007. The Danish fish fauna during the warm Atlantic period (ca. 7000–3900bc): Forerunner of future changes? Fisheries Research, 87: 167–180.

Fahay, M.P. 1983. Guide to the early stages of marine fishes occurring in the western North Atlantic Ocean, Cape Hatteras to the southern Scotian Shelf. Journal of Northwest Atlantic Fishery Science, 4: 3–423.

Fairweather, R., Bradbury, I.R., Helyar, S.J., et al. 2018. Range-wide genomic data synthesis reveals transatlantic vicariance and secondary contact in Atlantic cod. Ecology and Evolution, 8: 12140–12152.

Fauchald, P. 2010. Predator-prey reversal: A possible mechanism for ecosystem hysteresis in the North Sea? Ecology, 91: 2191–2197.

Fernandes, P.G., Coull, K., Davis, C., et al. 2011. Observations of discards in the Scottish mixed demersal trawl fishery. ICES Journal of Marine Science, 68: 1734–1742.

Fernández, J.E., Johansson, T.M., Skinner, J.A., et al. 2022. Fisheries and the Law in Europe: Regulation after Brexit. p. 171. Fernández, J.E., Johansson, T.M., Skinner, J.A., et al. (eds.). Taylor and Francis Group.

Fox, C.J., Armstrong, M., and Blanchard, J.L. 2005. Irish Sea cod. pp. 62–75. International Council for the Exploration of the Seas.

Fox, C.J., Taylor, M., Dickey-Collas, M., et al. 2008. Mapping the spawning grounds of North Sea cod (*Gadus morhua*) by direct and indirect means. Proceedings of the Royal Society B: Biological Sciences, 275: 1543–1548.

Freitas, C., Olsen, E.M., Knutsen, H., et al. 2016. Temperature-associated habitat selection in a cold-water marine fish. Journal of Animal Ecology, 85: 628–637.

Freitas, C., Olsen, E.M., Moland, E., et al. 2015. Behavioral responses of Atlantic cod to sea temperature changes. Ecology and Evolution, 5: 2070–2083.

Geffen, A., Fox, C., and Nash, R. 2006. Temperature-dependent development rates of cod *Gadus morhua* eggs. Journal of Fish Biology, 69: 1060–1080.

Gibb, F.M., Gibb, I.M., and Wright, P.J. 2007. Isolation of Atlantic cod (*Gadus morhua*) nursery areas. Marine Biology, 151: 1185–1194.

Godø, O.R., and Moksness, E. 1987. Growth and maturation of Norwegian coastal cod and Northeast Arctic cod under different conditions. Fisheries Research, 5: 235–242.

González-Irusta, J.M., and Wright, P.J. 2016. Spawning grounds of Atlantic cod (*Gadus morhua*) in the North Sea. ICES Journal of Marine Science, 73: 304–315.

González-Laxe, F. 2005. The precautionary principle in fisheries management. Marine Policy, 29: 495–505.

Graham, M. 1934. Report on the North Sea cod. ICES Document, 13: 2. 1–160.

Gulland, J.A. 1987. The management of North Sea fisheries: Looking towards the 21 st century. Marine Policy, 11: 259–272.

Hammond, P., and Harris, R. 2006. Grey seal diet composition and prey consumption off western Scotland and Shetland. Final report to Scottish Executive Environment and Rural Affairs Department and Scottish Natural Heritage: 41.

Harrald, M., Neat, F.C., Wright, P.J., et al. 2010. Population variation in thermal growth responses of juvenile Atlantic cod (*Gadus morhua* L.). Environmental Biology of Fishes, 87: 187–194.

Heath, M.R., Culling, M.A., Crozier, W.W., et al. 2014. Combination of genetics and spatial modelling highlights the sensitivity of cod (*Gadus morhua*) population diversity in the North Sea to distributions of fishing. ICES Journal of Marine Science, 71: 794–807.

Heath, M.R., Kunzlik, P.A., Gallego, A., et al. 2008. A model of meta-population dynamics for North Sea and West of Scotland cod—The dynamic consequences of natal fidelity. Fisheries Research, 93: 92–116.

Heath, M.R., and Lough, R.G. 2007. A synthesis of large-scale patterns in the planktonic prey of larval and juvenile cod (*Gadus morhua*). Fisheries Oceanography, 16: 169–185.

Heath, M.R., Rankine, P., and Cargill, L. 1994. Distribution of cod and haddock eggs in the North Sea in 1992 in relation to oceanographic features and compared with distributions in 1952–1957. ICES Marine Science Symposia, 198: 438–439.

Hedger, R., McKenzie, E., Heath, M., et al. 2004. Analysis of the spatial distributions of mature cod (*Gadus morhua*) and haddock (*Melanogrammus aeglefinus*) abundance in the North Sea (1980–1999) using generalised additive models. Fisheries Research, 70: 17–25.

Heessen, H.J. 1993. The distribution of cod (*Gadus morhua*) in the North Sea. NAFO Scientific Council Studies, 18: 7.

Heffernan, O.A., Danilowicz, B.S., and Milligan, S.P. 2004. Determination of species-specific spawning distributions of commercial finfish in the Irish Sea using a biochemical protein-based method. Marine Ecology Progress Series, 284: 279–291.

Heino, M., Dieckmann, U., and Godø, O.R. 2002. Estimating reaction norms for age and size at maturation with reconstructed immature size distributions: a new technique illustrated by application to Northeast Arctic cod. ICES Journal of Marine Science, 59: 562–575.

Heino, M.P., Díaz Pauli, B., and Dieckmann, U. 2015. Fisheries-induced evolution. Annual Review of Ecology, Evolution, and Systematics, 46: 461–480.

Hemmer-Hansen, J., Nielsen, E.E., Therkildsen, N.O., et al. 2013. A genomic island linked to ecotype divergence in Atlantic cod. Molecular Ecology, 22: 2653–2667.

Hill, A., Brown, J., and Fernand, L. 1997. The summer gyre in the western Irish Sea: shelf sea paradigms and management implications. Estuarine, Coastal and Shelf Science, 44: 83–95.

Hislop, J.R. 1996. Changes in North Sea gadoid stocks. ICES Journal of Marine Science, 53: 1146–1156.

Hoenig, J.M., Barrowman, N.J., Hearn, W.S., et al. 1998. Multiyear tagging studies incorporating fishing effort data. Canadian Journal of Fisheries and Aquatic Sciences, 55: 1466–1476.

Höffle, H., Van Damme, C.J.G., Fox, C., et al. 2017. Linking spawning ground extent to environmental factors—patterns and dispersal during the egg phase of four North Sea fishes. Dispersal during Early Life History of Fish, 01: 357–374.

Holden, M. 1981. The North Sea International O'group Gadoid Surveys 1969–1978. Conseil international pour l'exploration de la mer.

Holmes, S., Bailey, N., Campbell, N., et al. 2011. Using fishery-dependent data to inform the development and operation of a co-management initiative to reduce cod mortality and cut discards. ICES Journal of Marine Science, 68: 1679–1688.

Holmes, S.J., Millar, C.P., Fryer, R.J., et al. 2014. Gadoid dynamics: differing perceptions when contrasting stock vs. population trends and its implications to management. ICES Journal of Marine Science, 71: 1433–1442.

Holmes, S.J., Wright, P.J., and Fryer, R.J. 2008. Evidence from survey data for regional variability in cod dynamics in the North Sea and West of Scotland. ICES Journal of Marine Science, 65: 206–215.

Holt, E.W.L. 1893. North Sea Investigations. Journal of the Marine Biological Association of the United Kingdom, 3: 78–106.

Horsburgh, K., Hill, A., Brown, J., et al. 2000. Seasonal evolution of the cold pool gyre in the western Irish Sea. Progress in Oceanography, 46: 1–58.

Horwood, J.W., Nichols, J.H., and Milligan, S. 1998. Evaluation of closed areas for fish stock conservation. Journal of Applied Ecology, 35: 893–903.

Horwood, J.W., O'Brien, C., and Darby, C. 2006. North Sea cod recovery? ICES Journal of Marine Science, 63: 961–968.

Houghton, R.G., and Flatman, S. 1981. The exploitation pattern, density-dependent catchability, and growth of cod (*Gadus morhua*) in the west-central North Sea. ICES Journal of Marine Science, 39: 271–287.

Howarth, M. 1984. Currents in the eastern Irish Sea. Oceanography and Marine Biology: An Annual Review, 22: 11–54.

Huebert, K.B., Pätsch, J., Hufnagl, M., et al. 2018. Modeled larval fish prey fields and growth rates help predict recruitment success of cod and anchovy in the North Sea. Marine Ecology Progress Series, 600: 111–126.

Hunter, A., Speirs, D.C., and Heath, M.R. 2015. Fishery-induced changes to age and length dependent maturation schedules of three demersal fish species in the Firth of Clyde. Fisheries Research, 170: 14–23.

Huserbråten, M.B.O., Moland, E., and Albretsen, J. 2018. Cod at drift in the North Sea. Progress in Oceanography, 167: 116–124.

Hutchinson, W.F., van Oosterhout, C., Rogers, S.I., et al. 2003. Temporal analysis of archived samples indicates marked genetic changes in declining North Sea cod (*Gadus morhua*). Proceedings of the Royal Society of London. Series B: Biological Sciences, 270: 2125–2132.

ICES 1997. Report of the Working Group on the Assessment of Demersal Stocks in the North Sea and Skagerrak (WGNSSK), 7–15 October 1996, Copenhagen, Denmark. ICES CM 1997/Assess:6.

ICES 2005. Cod recovery plans.ICES 2017. Report of the Working Group on the Assessment of Demersal Stocks in the North Sea and Skagerrak (WGNSSK), 26 April–5 May 2016, Hamburg, Germany.. https://www.ices.dk/sites/pub/Publication%20Reports/Expert%20Group%20Report/acom/2016/WGNSSK/01%20WGNSSK%20report%202016.pdf.

ICES 2020a. Workshop on Stock Identification of North Sea Cod (WKNSCodID). https://doi.org/10.17895/ices.pub.7499.

ICES 2020b. Stock Annex: Cod (*Gadus morhua*) in divisions 7.e-k (eastern English Channel and southern Celtic Seas). https://doi.org/10.17895/ices.pub.18622229.v1

ICES2020c. Benchmark Workshop for Demersal Species (WKDEM). http://doi.org/10.17895/ices.pub.5548.

ICES 2020d. Working Group on the Assessment of Demersal Stocks in the North Sea and Skagerrak (WGNSSK). ICES Scientific Reports, 2: 61. 1140 pp. http://doi.org/10.17895/ices.pub.6092.

ICES 2020e. Working Group for the Celtic Seas Ecoregion (WGCSE). ICES Scientific Reports, 2:40. 924 pp. https://doi.org/10.17895/ices.pub.5978.ICES. 2020f. Benchmark Workshop on Celtic Sea Stocks (WKCELTIC). ICES Scientific Reports, 2:97. 166 pp. http://doi.org/10.17895/ices.pub.5983.

ICES 2021a. Working Group on Multispecies Assessment Methods (WGSAM; outputs from 2020 meeting). https://doi.org/10.17895/ices.pub.7695.

ICES 2021b. Benchmark Workshop in North Sea Stocks (WKNSEA). https://doi.org/10.17895/ices.pub.7922.

ICES. 2021c. Benchmark Workshop on North Sea Stocks (WKNSEA). ICES Scientific Reports, 3: 25. 756 pp. https://doi.org/10.17895/ices.pub.7922.

ICES 2022a. Working Group on the Assessment of Demersal Stocks in the North Sea and Skagerrak. Cod (Gadus morhua) in Subarea 4, Division 7.d and Subdivision 20 (North Sea, Eastern English Channel, Skagerrak).

ICES 2022b. Working Group for the Celtic Seas Ecoregion. https://doi.org/10.17895/ices.pub.19863796.v1.

ICES 2022c. Workshop on Stock Identification of west of Scotland Sea Cod (WK6aCodID; outputs from 2021 meeting). http://doi.org/10.17895/ices.pub.10031.

ICES 2022d. Benchmark Workshop for fish stocks in the North Sea and Celtic Sea.

ICES. 2023. Working Group on the Assessment of Demersal Stocks in the North Sea and Skagerrak (WGNSSK). ICES Scientific Reports, 5: 39. 1256 pp. https://doi.org/10.17895/ices.pub.22643143Editors.

Ikpewe, I.E., Baudron, A.R., Ponchon, A., et al. 2021. Bigger juveniles and smaller adults: Changes in fish size correlate with warming seas. Journal of Applied Ecology, 58: 847–856.

Jardim, E., Eero, M., Silva, A., et al. 2018. Testing spatial heterogeneity with stock assessment models. PLOS ONE, 13: e0190791.

Jaworski, A., and Penny, I. 2009. West of Four – Effectiveness of Windsock Area Closure. https://www.webarchive.org.uk/wayback/archive/20150219062454mp_/http://www.gov.scot/Uploads/Documents/SISP0209.pdf.

Jonsson, P.R., Corell, H., André, C., et al. 2016. Recent decline in cod stocks in the North Sea–Skagerrak–Kattegat shifts the sources of larval supply. Fisheries Oceanography, 25: 210–228.

Kamenos, N.A., Moore, P.G., and Hall-Spencer, J.M. 2004. Small-scale distribution of juvenile gadoids in shallow inshore waters; what role does maerl play? ICES Journal of Marine Science, 61: 422–429.

Kell, L.T., Nash, R.D.M., Dickey-Collas, M., et al. 2016. Is spawning stock biomass a robust proxy for reproductive potential? Fish and Fisheries, 17: 596–616.

Kell, L.T., Pilling, G.M., and O'Brien, C.M. 2005. Implications of climate change for the management of North Sea cod (*Gadus morhua*). ICES Journal of Marine Science, 62: 1483–1491.

Kelly, C.J., Codling, E.A., and Rogan, E. 2006. The Irish Sea cod recovery plan: some lessons learned. ICES Journal of Marine Science, 63: 600–610.

Kempf, A., Floeter, J., and Temming, A. 2009. Recruitment of North Sea cod (*Gadus morhua*) and Norway pout (*Trisopterus esmarkii*) between 1992 and 2006: the interplay between climate influence and predation. Canadian Journal of Fisheries and Aquatic Sciences, 66: 633–648.

Kempf, A., Stelzenmüller, V., Akimova, A., et al. 2013. Spatial assessment of predator–prey relationships in the North Sea: the influence of abiotic habitat properties on the spatial overlap between 0-group cod and grey gurnard. Fisheries Oceanography, 22: 174–192.

Keyl, F., Kempf, A.J., and Sell, A.F. 2015. Sexual size dimorphism in three North Sea gadoids. Journal of Fish Biology, 86: 261–275.

Kirby, R.R., Beaugrand, G., and Lindley, J.A. 2009. Synergistic effects of climate and fishing in a marine ecosystem. Ecosystems, 12: 548–61.

Knutsen, H., Jorde, P.E., André, C., et al. 2003. Fine-scaled geographical population structuring in a highly mobile marine species: the Atlantic cod. Molecular Ecology, 12: 385–394.

Knutsen, H., Olsen, E.M., Jorde, P.E., et al. 2011. Are low but statistically significant levels of genetic differentiation in marine fishes 'biologically meaningful'? A case study of coastal Atlantic cod. Molecular Ecology, 20: 768–783.

Koul, V., Sguotti, C., Årthun, M., et al. 2021. Skilful prediction of cod stocks in the North and Barents Sea a decade in advance. Communications Earth & Environment, 2: 140.

Kraak, S.B.M., Bailey, N., Cardinale, M., et al. 2013. Lessons for fisheries management from the EU cod recovery plan. Marine Policy, 37: 200–213.

Kristiansen, T., Drinkwater, K.F., Lough, R.G., et al. 2011. Recruitment variability in North Atlantic cod and match-mismatch dynamics. PLOS ONE, 6: e17456.

Kvile, K.Ø., Romagnoni, G., Dagestad, K.-F., et al. 2018. Sensitivity of modelled North Sea cod larvae transport to vertical behaviour, ocean model resolution and interannual variation in ocean dynamics. ICES Journal of Marine Science, 75: 2413–2424.

Lambert, R.A. 2002. The grey seal in Britain: a twentieth century history of a nature conservation success. Environment and History, 8: 449–474.

Lelievre, S., Vaz, S., Martin, C., et al. 2014. Delineating recurrent fish spawning habitats in the North Sea. Journal of Sea Research, 91: 1–14.

Lilley, R.J., and Unsworth, R.K. 2014. Atlantic Cod (*Gadus morhua*) benefits from the availability of seagrass (*Zostera marina*) nursery habitat. Global Ecology and Conservation, 2: 367–377.

Lorenzen, K. 1996. The relationship between body weight and natural mortality in juvenile and adult fish: a comparison of natural ecosystems and aquaculture. Journal of Fish Biology, 49: 627–642.

Needle, C.L. and Catarino, R. (2011). Evaluating the effect of real-time closures on cod targeting, ICES Journal of Marine Science 68(8): 1647–1655.

Mace, P., and Doonan, I. 1988. A generalised bioeconomic simulation model for fish population dynamics. New Zealand Fishery Assessment Research Document 88/4. Fisheries Research Centre, MAFFish, POB, 297.

Macer, C. 1983. Changes in growth of North Sea cod and their effect on yield assessments. ICES CM.

Madsen, N., Holst, R., Frandsen, R., et al. 2017. Development and test of selective sorting grids used in the Norway lobster (*Nephrops norvegicus*) fishery. Fisheries Research, 185: 26–33.

Madsen, N., and Valentinsson, D. 2010. Use of selective devices in trawls to support recovery of the Kattegat cod stock: a review of experiments and experience. ICES Journal of Marine Science, 67: 2042–2050.

Marty, L., Rochet, M.-J., and Ernande, B. 2014. Temporal trends in age and size at maturation of four North Sea gadid species: cod, haddock, whiting and Norway pout. Marine Ecology Progress Series, 497: 179–197.

McQueen, K., and Marshall, C.T. 2017. Shifts in spawning phenology of cod linked to rising sea temperatures. ICES Journal of Marine Science, 74: 1561–1573.

Melli, V., Krag, L.A., Herrmann, B., et al. 2019. Can active behaviour stimulators improve fish separation from Nephrops (*Nephrops norvegicus*) in a horizontally divided trawl codend? Fisheries Research, 211: 282–290.

Millner, R., Pilling, G., McCully, S., et al. 2011. Changes in the timing of otolith zone formation in North Sea cod from otolith records: an early indicator of climate-induced temperature stress? Marine Biology, 158: 21–30.

Minto, C., Mills Flemming, J., Britten, G.L., et al. 2014. Productivity dynamics of Atlantic cod. Canadian Journal of Fisheries and Aquatic Sciences, 71: 203–216.

Morgan, M.J., Wright, P.J., and Rideout, R.M. 2013. Effect of age and temperature on spawning time in two gadoid species. Fisheries Research, 138: 42–51.

Munk, P., Wright, P.J., and Pihl, N.J. 2002. Distribution of the early larval stages of cod, plaice and lesser sandeel across haline fronts in the North Sea. Estuarine, Coastal and Shelf Science, 55: 139–149.

Myers, R., Mertz, G., and Barrowman, N. 1995. Spatial scales of variability in cod recruitment in the North Atlantic. Canadian Journal of Fisheries and Aquatic Sciences, 52: 1849–1862.

Nash, R.D., Dickey-Collas, M., and Kell, L.T. 2009. Stock and recruitment in North Sea herring (*Clupea harengus*); compensation and depensation in the population dynamics. Fisheries Research, 95: 88–97.

Nash, R.D.M., Pilling, G.M., Kell, L.T., et al. 2010. Investment in maturity-at-age and -length in northeast Atlantic cod stocks. Fisheries Research, 104: 89–99.

Neat, F., and Righton, D. 2007. Warm water occupancy by North Sea cod. Proceedings of the Royal Society B: Biological Sciences, 274: 789–798.

Neat, F.C., Bendall, V., Berx, B., et al. 2014. Movement of Atlantic cod around the British Isles: implications for finer scale stock management. Journal of Applied Ecology, 51: 1564–1574.

Neat, F.C., Wright, P.J., Zuur, A.F., et al. 2006. Residency and depth movements of a coastal group of Atlantic cod (*Gadus morhua* L.). Marine Biology, 148: 643–654.

Neuenfeldt, S., Righton, D., Neat, F., et al. 2013. Analysing migrations of Atlantic cod *Gadus morhua* in the north-east Atlantic Ocean: then, now and the future. Journal of Fish Biology, 82: 741–763.

Nicolas, D., Rochette, S., Llope, M., et al. 2014. Spatio-temporal variability of the North Sea cod recruitment in relation to temperature and zooplankton. PLOS ONE, 9: e88447.

Nielsen, E.E., Wright, P.J., Hemmer-Hansen, J., et al. 2009. Microgeographical population structure of cod *Gadus morhua* in the North Sea and west of Scotland: the role of sampling loci and individuals. Marine Ecology Progress Series, 376: 213–225.

Nielsen, R., and Munk, P. 2004. Growth pattern and growth dependent mortality of larval and pelagic juvenile North Sea cod *Gadus morhua*. Marine Ecology Progress Series, 278: 261–270.

Núñez-Riboni, I., Taylor, M.H., Kempf, A., et al. 2019. Spatially resolved past and projected changes of the suitable thermal habitat of North Sea cod (*Gadus morhua*) under climate change. ICES Journal of Marine Science, 76: 2389–2403.

Ó Cuaig, M., and Officer, R. 2007. Evaluation of the benefits to sustainable management of seasonal closure of the greencastle codling (*Gadus morhua*) fishery. Final Report on Project 01.SM.TI.12, Irish Fisheries Bulletin, 27: 36.

O'Brien, C.M., Fox, C.J., Planque, B., et al. 2000. Climate variability and North Sea cod. Nature, 404: 142–142.

O'Leary, B.C., Smart, J.C.R., Neale, F.C., et al. 2011. Fisheries mismanagement. Marine Pollution Bulletin, 62: 2642–2648.

O'Neill, F.G., Feekings, J., Fryer, R.J., et al. 2019. Discard avoidance by improving fishing gear selectivity: helping the fishing industry help itself. pp. 279–296. *In:* Uhlmann, S.S., Ulrich, C., and Kennelly, S.J. (eds.). The European landing obligation: reducing discards in complex multi-species and multi-jurisdictional fisheries. Springer.

Olsen, E.M., Ottersen, G., Llope, M., et al. 2011. Spawning stock and recruitment in North Sea cod shaped by food and climate. Proceedings of the Royal Society B: Biological Sciences, 278: 504–510.

Oosthuizen, E., and Daan, N. 1974. Egg fecundity and maturity of North Sea cod, *Gadus morhua*. Netherlands Journal of Sea Research, 8: 378–397.

OSPAR-Commission 2015. 2014 Status Report on the OSPAR Network of Marine Protected Areas. https://www.ospar.org/documents?v=33572.

Pampoulie, C., Stefansson, M.Ö., Jörundsdóttir, T.D., et al. 2008. Recolonization history and large-scale dispersal in the open sea: the case study of the North Atlantic cod, *Gadus morhua* L. Biological Journal of the Linnean Society, 94: 315–329.

Patterson, K.R. 1998. Assessing fish stocks when catches are misreported: model, simulation tests, and application to cod, haddock, and whiting in the ICES area. ICES Journal of Marine Science, 55: 878–891.

Perry, A.L., Low, P.J., Ellis, J.R., et al. 2005. Climate change and distribution shifts in marine fishes. Science, 308: 1912–1915.

Pitois, S.G., and Armstrong, M. 2014. The growth of larval cod and haddock in the Irish Sea: a model with temperature, prey size and turbulence forcing. Fisheries Oceanography, 23: 417–435.

Pitois, S.G., and Fox, C.J. 2008. Empirically modelling the potential effects of changes in temperature and prey availability on the growth of cod larvae in UK shelf seas. ICES Journal of Marine Science, 65: 1559–1572.

Planque, B., and Frédou, T. 1999. Temperature and the recruitment of Atlantic cod (*Gadus morhua*). Canadian Journal of Fisheries and Aquatic Sciences, 56: 2069–2077.

Polet, H. 2002. Selectivity experiments with sorting grids in the North Sea brown shrimp (*Crangon crangon*) fishery. Fisheries Research, 54: 217–233.

Pope, J.G., and Macer, C.T. 1996. An evaluation of the stock structure of North Sea cod, haddock, and whiting since 1920, together with a consideration of the impacts of fisheries and predation effects on their biomass and recruitment. ICES Journal of Marine Science, 53: 1157–1069.

Poulsen, N.A., Hemmer-Hansen, J., Loeschcke, V., et al. 2011. Microgeographical population structure and adaptation in Atlantic cod *Gadus morhua*: spatio-temporal insights from gene-associated DNA markers. Marine Ecology Progress Series, 436: 231–243.

Raitt, D.F.S. 1967. Cod spawning in Scottish waters. Preliminary investigations. ICES CM 1967/F:29 (mimeo).

Rätz, H.-J., and Lloret, J. 2003. Variation in fish condition between Atlantic cod (*Gadus morhua*) stocks, the effect on their productivity and management implications. Fisheries Research, 60: 369–380.

Righton, D., Quayle, V.A., Hetherington, S., et al. 2007. Movements and distribution of cod (*Gadus morhua)* in the southern North Sea and English Channel: results from conventional and electronic tagging experiments. Journal of the Marine Biological Association of the United Kingdom, 87: 599–613.

Righton, D.A., Andersen, K.H., Neat, F., et al. 2010. Thermal niche of Atlantic cod *Gadus morhua*: limits, tolerance and optima. Marine Ecology Progress Series, 420: 1–13.

Rijnsdorp, A.D., Daan, N., van Beek, F.A., et al. 1991. Reproductive variability in North Sea plaice, sole, and cod. ICES Journal of Marine Science, 47: 352–375.

Riley, J.D., and Parnell, W. 1984. The distribution of young cod. *In*: The propagation of cod *Gadus morhua* L. an international symposium, Arendal, 14–17 June 1983.

Rindorf, A., Jensen, H., and Schrum, C. 2008. Growth, temperature, and density relationships of North Sea cod (*Gadus morhua*). Canadian Journal of Fisheries and Aquatic Sciences, 65: 456–470.

Rindorf, A., and Lewy, P. 2006. Warm, windy winters drive cod north and homing of spawners keeps them there. Journal of Applied Ecology, 43: 445–453.

Robichaud, D., and Rose, G.A. 2004. Migratory behaviour and range in Atlantic cod: inference from a century of tagging. Fish and Fisheries, 5: 185–214.

Romagnoni, G., Kvile, K.Ø., Dagestad, K.-F., et al. 2020. Influence of larval transport and temperature on recruitment dynamics of North Sea cod (*Gadus morhua*) across spatial scales of observation. Fisheries Oceanography, 29: 324–339.

Rowlands, W.L., Dickey-Collas, M., Geffen, A.J., et al. 2008. Diet overlap and prey selection through metamorphosis in Irish Sea cod (*Gadus morhua*), haddock (*Melanogrammus aeglefinus*), and whiting (*Merlangius merlangus*). Canadian Journal of Fisheries and Aquatic Sciences, 65: 1297–1306.

Ruddiman, W.F. 2003. The Anthropogenic Greenhouse Era Began Thousands of Years Ago. Climatic Change, 61: 261–293.

Russell, D., Jones, E., and Morris, C. 2017. Updated seal usage maps: the estimated at-sea distribution of grey and harbour seals. Scottish Marine and Freshwater Science, 8: 25.

SCOS. 2017. Scientific advice on matters related to the management of seal populations. 144 pp.

Serpetti, N., Baudron, A.R., Burrows, M., et al. 2017. Impact of ocean warming on sustainable fisheries management informs the Ecosystem Approach to Fisheries. Scientific Reports, 7: 1–15.

Sodeland, M., Jorde, P.E., Lien, S., et al. 2016. "Islands of Divergence" in the Atlantic Cod Genome Represent Polymorphic Chromosomal Rearrangements. Genome Biology and Evolution, 8: 1012–1022.

Sparholt, H., Hislop, J., Bergstad, O.A., et al. 2015. Cod fishes (Gadidae). *In:* Heessen, H.J.L., Daan, N., and Ellis, J.R. (eds.). Fish Atlas of the Celtic Sea, North Sea, and Baltic Sea., Edition 1st edn. Wageningen Academic Publishers.

Speirs, D.C., Guirey, E.J., Gurney, W.S.C., et al. 2010. A length-structured partial ecosystem model for cod in the North Sea. Fisheries Research, 106: 474–494.

Stearns, S., and Crandall, R. 1984. Plasticity for Age and Size at Sexual Maturity: A Life-history response to Unavidable Stress. pp. 13–34. *In:* Potts, G., and Wootton, R. (eds.). Fish reproduction: Strategies and Tactics. Academic Press, London.

STECF 2007. Evaluation of closed area schemes (SGMOS-07-03). http://stecf.jrc.ec.europa.eu/documents/43805/44876/07-09_SG-MOS+07-03+-+Evaluation+of+closed+areas+II.pdf.

Steinarsson, A., and Björnsson, B. 1999. The effects of temperature and size on growth and mortality of cod larvae. Journal of Fish Biology, 55: 100–109.

Stiasny, M.H., Mittermayer, F.H., Göttler, G., et al. 2018. Effects of parental acclimation and energy limitation in response to high CO_2 exposure in Atlantic cod. Scientific Reports, 8: 8348.

Stiasny, M.H., Mittermayer, F.H., Sswat, M., et al. 2016. Ocean acidification effects on Atlantic cod larval survival and recruitment to the fished population. PLOS ONE, 11: e0155448.

Stige, L.C., Ottersen, G., Brander, K., et al. 2006. Cod and climate: effect of the North Atlantic Oscillation on recruitment in the North Atlantic. Marine Ecology Progress Series, 325: 227–241.

Svåsand, T., Jørstad, K.E., Otterå, H., et al. 1996. Differences in growth performance between Arcto-Norwegian and Norwegian coastal cod reared under identical conditions. Journal of Fish Biology, 49: 108–119.

Temming, A., Floeter, J., and Ehrich, S. 2007. Predation hot spots: large scale impact of local aggregations. Ecosystems, 10: 865–876.

Thomas, L., Russell, D.J., Duck, C.D., et al. 2019. Modelling the population size and dynamics of the British grey seal. Aquatic Conservation: Marine and Freshwater Ecosystems, 29: 6–23.

Thompson, B., and Riley, J. 1981. Egg and larval studies in the North Sea cod (*Gadus morhua* L.). Rapport et Procès-Verbaux Des Réunions Conseil International pour l'Exploration de la Mer, 178: 553–559.

Thorsen, A., Witthames, P.R., Marteinsdóttir, G., et al. 2010. Fecundity and growth of Atlantic cod (*Gadus morhua* L.) along a latitudinal gradient. Fisheries Research, 104: 45–55.

Tobin, D., and Wright, P.J. 2011. Temperature effects on female maturation in a temperate marine fish. Journal of Experimental Marine Biology and Ecology, 403: 9–13.

Trijoulet, V., Holmes, S.J., and Cook, R.M. 2018. Grey seal predation mortality on three depleted stocks in the West of Scotland: What are the implications for stock assessments? Canadian Journal of Fisheries and Aquatic Sciences, 75: 723–732.

Trippel, E.A., Kjesbu, O.S., and Solemdal, P. 1997. Effects of adult age and size structure on reproductive output in marine fishes. pp. 31–62. *In:* Early life history and recruitment in fish populations, Springer.

Turner, K., Righton, D., and Metcalfe, J. 2002. The dispersal patterns and behaviour of North Sea cod (*Gadus morhua*) studied using electronic data storage tags. pp. 201–208. *In:* Aquatic Telemetry, Springer.

Van Alphen, J., and Heessen, H. 1984. Variations in length at age of North Sea cod. ICES CM.

van der Molen, J., Rogers, S.I., Ellis, J.R., et al. 2007. Dispersal patterns of the eggs and larvae of spring-spawning fish in the Irish Sea, UK. Journal of Sea Research, 58: 313–330.

van Overzee, H.M.J., and Rijnsdorp, A.D. 2015. Effects of fishing during the spawning period: implications for sustainable management. Reviews in Fish Biology and Fisheries, 25: 65–83.

Walther, B.D., and Thorrold, S.R. 2006. Water, not food, contributes the majority of strontium and barium deposited in the otoliths of a marine fish. Marine Ecology Progress Series, 311: 125–130.

West, W.Q.-B. 1970. The spawning biology and fecundity of cod in Scottish waters. Thesis. Department of Zoology. University of Aberdeen.

Wilson, L.J., and Hammond, P.S. 2019. The diet of harbour and grey seals around Britain: Examining the role of prey as a potential cause of harbour seal declines. Aquatic Conservation: Marine and Freshwater Ecosystems, 29: 71–85.

Wright, P.J. 2005. West of Scotland (ICES Division VIa). *In* Spawning and life history information for North Atlantic cod stocks, p. 89. Ed. by K. M. Brander. International Council for the Exploration of the Seas, Co-operative Research Reports.

Wright, P.J. 2014. Are there useful life history indicators of stock recovery rate in gadoids? ICES Journal of Marine Science, 71: 1393–1406.

Wright, P.J., Doyle, A., Taggart, J.B., et al. 2021. Linking Scales of Life-History Variation With Population Structure in Atlantic Cod. Frontiers in Marine Science, 8: e630515.

Wright, P.J., Galley, E., Gibb, I.M., et al. 2006b. Fidelity of adult cod to spawning grounds in Scottish waters. Fisheries Research, 77: 148–158.

Wright, P.J., Millar, C.P., and Gibb, F.M. 2011. Intrastock differences in maturation schedules of Atlantic cod, *Gadus morhua*. ICES Journal of Marine Science, 68: 1918–1927.

Wright, P.J., Neat, F.C., Gibb, F.M., et al. 2006a. Evidence for metapopulation structuring in cod from the west of Scotland and North Sea. Journal of Fish Biology, 69: 181–199.

Wright, P.J., Palmer, S.C.F., and Marshall, C.T. 2014. Maturation shifts in a temperate marine fish population cannot be explained by simulated changes in temperature-dependent growth and maturity. Marine Biology, 161: 2781–2790.

Wright, P.J., Régnier, T., Gibb, F.M., et al. 2018. Assessing the role of ontogenetic movement in maintaining population structure in fish using otolith microchemistry. Ecology and Evolution, 8: 7907–7920.

Wright, P.J., and Rowe, S. 2019. Reproduction and spawning. pp. 87–132. *In:* Rose, G. (ed.). Atlantic Cod. Wiley Blackwell.

Yoneda, M., and Wright, P.J. 2004. Temporal and spatial variation in reproductive investment of Atlantic cod *Gadus morhua* in the northern North Sea and Scottish west coast. Marine Ecology Progress Series, 276: 237–248.

CHAPTER 8

Kattegat and Baltic Sea Cod Stocks

*Karin Hüssy** and *Margit Eero*

Introduction

Atlantic cod (*Gadus morhua*) have traditionally been found in the North Sea, the Kattegat, Belt Sea and most of the western and eastern Baltic Sea. The cod inhabiting this region consists of at least two genetically distinct populations, the eastern and western Baltic cod stocks (Nielsen et al. 2003; Berg et al. 2015; Barth et al. 2019; Hemmer-Hansen et al. 2019). Cod spawning in the Kattegat and western Baltic Sea may belong to the same biological population or population complex, displaying the genetic signature of a hybrid zone between the differentiated North Sea and Baltic Sea populations (Nielsen et al. 2003). These populations are characterised by having distinct spawning areas and seasons (Brander 1994; Fox et al. 2008; Vitale et al. 2008; Hüssy, 2011a; Börjesson et al. 2013). Currently, three distinct cod management areas are specified by the ICES Subdivisions (SD): Kattegat in SD 21, the western Baltic in SD 22–24, and the eastern Baltic in SD 24–32 (Figure 1). Assessments for eastern and western Baltic cod are stock-specific and account for mixing the two populations in SD 24.

Geography and Hydrography

The Baltic Sea region was historically formed by processes of isostatic land uplift with concurrent sea-level rise following the latest deglaciation 15,000 years ago (Björck 1995, 2008). Today, the Baltic Sea is one of the largest brackish water ecosystems in the world, with consecutive deep basins and shallow connections to the North Sea and Atlantic Ocean through the Kattegat and the Belt Sea. The Kattegat is an environmental transition zone between the North Sea and the Baltic Sea, where the bathymetry becomes successively shallower from > 120 m in the North Sea and

National Institute of Aquatic Resources, Technical University of Denmark, Henrik Dams Allé, 201, 138, 2800 Kgs. Lyngby, Denmark.
* Corresponding author: kh@aqua.dtu.dk

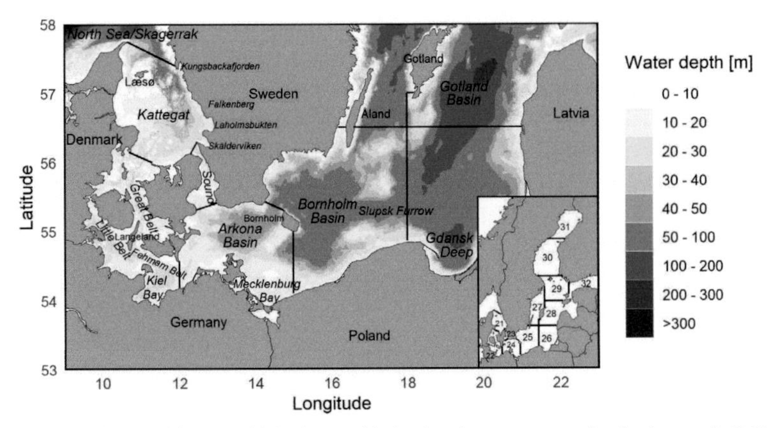

Figure 1. Map of the Baltic Sea with hydrographic landmark names occurring in the text in italics, and land areas in plain face. Inset provides an overview of the Baltic Sea with ICES subdivision outlines and numbers.

Skagerrak to 40–80 m in the northern Kattegat to depths of 20–40 m in the southern Kattegat. The Kattegat is connected to the western Baltic Sea through the Belt Sea, consisting of three narrow channels: the Little Belt, the Great Belt, and the Sound and the Fehmarn Belt. The western Baltic Sea consists of shallow bays (Kiel and Mecklenburg Bay) with a water depth of < 30 m, and to the east, the Arkona Basin with a maximum water depth of ca 45 m (Jakobsson et al. 2019). The eastern Baltic Sea contains a sequence of deep basins, the Bornholm Basin (max depth = 100 m), the Gdansk Deep (max depth = 110 m) and the Gotland Deep (max depth 200 m), and three bays towards the north (the Gulf of Riga, Gulf of Finland and Bothnian Bay) (Jakobsson et al. 2019).

The hydrography is characterised by deep-water inflow of saline water from the North Sea and outflow of freshwater from river runoff in the surface layer (Meier et al. 2006; Mohrholz et al. 2015). The inflow is driven by high air pressure associated with easterly winds over the Baltic, followed by longer periods of westerly winds over the North Atlantic and Europe (Matthäus and Franck 1992; Schinke and Matthäus 1998). These bathymetric and hydrographic conditions lead to a geographical salinity gradient, which ranges from close to fresh water in the Bothnian Bay in the north to around 8 PSU and in the central Baltic Proper, between 10 and 20 PSU in the southern Baltic, up to 34 PSU in the northern Kattegat and fully marine saltwater in the North Sea (Kullenberg and Jacobsen 1981; Matthäus and Franck 1992; Schinke and Matthäus 1998; Al-Hamdani et al. 2007). Freshwater input from river runoff creates a low salinity (~ 7‰) surface layer typically 45 to 65 m thick (Matthäus and Franck 1992; Matthäus and Schinke 1999; Meier et al. 2006; Naumann et al. 2020). Below this a 10 m layer, the upper deep-water pycnocline, overlies the deep, saline waters (10 to 18‰) of the Baltic Sea. This stable vertical salinity stratification and decades of eutrophication has caused the extent of hypoxic areas to spread in the form of a persistent layer in the deeper areas of the Bornholm and Gotland Basins (Helcom 2009; Viktorsson 2018; Hansson et al. 2019; Naumann et al. 2020). Today, the Baltic Sea contains the largest anthropogenic hypoxic area in the world with widely spread 'dead zones' (Conley et al. 2009; Lehmann et al. 2014; Carstensen and Conley 2019). Many coastal areas within the Baltic Sea and Kattegat suffer from restricted exchange of bottom water due to complex bottom topography that are prone to seasonal periods of hypoxia during summer (Conley et al. 2011; Carstensen and Conley 2019). A seasonal thermocline is found at 30 to 40 m in July to September, with a relatively homogeneous surface mixed layer down to the halocline during fall-winter (Matthäus and Franck 1992; Møller and Hansen 1994; Schinke and Matthäus 1998).

As a result of these complex bathymetric and hydrographic conditions, the hydrodynamic conditions within the Baltic Sea are extremely variable, particularly in the narrow Belt Sea, the Sound, and the Fehmarn Belt, through which all water passes between the Baltic Sea and the Kattegat (Matthäus and Franck 1992; Schinke and Matthäus 1998).

Kattegat Cod

Geographical Distribution and Migrations

Geographical Distribution

Historically, cod have been distributed across the whole Kattegat (ICES SD 21) throughout the feeding season, with a somewhat patchy distribution at depths > 40 m (Casini et al. 2005). Spawning cod could be found throughout the Kattegat decades ago, but the southern part was generally recognised as the main spawning area, especially Laholmsbukten and Skälderviken, the Kungsbackafjorden and the area north of Læsø island (Hagström et al. 1990; Svedäng and Bardon 2003) (Figure 2). Following the dramatic decline in stock size, spawning activity seems to have contracted into two areas in the south-eastern Kattegat, just north of the entrance to the Sound and off the coast at Falkenberg (Vitale et al. 2008; Börjesson et al. 2013), with some additional spawning in the deeper parts of the south-western Kattegat (Hagström et al. 1990). Spawning in the central and northern Kattegat seems to have been lost (Svedäng and Bardon 2003; Vitale et al. 2008). Concurrently, cod seem to have progressively disappeared from shallow coastal areas during the feeding season since the early 2000s (Støttrup et al. 2014; Dinesen et al. 2019).

Migrations

Cods in the Kattegat have traditionally been considered to consist of a mixture of resident and migratory ecotypes undertaking migrations as far as the North Sea (Svedäng et al. 2007). Over the years, the migrations of cod in the Kattegat have been inferred from tagging experiments from the 1950s, 1990s and 2000s and show patterns that seem to have been stable since the 1950s. Juvenile cod tagged as 1-year-old individuals in the Kattegat are fairly resident in shallow coastal waters until they reach an age of 2 years and 30–50 cm in length when they move offshore towards the spawning grounds in the southern Kattegat and the eastern North Sea (Pihl and Ulmestrand 1993). Adult cod tagged along the Swedish Kattegat coast migrated south towards the Sound and north towards the eastern North Sea, depending on the tagging location (Righton et al. 2010; Svedäng et al. 2010). Cod tagged in the

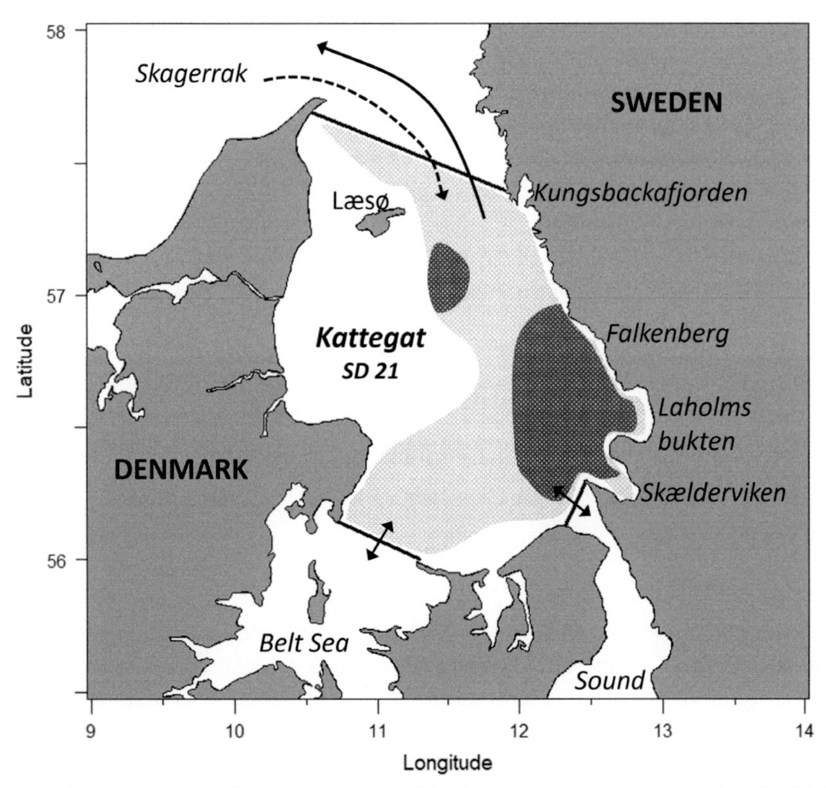

Figure 2. Map of the Kattegat (ICES SD 21), with all landmark names mentioned in the text. Solid black lines represent ICES SD boundaries, where the Kattegat cod management area consists of SD 21. Shaded areas represent the spawning areas (black) and feeding areas (light grey) of Kattegat cod. Solid arrows indicate known movements of adult cod between SDs, and the broken line shows the inflow of early life stages from the North Sea. Figure redrawn based on Hageström et al. (1990); Svedäng and Bardon (2003); Vitale et al. (2008); Börjesson et al. (2013).

northern Kattegat exhibited movements towards the Skagerrak or eastern North Sea, and the onset of these movements coincided with known spawning times of the North Sea stock (Svedäng et al. 2007), while cod tagged in the Sound and the southern-central Kattegat were primarily resident (Svedäng et al. 2010). André et al. (2016) suggested that migration direction may depend on the stock origin of individual cod rather than ecotype, with a strong link between population and philopatric migrations towards natal spawning grounds in the Skagerrak-Kattegat area.

Genetic studies of cod from the Norwegian Skagerrak coast to the central Kattegat found that a large proportion of juveniles can be genetically similar to cod from offshore spawning areas in the eastern North Sea, with substantial temporal variation in advection (Knutsen et al. 2004; Stenseth et al. 2006; André et al. 2016). Movement patterns derived from the chemical composition of otoliths and genetically assigned to either the endemic Kattegat population or the North Sea population revealed that the majority of individuals from both populations were distributed within the Kattegat throughout the first four years of life Hüssy et al. (2021b). Twenty-five per cent of all individuals remained resident within the Kattegat throughout their lives, while the majority of the individuals undertook frequent movements in and out of Kattegat, particularly towards the Belt Sea, but also the Skagerrak/North Sea. As these movements were not synchronous between individuals, they were not considered spawning migrations (Hüssy et al. 2021b). North Sea cod enter the Kattegat as larvae or pelagic juveniles (Hüssy et al. 2021b), presumably through passive advection by the strong Jutland current (Knutsen et al. 2004; Stenseth et al. 2006; André et al. 2016; Eero et al. 2016). This results in a geographical gradient in the distribution of cod within Kattegat, with higher frequencies of individuals with a North Sea population origin in the deeper northern areas (Hemmer-Hansen et al. 2020). This distribution pattern seems to persist across several age classes. No North Sea cod older than four years were found in the Kattegat, supporting the hypothesis of philopatric migrations towards natal spawning grounds once the North Sea cod become sexually mature, without a subsequent return to the Kattegat (Hemmer-Hansen et al. 2020).

Life History

Maturation

Cods in the Kattegat start maturing in November, with peak spawning in February–March (Vitale et al. 2005). Maturity ogives are estimated from the visual (macroscopic) staging of individual fish's gonads during the first quarter of International Bottom Trawl Surveys (IBTS). The age at which 50% of the population is mature (A_{50}) is two years (ICES 2021a). Due to the low numbers of cod in the survey, the maturities in recent years are based on a running mean of three years. It is unclear what influence the extensive mixing of stocks in Kattegat has on the estimation of maturity. No spawning North Sea cod were found in the Kattegat despite extensive sampling (Hemmer-Hansen et al. 2020; Hüssy et al. 2021b), suggesting that the values on maturity reported here most likely represent the endemic Kattegat population (Marty et al. 2014).

Growth and Condition

Growth of Kattegat cod varies somewhat between years but was fairly stable from 1980 to 2000, increasing steadily by 1–2% per year for adult cod over the last two decades (ICES 2021a). Size-at-age averaged over the last 5 years (Figure 3D) is described by the von Bertalanffy growth parameters $L_\infty = 237.3$ cm, $k = 0.06$ and $t_0 = -0.59$. Since no differences in growth were detected between North Sea and Kattegat cod at ages < 4 years (Hüssy et al. 2021b), these growth parameters adequately describe the size-at-age of the Kattegat cod population.

From individual data on weight and length over the entire time series available from the ICES survey data base DATRAS (Appendix 1), it is evident that Fulton's *K* in the first quarter of the IBTS, calculated based on whole body weight, has steadily decreased from a mean of 0.97 in 1996 to 0.92 in 2021 for cod < 30 cm and from 1.05 to 0.95 in cod > 30 cm. Compared to the two Baltic cod stocks, this is a considerably smaller decline in condition for all size classes.

Recruitment

Recruitment of cod in the Kattegat (abundance of a year class at age 1) varies inter-annually but has generally declined by a factor of ten since 1997 (the first year in the stock assessment time series), with occasional strong year-classes (ICES 2021a). There has not been a recruitment above the average since 2013, the year-classes of 2018 and 2021 are the lowest in the times-series (Figure 3B). The nursery areas for juvenile cod in the Kattegat are habitats such as eelgrass beds and shallow rocky areas with macroalgae (Pihl and Wennhage 2002). In both the historic and the present spawning areas within Kattegat (Vitale et al. 2008; Börjesson et al. 2013), the environmental conditions ensure that the spawning habitats are generally well suited for development of eggs with respect to salinity conditions and temperature (Hüssy et al. 2012) and also support survival and development of eggs and larvae (Hinrichsen et al. 2012). In the Kattegat and in the Sound, cod eggs are neutrally buoyant at 21.1 PSU (Nissling and Westin 1997) and are found at salinities of 18–20 PSU, corresponding to depths of 11–14 and 1–4 m, respectively (Westerberg 1994). This position in the water column leaves eggs and larvae exposed to hydrodynamic conditions.

Advection and retention of early life stages are essential factors for recruiting fish stocks and often depend on local meteorological and hydrographical conditions. Drift patterns of early life stages have been studied using different approaches based on Lagrangian particle-tracking models with hydrodynamic ocean circulation models (Lehmann 1995; Hinrichsen et al. 2012; Pacariz et al. 2014a; Jonsson et al. 2016; Köster et al. 2017). These approaches track simulated eggs and larvae, released as virtual drifters into the model velocity fields through space and time,

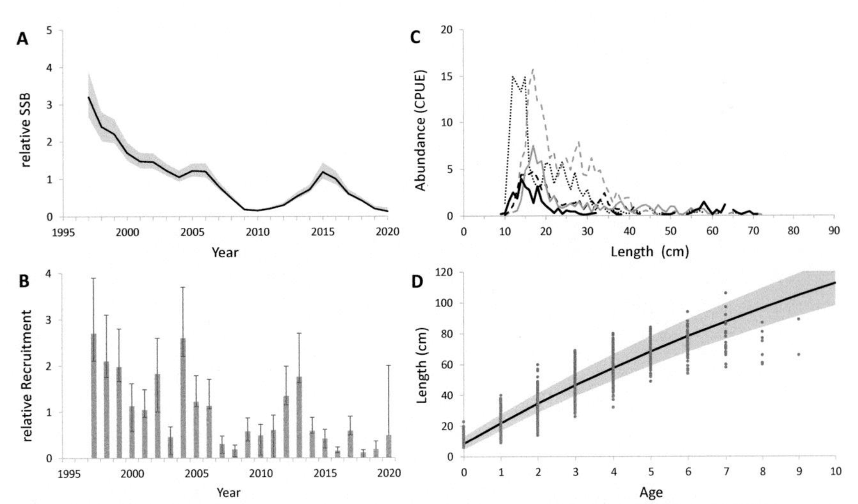

Figure 3. Cod in the Kattegat: (A) Time series of spawning stock biomass (SSB), and (B) Recruitment at age 1, both represented as relative values scaled to the average of the time series, as absolute values are uncertain. The figures on the right show (C) Length distribution (CPUE = catch per 30 min haul), and (D) Fish length (cm) at age, representing mean values over the years 2017 – 2021. Figures A, B and D show mean and 95% confidence intervals.

and allow us to draw inferences on drift directions and causes of mortality under prevailing meteorological conditions. These studies agree that the majority of the released drifters, representing eggs and larvae, are retained within the Kattegat (Hinrichsen et al. 2012; Pacariz et al. 2014a; Huwer et al. 2016; Jonsson et al. 2016), with substantial northward advection of larvae which then contribute to recruitment in the Skagerrak (Huwer et al. 2016; Jonsson et al. 2016). Southward advection to the Belt Sea and Sound, on the other hand, seems generally to be limited (Huwer et al. 2016; Jonsson et al. 2016). In recent years, following the stock decline, the protected area in the Sound and the Belt Sea may have provided most of the larvae recruiting in the Kattegat (Hinrichsen et al. 2012; Jonsson et al. 2016) and may always have been of importance for the Kattegat stock (Huwer et al. 2016). There is also recent evidence that the inflow of cod early life stages from the North Sea can be significant although highly variable between years (Jonsson et al. 2016; Hemmer-Hansen et al. 2020; Hüssy et al. 2021b).

Advection and retention of early life stages (eggs, larvae and pelagic juveniles) are strongly influenced by prevailing meteorological conditions, where the majority of early life stages are retained in the area where they were spawned during variable, weak wind directions and a strong outflow of surface water from the Baltic Sea (Pacariz et al. 2014b). On the other hand, strong westerly winds cause higher advection of early life stages towards the Belt Sea and Sound. These westerly wind conditions have increased in recent decades due to a change in the large-scale atmospheric circulation pattern, thereby potentially affecting the recruitment of cod in the Kattegat (Pacariz et al. 2014b). Adding further complexity to the interpretation of these results is the suggestion that advection and retention also depend on egg density, where larger and heavier eggs seem to be retained in the Kattegat or transported southward, while lighter eggs are transported towards the Skagerrak (Pacariz et al. 2014a). A change in source population dynamics has thus possibly occurred since the 1970s, where the importance of locally retained Kattegat larvae has decreased dramatically concurrent with the decline in stock size. While recruitment in the 1970s predominantly originated from the local contribution of larvae, it is now suggested that the stock relies heavily on larvae from the Sound and the Belt Sea to uphold recruitment in the Kattegat (Jonsson et al. 2016). It remains unclear whether these patterns are related to stock size or changes in ocean circulation.

Despite these highly variably advection and retention patterns, and to some extent contradictory drift patterns, juvenile cod in the North Sea, Kattegat/Skagerrak, Belt Sea, and eastern Baltic are genetically related to adults in the same areas (Nielsen et al. 2005). This suggests that the majority of the early life stages are retained within the area in which they hatched or, alternatively, die during their advection to other areas.

Stock Size and Structure

Historical Changes in Abundance

The stock size of cod in the Kattegat has been declining since the early 1990s (3A) to historical low levels in 2010, and following a small recovery in the following decade, it decreased again to similar low levels (ICES 2021a). This stock decline is presumably caused by the combination of low recruitment and high removals from the stock, which besides mortality (Figure 4A), include migration. A notable increase in stock size was recorded from 2012–2015 (Figure 3A), which was dominated by the influx of large numbers of North Sea cod (Hemmer-Hansen et al. 2020; Hüssy et al. 2021b). Since then the stock size has diminished again and is currently at a historical low size.

Stock Structure

The size distribution of cod in the Kattegat has become increasingly truncated towards sizes < 50 cm (Figure 3C), with lengths from 50–80 cm making up a progressively smaller fraction of the stock (ICES 2021a). Size distributions vary somewhat between years, but the general pattern has persisted over the last decade. Interannual differences in the smaller part of the size range, representing age classes 0 and 1, occur presumably as a result of variable inflow of North Sea cod. Cod in the Kattegat is evidently a mixture of different populations and stock components. Particularly in the northern part of the Kattegat, the local cod population mix with the adjacent North Sea population (Knutsen et al. 2004; ICES, 2015; André et al. 2016; Hemmer-Hansen et al. 2020). The connectivity between these populations exists primarily through the advection of early life stages from the North Sea (André et al.

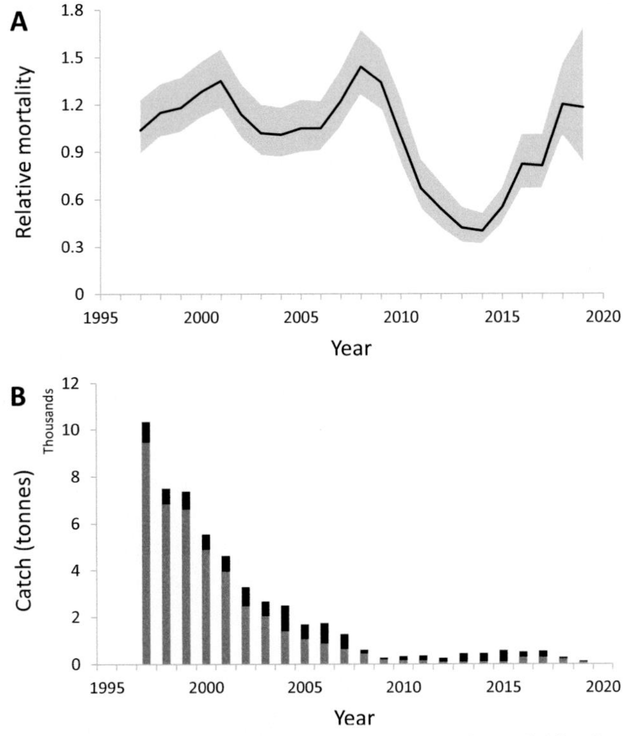

Figure 4. Cod in the Kattegat: (A) Relative mortality (total mortality minus 0.2) (showing mean and 95% confidence intervals) and (B) Catch in thousand tonnes (B, colour represents landings: grey, and discards: black) for the available time series.

2016; Eero et al. 2016; Jonsson et al. 2016; Hüssy et al. 2021b), but possibly also via migration of adult individuals (Svedäng et al. 2007). Owing to the low stock size of Kattegat cod, Jonsson et al. (2016) estimated that currently, only approximately 34% of the juveniles in Kattegat are from locally retained spawning areas compared to 83% in the 1970s when the stock was larger. The perceived stock size is thus strongly influenced by the influx of North Sea cod.

The North Sea and Kattegat populations are genetically distinct, and they mix purely through admixture but not through interbreeding (Hemmer-Hansen et al. 2020). In the southern Kattegat and the Sound, on the other hand, genetic stock identification based on microsatellite analysis could not identify reproductively isolated populations (Svedäng et al. 2010). However, information from tagging and the chemical composition of otoliths suggests that these stock components are separated to some extent, as they indicate philopatric behaviour by dispersing from mixed feeding areas and returning to their natal spawning grounds (Svedäng et al. 2010). Natal homing thus seems to be an important stock-separating mechanism even over distances of < 100 km in the transition zone between the North Sea and the Baltic Sea (Svedäng and Svenson 2006).

Environmental Challenges

One of the drivers behind the observed decline in the Kattegat cod stock may be related to temperature. Bottom temperatures have increased by 1–2°C. in the deeper parts of the area over the last 30 years, while they have increased by as much as 3–4°C in shallow coastal waters during summer (Mariani et al. 2020). These areas are known nursery areas of juvenile cod, where they remain without undertaking extensive migrations during their first two years of life (Pihl and Ulmestrand 1993). While an increase in environmental temperature may not have a negative effect on fish's growth and survival, it does have a direct effect on fish by increasing their metabolic rate (Schurmann and Steffensen 1997; Claireaux et al. 2000). In addition to warming waters, the Kattegat is suffering from continuously intensifying hypoxia, which is exacerbated by both eutrophication and increased water temperature; the latter leading to the reduced solubility of oxygen (Carstensen et al. 2014). If the temperature-related increase in energy demand is not met through increased consumption, survival will evidently be affected directly or, alternatively, lead to fish avoiding the warming areas. Moving to more suitable areas may, in turn, lead to an increase in spatial overlap with predators with a concurrent increase in mortality. Cod catches in shallow water have decreased since 2000 (Støttrup et al. 2014), which seems to be attributable to rising water temperatures that are particularly evident in these shallow areas and indicate shrinkage of suitable habitat for particularly smaller fish (Dinesen et al. 2019).

Stock Management

Fishery

Cod catches in Kattegat are almost exclusively taken by Denmark and Sweden. Cod have historically been captured primarily by trawls, in addition to Danish seines and gillnets. The reported landings of cod in Kattegat have declined from more than 15,000 tonnes in the 1970s to less than 10,000 tonnes in the late 1990s and further to below 100 tonnes in 2019 (ICES 2021a). Kattegat cod were traditionally caught as target species during the spawning season at the beginning of the year. At present, there is no targeted cod fishery in Kattegat, and cod is mainly taken as by-catch in the trawl fishery for Norway lobster (*Nephrops norvegicus*) throughout the year (ICES 2021a).

Management

Total Allowable Catch (TAC) is the primary management measure regulating cod fishing in Kattegat and is presently restricted to by-catch (mainly in the *Nephrops* fishery). Since 2017, the cod in Kattegat is under the EU landing obligation. For a period (2004–2009), the fishery in the Kattegat was additionally regulated by an effort system introduced in the first cod recovery plan (EC No. 423/2004). Under this system, effort was regulated through limitation of number of fishing days allowed for individual vessels. This effort regulation was changed to a system where effort was regulated by kW days for different gear groups as part of the new cod management plan for the North Sea (incl. Kattegat) in 2009 (EC No. 1342/2008). The effort regulation was abandoned in 2016. Additional measures introduced in the Kattegat include selective gears, e.g., mandatory use of SELTRA 270 trawl (bekendtgørelse om regulering af fiskeriet (BEK) nr. 1,461 af 15/12 2019 § 31) with at least 180 mm panel in the Danish fishery. The Swedish sorting grid was intensively used in years when the effort system generated incentives for it. In 2009, Denmark and Sweden introduced marine protected areas on the historically most important spawning grounds in the south-eastern Kattegat, with the aim to contribute to rebuilding of the cod stock. The protected zone consists of three areas in which fishing is either completely forbidden or limited to certain selective gears (Swedish grid and Danish SELTRA 300 trawl) during all or different periods of the year. Further details relating to the management of cod in the Kattegat, may be found in the ICES Stock Annex and Advice Sheet (Appendix 1).

Challenges

Stock assessment for cod in the Kattegat is only indicative of trends. The current absolute level of fishing mortality is unknown because the estimated total removals from the stock represent a combination of fishing mortality, unaccounted natural mortality, and migration out from the Kattegat area (Figure 4A). It has so far not been possible to disentangle these processes. Recent evidence of stock mixing (emigration of North Sea cod upon becoming sexually mature) seems to be the key issue to resolve (ICES 2017). Due to the lack of a quantitative stock assessment, MSY or precautionary approach reference points are not available either, and it is not possible to assess the stock status relative to these. However, the spawning stock biomass is presently historic low (Figure 3A) and thus can be considered below possible reference points. The management measures taken so far (effort restrictions, area closures, and by-catch quota) have not been sufficient to ensure the recovery of the stock, and the discards remain high (Figure 4B). Selective gears to reduce cod by-catch are available but are not fully utilised.

Western Baltic Cod

Geographical Distribution and Migrations

Geographical Distribution

Cod have historically been occupying all areas from shallow to deep water within the western Baltic Sea (ICES SDs 22, 23 and 24), but primarily at depths > 20 m (Oeberst 2008), with some seasonal movements between greater depths during winter and shallower areas in summer (Thurow 1970; Oeberst 2008) (Figure 5). In addition, depth distribution seems to be related to size, where smaller/younger cod occupy shallower areas throughout the year (Oeberst 2008). The most important spawning areas of the western Baltic cod have historically been identified as the deep parts of the Kiel Bay and the Belt Sea (Kändler 1950; Thurow 1970; Bagge et al. 1994), in particular the areas south-southwest of Langeland and northwest of Fehmarn. These spawning areas have persisted over the years (Bleil and Oeberst 2000, 2002). More recent surveys have also shown that the spawning activity seems to be limited in the Fehmarn Belt and Mecklenburg Bay, where only a few spawning individuals have been observed over the years (Thurow 1970; von Westernhagen et al. 1988) and restricted to depths > 20 m (Bleil and Oeberst 2000, 2002). The deepest part of the Arkona Basin (> 40 m) is considered to be used for spawning by both western (Bleil and Oeberst 2000, 2002) and eastern Baltic cod (Bleil and Oeberst 2004; Bleil et al. 2009). Owing to this overlap in stock distribution, it is important to note that while the biological western cod population is distributed across the Belt Sea, Sound and Arkona Basin (SDs 22, 23 and 24), both the eastern and western populations co-occur in SD 24. Cod management in this area is carried out for the entire stock in SDs 22–24, while the stock assessment is carried out for the 'true' western Baltic cod stock, i.e., including SDs 22–23 and the fraction of the cod in SD 24, which belong to the western population. All biological information (size distribution, maturation, and growth) provided in the following text relate to the true western Baltic cod in SDs 22 and 23 only, while the information originating from stock assessment (stock size and recruitment) is for the entire western population in SDs 22–24.

Migrations

Extensive tagging programmes have shown that adult cod undertake distinct and highly complex migrations after the onset of maturation between coastal feeding areas and offshore spawning areas (Aro 1989; Bagge et al. 1994). These migrations are pronounced but suggest sub-structuring of the stock as they vary between eastern and western areas (Otterlind 1985). The general direction of cod tagged in the western part of the western Baltic Sea is towards the southern Kattegat and Danish Belts (Bagge 1969; Otterlind 1985), whereas cod tagged in the Mecklenburg Bay move towards Kiel Bay, Little Belt, and particularly Great Belt and the southern Kattegat in early winter (Berner 1967). Cods in the Arkona Basin, the easternmost part of the area, follow a quite different migration pattern. Adult cod tagged in the Arkona Basin migrate both westward toward the Sound and the Belts and eastward as far as the Bornholm Basin (Berner 1967, 1971, 1974; Otterlind 1985). Within the Arkona Basin, there seem to be stock-specific geographic patterns. Cod from the eastern Arkona Basin move into the Bornholm Basin, whereas the main destination

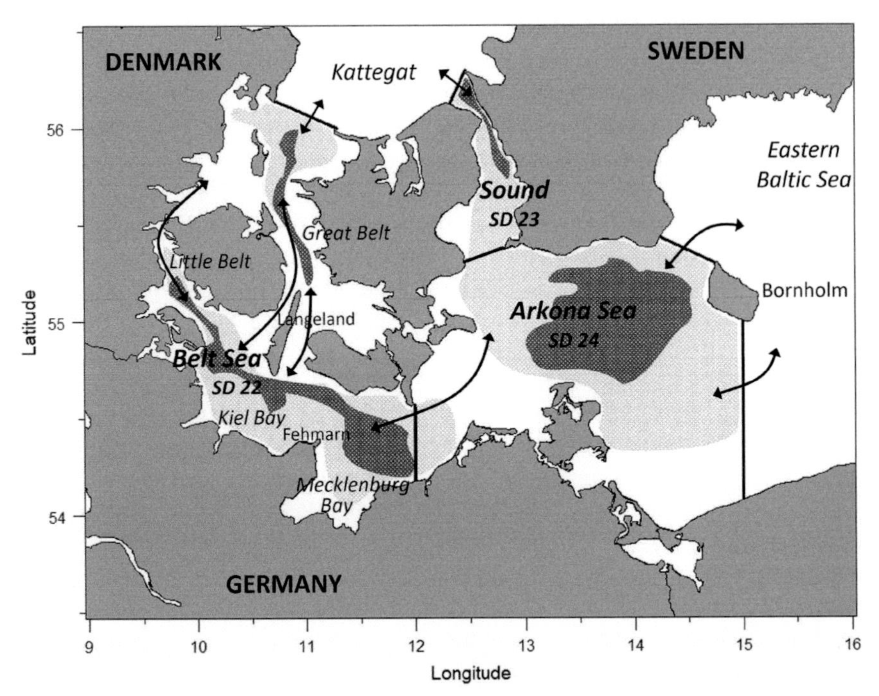

Figure 5. Map of the western Baltic Sea, consisting of the Belt Sea (ICES SD 22), the Sound (ICES SD 23), and the Arkona Sea (ICES SD 24), with all landmark names mentioned in the text. Solid black lines represent ICES SD boundaries, where the western Baltic cod management area consists of SDs 22, 23 and 24. Shaded areas represent the spawning areas (black) and feeding areas (light grey) of western Baltic cod. Solid arrows indicate known movements of adult cod between SDs. The figure is redrawn based on Aaro (1989), Otterlind (1985), Bagge et al. (1994), Bleil and Oeberst (2000, 2002, 2004), Bleil et al. (2009), and Hüssy (2011).

of cod from the western Arkona Basin is towards the Belt Sea in what appear to be spawning migrations by the two cod populations in the area (Bagge 1969; Otterlind 1985). In addition to these horizontal movements, cod in the western Baltic also exhibit pronounced seasonality in their depth distribution. In this area, cod predominantly stay at a depth > 15 m during December–March and July–August, apparently avoiding the coldest and hottest temperatures (Funk et al. 2020). During the rest of the year, the western Baltic cod occupy shallow water areas during their feeding season. The vertical movements of western Baltic cod thus seem strongly linked to sea surface temperature (Funk et al. 2020).

Life History

Maturation

Cod in the western Baltic Sea start maturing in November, with peak spawning restricted to 1–2 months (Kändler 1949; Thurow 1970; Bleil and Oeberst 1997, 2004; Bleil et al. 2009), and with a progressive delay from west to east (Hüssy 2011b). The peak spawning of cod in the Belt Sea and Sound occurs in January/February (Vitale et al. 2008) and in Kiel Bay and Mecklenburg Bay in March/April (Bleil and

Oeberst 2004; Bleil et al. 2009). In the Arkona Basin, spawning occurs from February to July (Kändler 1949; Bleil et al. 2009), thereby overlapping considerably with the spawning times of cod in the Belt Sea and the eastern Baltic cod in the Bornholm Basin that peak in July/August (Wieland et al. 2000; Bleil et al. 2009). Males generally mature at a younger age and smaller size and stay longer on the spawning grounds than females (Thurow 1970; Bleil and Oeberst 1997, 2004; Tomkiewicz et al. 1997; Bleil et al. 2009). This exposes males to prolonged intense fishing pressure (Thurow 1970; Tomkiewicz et al. 1997), resulting in a skewed sex ratio with fish age (Thurow 1970; Berner, 1985). In addition to these sex-specific patterns, maturation and fecundity also depend on fish size. Large females arrive at the spawning grounds earlier in the season than their smaller conspecifics and produce a larger number of egg batches (Thurow 1970; Bleil and Oeberst 1997, 1998, 2004; Tomkiewicz et al. 2003; Bleil et al. 2009).

For western Baltic cod, maturity ogives are estimated from the visual (macroscopic) staging of individual fish's gonads during the first quarter Baltic International Trawl Surveys (BITS). The age where 50% of the population is mature (A_{50}) ranges between 1.5 to 2.2 years (ICES 2021a). According to survey data, A_{50} has gradually decreased since the early 2000 (ICES 2021b). However, this pattern may be associated with biased sampling design in relation to observed depth movements of cod (Funk et al. 2020) and so may not be representative of an actual change in maturity (ICES 2021b).

Growth and Condition

From individual fish data on size-at-age over the entire time series available from ICES survey database DATRAS (Appendix 1), it is evident that while there are considerable interannual differences in size-at-age, there are no consistent trends across time series suggesting that the growth rate of western Baltic cod has not changed. The average size-at-age of the western Baltic cod (data from SDs 22–23) in the last five years (2015–2020; Figure 6D) is described by the von Bertalanffy growth parameters $L_\infty = 165.8$ cm, $k = 0.10$ and $t_0 = -0.36$. Growth estimates from an extensive tagging programme (2007–2015) are very similar ($L_\infty = 154.6$ cm, $k = 0.11$ and $t_0 = -0.13$) (McQueen et al. 2019).

The average Fulton's K in SD 22 has steadily decreased for both large and small cod over the years 1977 to 2020, from approximately 1.1 to 0.9 in large cod and from 1.0 to 0.85 in small cod (during the first quarter of the year) (Receveur et al. 2022). This decline in condition results from decreasing weight at a given length for all age groups and is particularly pronounced in the large 2016 year class (ICES 2021b). Concurrently, the hepatosomatic index, also assessed in the first quarter of the year, decreased by 50%. Receveur et al. (2022) identified this decline in energy reserves to be associated with an increase in water temperature, expansion of hypoxic areas during late summer/autumn, and changes in the diet composition of the cod. Spatial patterns in the condition in SDs 22–24 are evident, with relatively lower Fulton's K in SD 24 (Eero et al. 2014), which is likely an effect of larger proportions of immigrating individuals from the eastern Baltic cod stock in this area. The condition of the eastern Baltic cod has substantially declined over the last decades (see below), which is therefore expected to also influence the condition in SD 24.

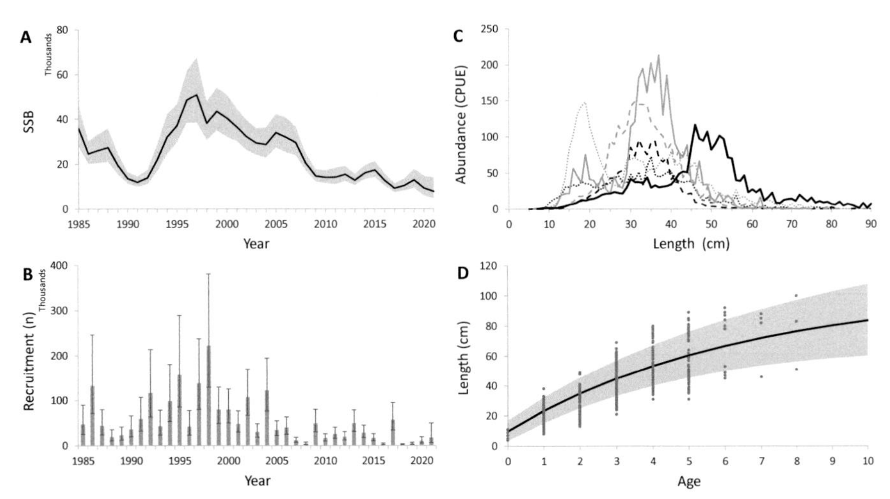

Figure 6. Cod in the western Baltic Sea (true western stock in SD 22 and 23): (A) Time series of spawning stock biomass (SSB, in thousand tonnes), and (B) Recruitment in numbers, where both figures cover the entire western Baltic cod stock in SDs 22, 23 and the fraction of western cod in SD 24, (C) Length distribution (CPUE = catch per 30 min haul) for six years at five-year time intervals, and (D) Fish length (cm) at age representing mean values over the years 2017–2021. Figures C and D show only data from SD 22 and 23. Figures A, B and D show mean and 95% confidence intervals.

Recruitment

Recruitment of western Baltic cod (abundance of a year class at age 1) varies interannually by a factor of 10 but has generally declined from the highest levels in the mid-1990s to low levels since the mid-2000s (ICES 2021a). Recruitment has remained low since then, with occasional strong year-classes, the last of which occurred in 2016 (Figure 64B). Hüssy (2011b) provided a complete overview of processes regulating the recruitment of western Baltic cod. Here, only the most important issues are highlighted. One of the key factors that affect egg buoyancy, its ability to float in the water column, is environmental salinity. Cod in the Baltic Sea are adapted to the prevailing brackish conditions in that eggs are neutrally buoyant at salinities as low as 19 PSU in the Sound (Westerberg 1994), Belt Sea, Kiel Bay and Mecklenburg Bay (von Westernhagen et al. 1988; Nissling and Westin 1997) and at salinities down to 13.7 PSU in the Arkona Basin (Nissling and Westin 1997). At lower salinity, eggs will sink to the bottom and fail to develop (von Westernhagen 1970). Temperature also plays a crucial role in the survival and development of eggs and larvae. Reported minimum temperatures for egg development in the laboratory range from 4–8°C in cod from the Fehmarn Belt (von Westernhagen 1970) to 5.5–8.5°C in cod from the Mecklenburg Bay (Bleil 1995). At temperatures above or below these conditions, there is higher egg mortality and more deformities (von Westernhagen 1970). Total egg mortality may be as high as 96–99% (von Westernhagen et al. 1988).

The suitability of the environmental conditions for supporting viable egg production in the western Baltic varies geographically and with season. Due to the salinity and temperature thresholds for egg development, conditions are suboptimal from December through to mid-March in all areas and increase towards the end of the spawning season (Hüssy et al. 2012). Survival of eggs is generally limited by salinities that are too low, while temperature was limiting in the 1980s only. The Great Belt generally has the best conditions for spawning, followed by the Sound, and with severely reduced conditions in the Kiel Bay and Mecklenburg Bay (Hüssy et al. 2012). The suitability of the Arkona Basin for spawning of the western Baltic cod is somewhat unclear, but given that salinities are even lower than in the aforementioned areas, conditions are presumably even less suitable. Similar patterns in geographic and seasonal differences in the environmental conditions suitable for larval development and survival also exist (Hinrichsen et al. 2012). Overall, conditions for successful recruitment of cod eggs and larvae are thus highest in the Great Belt, decreasing further towards the Sound, Kiel Bay and Mecklenburg Bay. However, western Baltic cod recruitment is not correlated with habitat suitability for spawning and early life stage survival, indicating that other processes contribute to modulating the survival of these stages (Hüssy et al. 2012).

Hydrodynamic conditions in the narrow Belt Sea, the Sound and the Fehmarn Belt, and shallow German bays are extremely dynamic (Matthäus and Franck 1992; Schinke and Matthäus 1998), depending on meteorological conditions. The prevailing wind speeds and directions profoundly impact the advection of cod eggs and larvae as they drift passively with the currents. Drift modelling suggests a strong general advection of cod eggs and larvae from east to northwest, depending on wind direction and strength (Huwer et al. 2016). Of the early life stages from the Great Belt, approximately 60% are advected towards the Kattegat and Skagerrak, while 10–30% are retained. From the Sound, virtually all individuals are advected northward (ca 80%), 10–20% drift into the western Baltic, while only a few per cent of eggs are retained within the area. Eggs and larvae from Kiel Bay and Mecklenburg Bay, on the other hand, predominantly remain in the area, with some northward and eastward advection, depending on prevailing wind conditions (Huwer et al. 2016). Advection and retention thus vary both within and between years as a result of these conditions, where low and variable wind forcing leads to a higher degree of early life stage retention, while sustained strong, mainly westerly, winds that are also responsible for the inflow of Atlantic water, result in a large degree of advection (Hinrichsen et al. 2001). For eggs transported to the Bornholm Basin, survival is limited because their density causes them to sink to depths with insufficient oxygen (Nissling et al. 1994; Petereit et al. 2014). Overlap of modelled early life stage destination corresponds well with observed distributions of juvenile cod in bottom trawl surveys, suggesting that drift simulations provide reasonable estimates of early life stage connectivity between cod populations (Huwer et al. 2016). Juvenile cod from the western Baltic Sea are genetically most closely related to spawning adults from the Belt Sea (Nielsen et al. 2005), which suggests that cod from this stock also undertake natal homing despite the extensive dispersion of early life stages.

Stock Size and Structure

Historical Changes in Abundance

The western Baltic cod stock has seen considerable changes in stock size since the 1980s (Figure 6A). Like the other cod stocks from the North Sea to the eastern Baltic Sea, the spawning stock biomass of the western Baltic cod has declined continuously since the mid-1990s. In recent years, the biomass has exclusively depended on the large 2016-year class, which was preceded by some of the weakest year-classes in the time series (Figure 64A). The stock is currently estimated to be at its historically lowest level and below biomass reference points (ICES 2021a; Möllmann et al. 2021).

Stock Structure

The size distribution of the stock has not varied considerably over the years (Figure 6C). The shape of these size distributions depends largely on the age composition of the stock. The year 2020, for example, stands out because the size distribution in the stock is exclusively driven by the large 2016-year class. This year's class has been dominating the size distributions from 2016 onward, and considering 2020 implies only a larger fraction of larger cod in 2020 compared to earlier years. Historically, a number of studies have suggested that the eastern and western Baltic cod stocks overlap in their distribution based on tagging-derived migration patterns (see above), phenotypic differences such as the number of fin rays in the dorsal fin, body height, otolith size, head characteristics, and length-weight ratios (Birjukov 1969; Berner and Vaske 1985; Berner and Müller 1989, 1990; Müller 2002), and genetics (Nielsen et al. 2003, 2005). The Arkona Basin, particularly, has been considered the centre of the mixing area (Aro 1989; Müller 2002; Nielsen et al. 2005). A time series of stock mixing based on genetically validated shape analysis of archived otoliths (Hüssy et al. 2016a) documented pronounced spatial and temporal patterns in stock mixing within the Arkona Basin (Hüssy et al. 2016b). Eastern Baltic cod have thus throughout the time series available (since 1977), been present in the Arkona Basin (ICES 2019a, 2021a). However, the percentage of eastern Baltic cod in the eastern Arkona Basin increased from ca. 50% before 2005 to 80% in the following years. Immigration occurred north of Bornholm but not as spawning migrations as there was no seasonal trend in mixing proportions. A pronounced geographic east-west trend in stock mixing throughout the time series suggests that immigration is presumably due to spill-over from the eastern stock (Eero et al. 2014).

In a recent study using single nucleotide polymorphisms, Hemmer-Hansen et al. (2019) confirmed that stock mixing does indeed occur in the Arkona Basin and that this mixing occurs as mechanical mixing rather than hybridisation and introgression. This mechanical mixing is why the geographical gradient with higher proportions of eastern Baltic cod in the eastern parts of the Arkona Basin was observed by Hüssy et al. (2016b). Hemmer-Hansen et al. (2019) provided support for the earlier hypotheses that the Arkona Basin is used as a spawning area by both stocks in that spawning individuals from both stocks occurred in the area during the first quarter of the year but exclusively in eastern Baltic cod during the second and third quarter. Juveniles of age 0 and 1 of both stocks were also found in the area. Stock mixing has been taken into account in stock assessments of eastern and western Baltic cod since 2015.

Environmental Challenges

Climate change is expected to affect the western Baltic Sea substantially, with a predicted temperature increase of 3°C by 2100 (Friedland et al. 2012; Gräwe et al. 2013) with a concurrent decline in salinity of 1.5 PSU (Friedland et al. 2012). Additionally, the ecosystem is predicted to shift from a nitrogen-limited system towards a phosphorus-limited system, increasing oxygen saturation (Friedland et al. 2012). These changes will significantly impact all life stages of the cod stock. Adult cod already respond to higher temperatures and thermal stratification by adjusting their preferred depth distribution (Funk et al. 2020). Shallow water and structured habitats are highly important feeding areas for cod in the area (Funk et al. 2020), and heating surface temperatures will, therefore, limit the cods' ability to acquire adequate energy. Decreasing salinity, on the other hand, may influence the production of offspring by limiting both egg production and survival. Suboptimal salinity has been shown to cause cessation of spawning in the laboratory (Bleil 1995) and in the wild (Kändler 1950). Environmental salinity is furthermore the limiting factor for the survival of eggs and larvae, and lower salinities will cause eggs to sink, and they are expected to die upon contact with the bottom (Hinrichsen et al. 2012; Hüssy et al. 2012). The western Baltic cod stocks' ability to produce viable recruitment for sustaining the stock seems thus limited under the predicted climate-change scenarios, and recovery of the stock is questionable (Möllmann et al. 2021).

Stock Management

Fishery

The commercial catches of the western Baltic cod are taken primarily by Danish, Swedish and German trawlers, gillnetters and to a less extent, Danish seiners. Small vessels < 15 metres are a large fraction of the fleet operating in the western Baltic. Commercial landings of western Baltic cod have declined over the last decades from a maximum of 30,000–40,000 t in the mid-1990s to below 3,000 t in 2020 (ICES 2021a) (Figure 7B). Historically, commercial catches originate primarily from SDs 22 and 24, because a trawling ban was imposed on the Sound (SD 23) in 1932, where commercial catches have been taken by gillnetters since then. The fraction of catches in SD 24 has reduced since the early 2010s due to management regulations imposed to protect the eastern Baltic cod. Western Baltic cod have historically been caught primarily during their spawning period in the first quarter of the year (50–80%). Cods in this area are mainly caught in a targeted fishery and, to a lesser degree, by-catch in other fisheries. In the western Baltic Sea, considerable cod catches are also taken by the recreational fishery, mainly by private and charter boats and to a small degree by land-based fishing methods. In the last 10 years (2010–2020), recreational catches have, on average, amounted to close to 30% of the total catches from the stock. Fishing mortality has throughout the time series been exceptionally high and continues to be at a high level despite continuously decreasing catches (Figure 7A). Since catches have decrease according to TAC levels, the fishing mortality is now considered unreliable (ICES 2023).

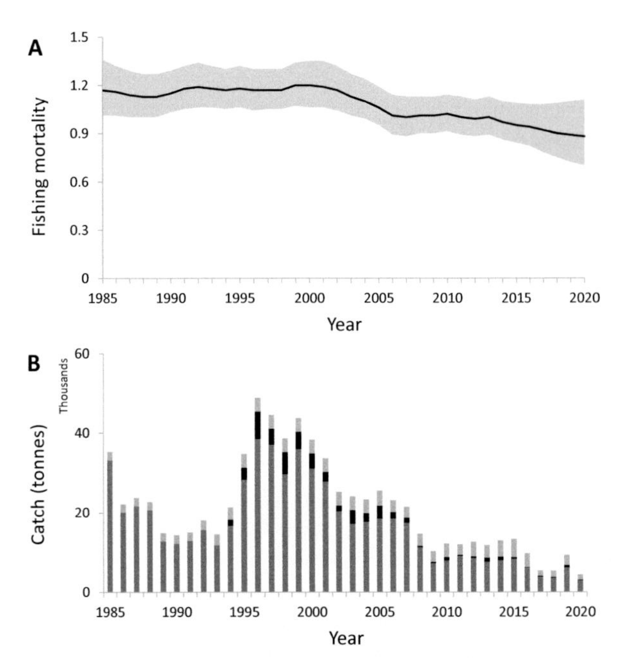

Figure 7. Cod in the western Baltic Sea (entire stock in SDs 22–24): (A) Fishing mortality Fbar (showing mean and 95% confidence intervals) and (B) Catch in thousand tonnes (B, colours represent landings: dark grey, discards: black, recreational catch: light grey) for the available time series.

Management

Cod in the Baltic Sea was managed as one stock until 2002. Since 2003, cod management has been conducted separately in the western (SDs 22–24) and eastern (SDs 25–32) Baltic Sea. Cod commercial catch amounts are primarily regulated by an annual TAC. Historically, the fishery in the western Baltic Sea has additionally been regulated by effort limitations (EC No. 52/2006). Regulations on mesh sizes in fishing gear have varied over time (see Stock Annex (ICES 2021c)). Additionally, seasonal fisheries closures are applied in the spawning time of the western Baltic cod (ref to Stock Annex; Eero et al. 2019). Since 2015, an EU landing obligation has been in place, whereas cod at or above a conservation reference size of 35 cm can commercially be landed. Since 2016, the western Baltic cod fishery has been regulated according to the multi-annual management plan for the Baltic Sea (2016/1139). In the recreational fishery, bag limits have been in place since 2017 (see Stock Annex). Since June 2019, directed fishery for cod in SD 24 has been prohibited as a measure to protect the eastern Baltic cod stock (EU 2019/1248), with special regulations for small vessels using passive gears. Further details relating to managing the western Baltic cod stock may be found in the ICES Stock Annex and Advice Sheet (Appendix 1).

Challenges

The cod stock in the western Baltic has historically been much smaller than the neighbouring eastern Baltic stock, and eastern Baltic cod have, throughout the time series available (since 1977), been present in the Arkona Basin (Hüssy et al. 2016b; ICES 2019a, 2021a). The presence of eastern cod in SD 24 poses a number of challenges for fisheries management related to the potential depletion of one of the stocks fished in a mixed fishery (Eero et al. 2014). Stock assessment is currently taking into account mixing between eastern and western Baltic cod, so the resulting stock estimates and management advice correspond to biological populations rather than management areas (ICES 2014, 2015). This, however, complicates setting TAC by management areas, as part of the eastern cod TAC is utilised in the western Baltic cod management area.

In 2016, ICES called for the imminent collapse of the western Baltic cod stock (ICES Advice 2016). While the strong 2016-year class led to an intermittent stock recovery, the stock size is now again at an all-time low biomass. This is due to a lack of recruitment and too high fishing pressure, i.e., substantially above the Fmsy target. Möllmann et al. (2021) further highlight a failure to incorporate changing environmental conditions into catch predictions, leading to large uncertainty in the assessment of the state of the stock, as a reason why the western Baltic cod stock has declined.

Eastern Baltic Cod

Geographical Distribution and Migrations

Geographical Distribution

Cod have historically been distributed throughout the eastern Baltic Sea, from the south-eastern Baltic to the Bothnian Bay and the eastern parts of the Gulf of Finland (ICES SDs 24–32) (Aro 1989; Sparholt et al. 1991) (Figure 8). The highest abundances of cod generally occurred in the Bornholm Basin (SD 25), followed by the Gotland Basin (SD 28) and Gdansk Deep (SD 26) (Sparholt et al. 1991). While virtually the entire Baltic Sea was considered a potential feeding area (Aro 1989; Bagge et al. 1994), the historical spawning areas of the eastern Baltic cod were located in the deeper areas of the Bornholm Basin, the Slupsk Furrow, the Gdansk Deep and the Gotland Basin (Aro 1989). The geographical distribution of adult cod was historically regulated by population size and only to a minor degree by hydrological conditions (Bartolino et al. 2017). Concurrent with the stock decline in the 1990s (see below), the distribution range of the stock contracted to the present main distribution area in the southern Baltic Sea (SDs 25, 26 and 28) (Hjelm et al. 2004; Eero et al. 2012b, 2015; Bartolino et al. 2017; Mion et al. 2022). Despite a small stock recovery in the late 2000s (Eero et al. 2012a), the cod did not expand its distribution range to the northeast, in particular the Gotland Basin, to re-occupy its former wide habitat area (Eero et al. 2012b; Mion et al. 2022). This contraction of the distribution range, in combination with deteriorating environmental conditions in the northern spawning areas (Hinrichsen et al. 2016), is the reason why only spawning in the Bornholm Basin currently results in viable offspring (Bagge et al. 1994; Wieland

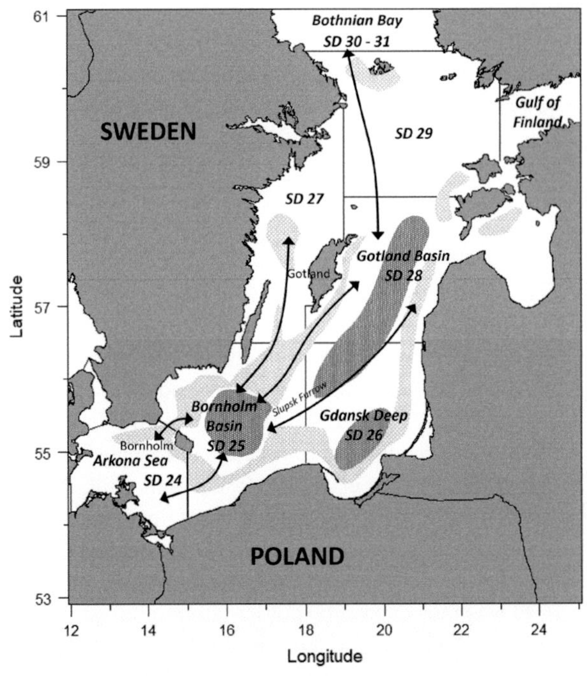

Figure 8. Map of the eastern Baltic Sea, where the eastern Baltic cod management area consists of ICES SDs 24–32, including all landmark names mentioned in the text. Solid black lines represent ICES SD boundaries. The main distribution area of eastern Baltic cod is almost exclusively restricted to the Bornholm Basin (ICES SD 25), the Gdansk Deep (ICES SD 26), and the Gotland Basin (ICES SD 28). Shaded areas represent the spawning areas (black) and feeding areas (light grey) of Kattegat cod. Solid arrows indicate known movements of adult cod between SDs. Figure redrawn based on Aro (1989); Otterlind (1985); Bagge et al. (1994); Bleil et al. (2009); Eero et al. (2012a); Wieland et al. (2000); Mion et al. (2022).

et al. 2000; Bleil et al. 2009; Köster et al. 2009, 2017). In addition to the geographical distribution, cod also exhibit distinct vertical depth distribution. Cods older than three years of age are generally distributed offshore in water deeper than 40 m, while juveniles occupy shallower coastal habitats (Sparholt et al. 1991; Hjelm et al. 2004; Hinrichsen et al. 2017). Horizontal and vertical distribution further depends on the salinity and oxygen concentration of the environment, where particularly low oxygen concentrations limit the distribution of cod (Hjelm et al. 2004; Casini et al. 2016).

Migrations

Decades of tagging data have shown that migrations of cod from the southern feeding areas in the Gdansk Deep and Slupsk Furrow toward the spawning area in the Bornholm Basin used to commence in December–February (Netzel 1963, 1968, 1974a, 1974b; Aro 1989). Cod from the feeding areas along the Swedish west coast, Åland (SD 27), the Gulf of Finland (SD 32) and the southern Bothnian Bay (SD 30), on the other hand, used all the historically active spawning areas (Bornholm Basin, Gdansk Deep and Gotland Basin) and commenced their spawning migrations in December–March (Otterlind 1985; Aro 1989). Migrations to the Arkona Basin also occurred frequently, but apparently not as spawning migrations (Aro 1989). Archival tags recording depth movements of individual fish provided a higher temporal resolution of movements and documented that the movement towards the spawning area in the deepest parts of the Bornholm Deep was highly targeted and completed within 11 days on average (Nielsen et al. 2013).

Life History

Maturation

Gonadal maturation of cod in the eastern Baltic Sea begins in winter and lasts about four months (Baranova and Uzars 1986; Uzars et al. 2001; Baranova et al. 2011). Eastern Baltic cod have an exceptionally extended spawning season, with peak spawning much later in the year than other cod stocks, historically ranging from March to August (Bagge et al. 1994). Peak spawning has changed over time, from the end of April to the beginning of June during the 1970s, from mid-May to mid-June in the late 1980s (Bagge et al. 1994; Wieland et al. 2000), to the end of July in the 1990s (Wieland et al. 2000), and in most years since 2010, highest egg abundance has been recorded in June (Eero et al. 2019). Temperature during gonadal maturation is the key driver of peak maturation time. The shift in peak spawning time towards June in 1990 was thus caused by colder temperatures together with a higher proportion of first-time spawners, who tend to spawn earlier in the year (Wieland et al. 2000). Males mature earlier and at a smaller size than females, and arrive in the spawning areas earlier, and larger individuals of both sexes start spawning earlier and spawn over a longer period than smaller individuals (Tomkiewicz and Köster 1999). Consequently, the sex ratio varies over the season, becoming dominated by females towards the end of the season. Fecundity also varies over the life of the fish, with higher potential fecundity in larger fish and fish with higher somatic conditions (Mion et al. 2018).

For eastern Baltic cod, maturity ogives are also estimated from the macroscopic staging of individual fish's gonads during the first quarter of Baltic International Trawl Surveys (BITS). The size where 50% of the population is mature has progressively declined from 38 cm in 1991–1997 to 21 cm in 2015–2020 (Eero et al. 2015; ICES 2021a). Vainikka et al. (2009) estimated probabilistic maturation reaction norms to remove the impact of growth and demography from maturation and found that the length and condition at 50% probability of maturing decreased by 15–20% (most pronounced in females), which indicates an evolutionary change in the eastern Baltic cod maturation.

Growth and Condition

In fisheries, growth has traditionally been described as size-at-age based on age estimation from counting seasonally recurring growth patterns in the otoliths of the fish (Campana 2001). In eastern Baltic cod, age readings have notoriously been an issue of concern (Hüssy et al. 2016c). Over the last 40 years, various expert groups have struggled to document and improve the agreement of age estimation between national otolith readers, standardise methods and age estimations through repeated exchanges and reference collections as well as an internationally agreed manual.

Despite these initiatives, the precision of the age estimations based on traditional ageing did not improve, with significant bias persisting among and within readers (Hüssy et al. 2016c). The reason for these problems is linked to a combination of the hydrographic conditions prevailing in the Baltic Sea proper, coupled with the movements of the cod through complex thermal and salinity gradients and the extended spawning season and peak spawning in summer, which results in low contrast between otolith growth zones (Hüssy et al. 2009, 2010, 2016c; Hüssy 2010; Baranova et al. 2011). A complete overview of age reading exchange results and studies addressing the biological reasons for the difficulties encountered and alternative age estimation methods that have been tested over the years may be found in Hüssy et al. (2016c). To date, the most reliable growth estimates are from tag/recapture programmes spanning the seven decades Mion et al. (2020, 2021), details of which may be found in Box 1. First, empirical evidence using an age estimation technique based on daily growth increments rather than seasonal growth patterns coupled with length frequency analysis suggested that growth decreased by > 30% from the early 2000s onward (Hüssy et al. 2018). Later analyses of tagging data confirmed that the growth of eastern Baltic cod declined dramatically from an all-time high in the 1980s to recent years. To be able to address growth changes in the future, extensive efforts are currently invested into the development of a new age estimation approach based on seasonally varying concentrations in the chemical composition of otoliths (Heimbrand et al. 2020; Hüssy et al. 2021a).

Concurrent with the decline in growth, the average somatic condition of adult cod (Fulton's *K*) started to decline in the late 1990s from a value of 1.1 to < 0.9 in 2014 (Eero et al. 2015; Casini et al. 2016) and further to 0.85 in 2018 (Casini et al. 2021). This decline in condition was evident throughout all areas of the eastern Baltic Sea and progressively more pronounced with the increasing size of the fish (Casini et al. 2016, 2021). At the same time, the proportion of cod with a Fulton's *K* below 0.8, which can be considered a critical threshold level (Marteinsdottir and Begg 2002), increased rapidly to 30–40% of large cod and 20% in smaller cod during the main growth season (October–December) (Casini et al. 2021). The driving force behind the declining condition and growth is presumably a combination of several factors. These include habitat compression due to deteriorating environmental conditions (increasing temperatures and hypoxic areas) (Conley et al. 2009; Carstensen et al. 2014; Carstensen and Conley 2019) and increased hypoxia exposure due to changes in cods' depth distribution (Casini et al. 2021), potentially with a direct impact of hypoxia exposure on cods' metabolism (Plambech et al. 2013). Together with decreasing pelagic fish prey availability in the main distribution area (Eero et al. 2012b; Casini et al. 2016), these conditions have led to changes in feeding patterns over time, such as a decrease in sprat (*Sprattus sprattus*) and the benthic *Saduria entomon* in the diet of cod (Neuenfeldt et al. 2020). These conditions may have led to periods of density-dependent competition for increasingly scarce prey (Eero et al. 2012b, 2015; Casini et al. 2016, 2021). Further negative effects on growth and condition are related to seal predation and infection from seal-borne parasites (Plambech et al. 2013; Sokolova et al. 2018; Ryberg et al. 2020). This suite of drivers can potentially affect the somatic growth and condition of the eastern Baltic cod and cause a pronounced increase in natural mortality (Eero et al. 2015; ICES, 2021a).

BOX 1: Multi-Decadal Changes in Growth of Eastern Baltic Cod

Tag/recapture information of individually tagged fish provides an alternative to estimating growth from otolith age readings. Length measurements at release and recapture, together with the days spent at liberty (DAL), provide a measure of growth rate that allows the estimation of von Bertalanffy growth parameters. Since the 1950s, a number of tagging programmes have been carried out in the Baltic Sea. During the "Tagging Baltic Cod" (TABACOD) project (Hüssy et al. 2020), data from these historic programmes were digitised and combined with contemporary data to span the years from 1955 to 2019 (Mion et al. 2020, 2021). Using different complementary analytical techniques, Mion et al. (2020, 2021) documented strong temporal changes in growth, where growth from 1955 to the end of the 1980s increased by 8.6–10.6 cm year^{-1} for a 40 cm long cod, with the largest changes from the 1970s onward (Figure BOX 1). Following this period, growth decreased sharply to currently 4.3–5.1 cm year^{-1} for a 40 cm long cod, the lowest growth rate in the past seven decades (1950s, 1960s, 1970s, 1980s, and 2010s). Growth in the 1980s was significantly faster than during previous decades and significantly reduced in the 2010s compared to all previous decades. The amplitude in growth rate changes, including the sharp decline from the 1990s to the present, was most pronounced in size classes > 30 cm but relatively limited for smaller individuals (Mion et al. 2021).

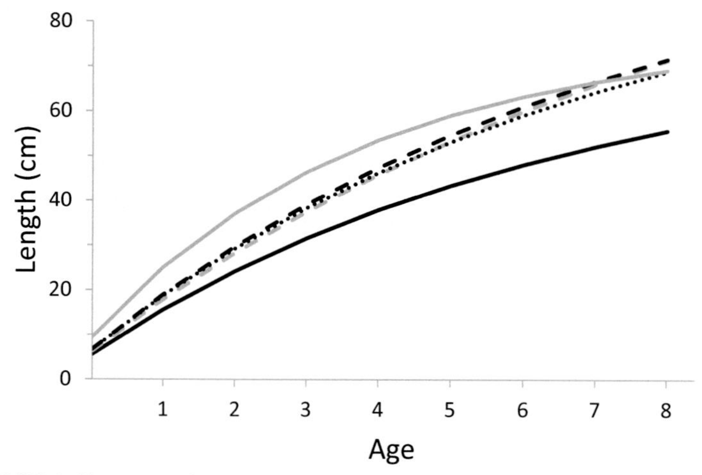

Figure BOX 1. Size-at-age of eastern Baltic cod estimated from tag/recapture data, where colours and line types represent different decades. 1955–1964: dotted black, 1965–1970: broken grey, 1971–1980: broken black, 1981–1990: solid grey, 2016–2019: solid black. Growth curves reproduced from parameter estimates in (Mion et al.,al. 2021) (see Table 1).

Growth parameter estimates of the 1950s and 1960s presented by Bagge et al. (1994) showed that estimates might vary somewhat between years. However, the overall trend is similar to the estimates by Mion et al. (2020, 2021). The most notable difference between these growth estimates and those derived from age readings from the DATRAS survey database is the lower maximum attainable length for the same period (survey: 132.2 cm, tagging: 77.2 cm), while the age reading derived growth coefficient k is within the confidence interval of the tagging-derived estimates. Since mortality related to tagging is not an issue, particularly for larger individuals (Haase et al. 2021), these results highlight that growth rates estimated from age readings may be biased towards faster growth.

Table 1. von Bertalanffy growth parameters estimated from tag/recapture data of eastern Baltic cod. Data provided are mean estimates of the growth coefficient k and the maximum attainable length L_∞, including 95% confidence intervals (CI).

Years	k	95% CI	L_∞	95% CI
1955–1964	0.14	0.09 – 0.19	98. 9	84.2 – 128.0
1965–1970	0.11	0.07 – 0.15	117.4	92.8 – 155.0
1971–1980	0.13	0.10 – 0.16	107.0	96.1 – 122.0
1981–1990	0.26	0.16 – 0.38	77.6	68.4 – 97.4
2016–2019	0.15	0.05 – 0.33	77.2	57.0 – 141.0

In a complementary study, McQueen et al. (2020) investigated to what extent growth rates in the Baltic Sea were regulated by geography or genotype, using tag/recapture data of individuals genetically identified as being of the western or the eastern Baltic stock during 2007–2019. An average-sized western Baltic cod caught in the western Baltic Sea grew more than double the rate of a similar-sized eastern Baltic cod caught in the eastern Baltic Sea (western: 14.5 cm year^{-1}, eastern: 5.8 cm year^{-1}). The difference in growth rate between geographic areas (eastern and western Baltic Sea) was more than twice as large as the difference between stocks (geography: 6.3 cm year^{-1}, stock: 2.4 cm year^{-1}). In other words, differences in growth rate between the two stocks were primarily driven by environmental conditions experienced within their respective geographical distribution range rather than by some size- or growth-selective mechanism. These growth estimates are similar to the historic values reported by Bagge et al. (1994), who also found that growth decreased from the Belt Sea in the west to the Gdansk Deep in the east.

Recruitment

Since the beginning of the time series in 1946, recruitment (abundance of a year class at age 0) has seen considerable changes, being relatively stable up to the mid-1960s, when recruitment more than doubled to peak values of 12 million in the late 1970s followed by a rapid decline back to a pre-1970s level (Figure 9A). During the 2010s, recruitment declined to an all-time low of < 1 mil, with the last two year-classes being estimated as the lowest since 1946 (Figure 9B). Recruitment of eastern Baltic cod has been studied intensively over the years, and a number of interacting conditions have been identified as key processes regulating year-class strength and have been summarised by Köster et al. (2005, 2017). Here, only the most prominent are highlighted. The hydrography of the eastern Baltic Sea substantially challenges the adaptability of marine species. Salinities throughout the water column are much lower than in the habitats of all other cod stocks, and the strong stratification combined with a high nutrient load from river runoff has led to the build-up of large hypoxic areas. Cod spawn pelagic eggs must remain buoyant to float above the hypoxic deep water to survive and develop. Several environmental threshold values have been identified for the successful development of early life stages: The salinity requirements of this stock are 11–12 psu for spermatozoa activity and 14.5±1.2 psu for neutral egg buoyancy (Nissling et al. 1994; Nissling and Westin 1997; Vallin and Nissling 2000). Egg survival declines when oxygen concentrations are < 5 ml l^{-1} (Köster et al. 2005), and no eggs survive at concentrations < 2 ml l^{-1} (Wieland et al. 1994). Finally, cod eggs need temperatures between 2–10°C for hatching (Wieland and Jarre-Teichmann, 1997; Petereit et al. 2004). The volume of water within these environmental threshold values, limited by salinity at the upper end and oxygen at the lower, is called the reproductive volume (MacKenzie et al. 2000).

Eastern Baltic cod have adapted to this environmental challenge through larger egg sizes that are neutrally buoyant at lower salinities (Nissling et al. 1994; Nissling and Westin 1997). In addition to the reproductive volume, biological features associated with fish size are also important. Egg size depends on the size of the female (Nissling et al. 1994). Larger females have a higher fecundity and produce larger and more viable numbers of eggs (Vallin and Nissling 2000). Larvae hatching from larger eggs are larger and have a higher probability of migrating from the spawning depth to the sub-surface layer (Grønkjær and Schytte 1999), where they additionally need to encounter suitable prey for particularly first-feeding larvae (Hinrichsen et al. 2002; Huwer et al. 2011). Additional influence on recruitment was attributed to predation by sprat on cod eggs (Hinrichsen et al. 2002; Köster et al. 2005; Möllmann et al. 2005), fecundity related to prey availability (Kraus et al. 2000), transport of larvae to juvenile nursery areas (Hinrichsen et al. 2003) as well as cannibalism depending on the spatial overlap between size classes and the abundance of prey for adults (Neuenfeldt and Köster 2000; Uzars and Plikshs 2000). Since the size of the reproductive volume was correlated with recruitment, the key controlling factors for cod recruitment were thought to be major Baltic inflows of saline and oxygen-

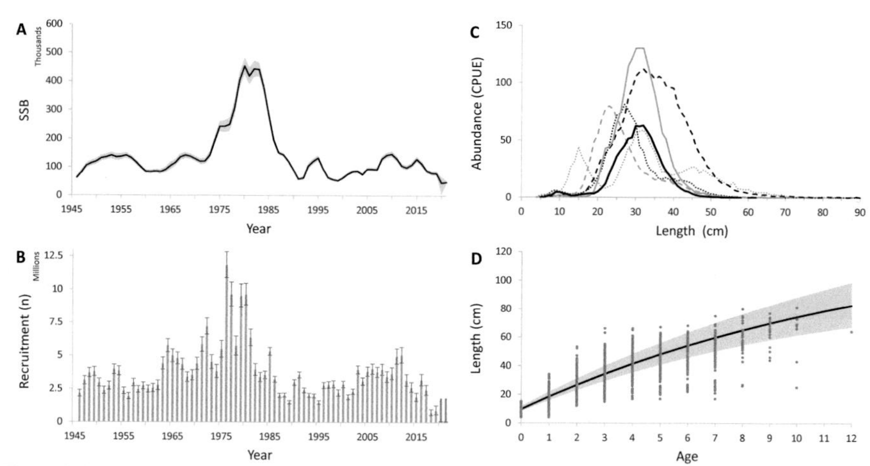

Figure 9. Eastern Baltic cod: (A) Time series of biomass of cod > 35 cm (dotted line) and spawning stock biomass (solid line) (both in thousand tonnes), (B) Recruitment in numbers, and (C) Length distribution (CPUE = catch per 30 min haul in quarter 1) for six years at five year time intervals, and (D) Fish length (cm) at age representing mean values over the years 1991–2005. Figures A, B and D show mean and 95% confidence intervals.

rich water from the North Sea in combination with oxygen consumption in the deep basins (Köster et al. 2005).

However, the increase in recruitment from the 2000s coincided with prolonged stagnation periods without any major inflow events (Lehmann et al. 2014). Köster et al. (2017) pointed towards a number of different conditions that may be of importance: Minor inflows of saline waters during winter and summer are sufficient to sustain successful egg development in the Bornholm Basin, and the climate-change-related change in hydrographical conditions may promote higher egg survival in the main spawning season of the stock. Predation pressure by sprat on eggs may have been reduced (Neumann et al. 2017) due to reduced vertical overlap in cod eggs and clupeid predators (Neumann et al. 2018) and a shift in the horizontal distribution of sprat to the east and north of the Baltic Sea (Eero et al. 2012b). Huwer et al. (2014) further showed that larval growth and survival are less linked with the spatial distribution of egg production than with the availability of suitable larval prey. Finally, it seems that not only the transport of larvae to juvenile nursery areas but also environmental conditions in these areas (Hinrichsen et al. 2009, 2011) may influence the survival of early life stages and suitable habitats are currently restricted to the Bornholm Basin (Hinrichsen et al. 2009, 2017). Recruitment in recent years, therefore, is more related to spawning stock biomass, egg production, prey availability, and predation-related egg survival (Köster et al. 2017, 2020). The deteriorating somatic condition of adults combined with the progressively smaller size at first maturity may thus have a negative impact on recruitment.

Stock Size and Structure

Historical Changes in Abundance

The eastern Baltic cod stock has been assessed annually since 1973, and the current time series of stock dynamics started in 1946 (Figure 9A). Stock size has varied considerably over the years but was estimated to be relatively low after World War II, then increased during the first half of the 1950s (Eero et al. 2007). During the 1970s, the spawning stock biomass saw a dramatic increase from around 100,000 tonnes to the record high levels of 450,000 tonnes in the early 1980s, followed by a rapid decline towards the mid-1990s as a result of increasing fishing pressure and failing recruitment (Eero et al. 2007, 2015; Köster et al. 2009, 2017). This period saw some major restructuring of the ecosystem with a pronounced regime shift towards a sprat-dominated system (Casini et al. 2009; Möllmann et al. 2009). While some stronger year-classes, in combination with effective management measures, temporarily led to an increase in stock biomass in the mid-2000s (Eero et al. 2012b), the stock has since deteriorated to historic low levels (ICES 2021a). As a result of the decreasing size at maturity, implying that a larger fraction of smaller cod is included in spawning stock biomass in recent years, the decline in total stock biomass of larger cod (> 35 cm in length) is even more pronounced (Eero et al. 2015).

Stock structure

The size distribution of the eastern Baltic cod stock has undergone a dramatic change since the 1990s (Figure 9C). The size distribution has become increasingly truncated towards smaller sizes (Svedäng and Hornborg 2017), except for a temporary increase in the abundance of market-sized cod (> 38 cm) in 2007–2012 (Eero et al. 2015). The reasons for this truncation were thought to be associated with a combination of decreasing growth rates and increased natural mortality (Eero et al. 2015; Svedäng and Hornborg 2017). Vainikka et al. (2009) also suggested that the combined evidence of a decrease in growth, somatic condition and size at maturity indicates that a fishing pressure-induced genetic change has occurred in eastern Baltic cod, which would favour a shift towards smaller sizes in the population (Svedäng and Hornborg 2017).

Environmental Challenges

Over the last decades, the Baltic Sea has experienced environmental changes that seem to be accelerating. The time between major Atlantic inflow events that bring saline and oxygenated water to the Baltic Sea has been increasing. Where they occurred regularly every few years in the past, there have only been two strong inflows in the last 30 years (Mohrholz et al. 2015). Overall, salinity has not changed considerably in all areas, but the stratification seems to have intensified, strengthening the barrier of vertical mixing between water layers (Carstensen et al. 2014; Lehmann et al. 2014; Carstensen and Conley 2019; Schmidt et al. 2021). Together with the continued high nutrient load from agriculture, this stratification and lack of renewal of the bottom water have caused the oxygen concentration at the bottom to decrease (Schmidt et al. 2021), leading to the spreading of hypoxic areas (Carstensen et al. 2014) that extend further up into the water column (Hinrichsen et al. 2011). The deeper basins are even anoxic, resulting from increased oxygen consumption at depth (Carstensen et al. 2014). The Baltic Sea has also experienced climate-change-related increases in temperature, not only at the surface (Lehmann et al. 2014) but also below the halocline (Carstensen et al. 2014; Lehmann et al. 2014; Schmidt et al. 2021). The hypoxic state is continuously intensifying, exacerbated by both eutrophication and increased water temperature, the latter leading to reduced solubility of oxygen (Carstensen et al. 2014b). These environmental changes are increasingly limiting the suitability of the habitat for all life stages of cod.

The biological changes observed in the eastern Baltic cod outlined above (concurrent reduction in growth, somatic condition, size at maturity, and truncated size distribution) are strong evidence for the stock being under severe negative pressure. While the studies mentioned above address causal relationships between specific environmental parameters and biological responses, each alone does not answer why the eastern Baltic cod health status has been deteriorating so quickly. However, they provide strong evidence indicating that it is the sum of adverse conditions that are driving the stock collapse. While we now know that there are multiple pressures targeting the different life stages of cod, we do not know how these drivers interact or if we have overlooked a crucial issue. In addition to the pressure imposed by the environment, the eastern Baltic cod stock may also suffer from evolutionary changes caused by size-selective fishing pressure, which causes individuals to mature at earlier ages and smaller sizes and is accompanied by decreasing growth rates (Vainikka et al. 2009; Svedäng and Hornborg 2017). It is unclear to what extent the reason for the stock collapse can be solely attributed to changing environmental conditions, seal-related effects or fishery-related evolution. The most likely scenario is an interaction of a whole suite of unfavourable conditions that leave only a little optimism for a rebound to historic stock biomass.

Stock Management

Fishery

Cods in the eastern Baltic Sea have traditionally been caught in a directed fishery, using mainly demersal and semi-pelagic trawls and gillnets. The by-catch of cod in pelagic fisheries has been very limited. Fishing mortality was relatively low after World War II but increased rapidly from the 1950s to the 1970s (Eero et al. 2007). After a period of reduced fishing pressure in the late 1970s-early 1980s, high levels of fishing intensity were observed in the late 1980s to early 2000s. Since the 2000s, fishing mortality has been declining to the present very low level (ICES 2021) (Figure 10B). The highest landings (350–400,000 tonnes) occurred in the early 1980s (Figure 10A), corresponding to the record high cod biomass (Figure 9A). Since the second half of 2019, targeted fishing for the eastern Baltic cod has been prohibited, and cod catch in 2020 amounted to less than 3,000 t. In accordance with stock distribution, in the recent two decades, the eastern Baltic cod catches were almost exclusively taken in ICES SDs 24–26.

Management

Since 2003, there have been two management areas for cod in the Baltic Sea, Western (ICES SD 22–24) and Eastern (ICES SD 25–32), although both stocks occur in ICES SD 24. Catch amounts of cod in the eastern Baltic Sea are mainly regulated by Total Allowable Catch (TAC). Additionally, various regulations on fishing gear and effort have been introduced over time (ICES 2019b), including seasonal and area closures related to cod spawning (Eero et al. 2019). A first EU multi-annual management plan aimed at the recovery and sustainable exploitation of Baltic cod stocks was implemented in 2008 (EC No. 1098/2007), later replaced by the EU Baltic multi-annual management plan (EU 2016). Since 2015, there has been a landing obligation in place for cod in the Baltic Sea, with a minimum conservation reference size of 35 cm, implying that smaller cod cannot be sold for human consumption. Due to the poor state of the stock, all fishing targeting cod has been prohibited since July 2019 (ICES 2021a). By-catch of cod is still allowed in pelagic fisheries and demersal fisheries targeting other species than cod. Further details relating to the management of the eastern Baltic cod stock, including a historical overview of management measures, may be found in the ICES Advice Sheet (Appendix 1).

Challenges

Management of the eastern Baltic cod stock was affected by uncertainties in stock assessment and lack of quantitative stock estimates in 2014–2019, which was due to a failure to account for the large changes in cod productivity (growth and natural mortality) in stock assessment (Eero et al. 2015). Quantitative stock assessment was re-established in 2019 after a period of intensified research to understand the

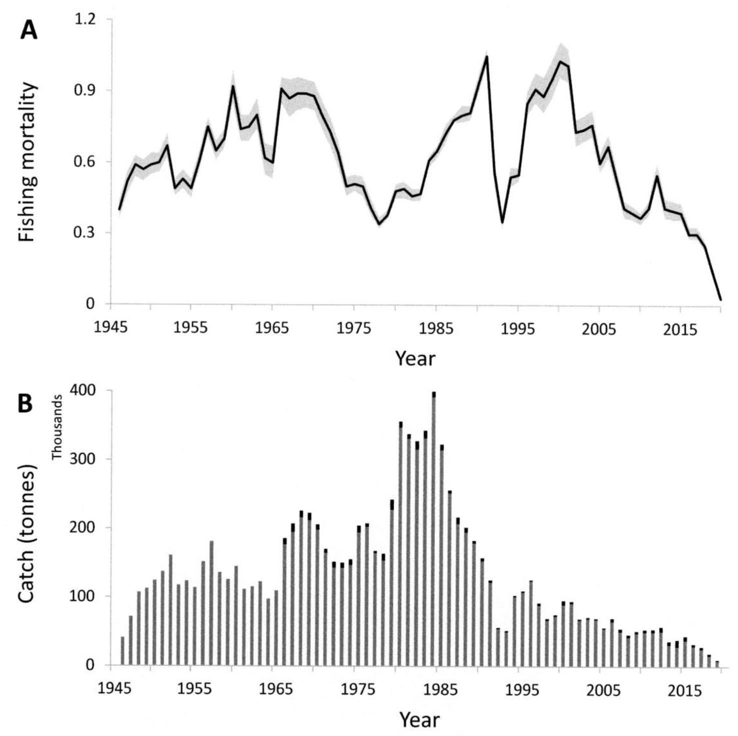

Figure 10. Eastern Baltic cod: (A) Fishing mortality Fbar (showing mean and 90% confidence intervals) and (B) Catch in thousand tonnes (lower panel, colour represents landings: dark grey, discards: black) for the available time series.

rapidly deteriorating stock status and incorporate relevant biological changes in stock assessment (ICES 2019a). Due to a pronounced decline in productivity, there is currently no surplus in the stock to support fisheries; thus, no Fmsy target could be considered sustainable in the long term. Also, estimating biomass reference points is challenging, as the standard approaches are unsuitable for dealing with such changes in stock productivity, as seen for the eastern Baltic cod (ICES 2019). Presently, fishing mortality of cod is estimated to be much lower than natural mortality, which has increased to an exceptionally high level since the 2000s, owing to the environmental and biological pressures summarised above. Thus, managing cod fisheries alone is unlikely to improve the stock status. Future development in eastern Baltic cod is very dependent on the state of the ecosystem, affecting its productivity. This has drawn increasing attention to possible ecosystem-based management measures, although the relevant aspects of such an approach are difficult to implement (Eero et al. 2020).

References

Al-Hamdani, Z.K., Reker, J., Leth, J.O., et al. 2007. Development of marine landscape maps for the Baltic Sea and the Kattegat using geophysical and hydrographical parameters. Geological Survey of Denmark and Greenland Bulletin, 13: 61–64.

André, C., Svedäng, H., Knutsen, H., et al. 2016. Population structure in Atlantic cod in the eastern North Sea-Skagerrak-Kattegat: early life stage dispersal and adult migration. BMC research notes, 9: 63.

Aro, E. 1989. A review of fish migration patterns in the Baltic. Rapports et Proces-Verbaux des Réunions du Conseil international pour l'Exploration de la Mer, 109: 72–96.

Bagge, O. 1969. Preliminary results of the cod tagging experiments in the Western Baltic. ICES CM, 1969/F:29, 7 p.

Bagge, O., Thurow, F., Steffensen, E., et al. 1994. The Baltic cod. Dana, 10: 1–28.

Baranova, T., and Uzars, D. 1986. Growth and maturation of cod (*Gadus morhua callarias* L.) in the Eastern Baltic. ICES CM, 1986/J:07.

Baranova, T., Müller-Karulis, B., Sics, I., et al. 2011. Changes in the annual life cycle of eastern Baltic cod during 1950–2010. ICES C.M. 2011/R:10, 22 p.

Barth, J.M.I., Villegas-Ríos, D., Freitas, C., et al. 2019. Disentangling structural genomic and behavioural barriers in a sea of connectivity. Molecular Ecology, 28: 1394–1411.

Bartolino, V., Tian, H., Bergström, U., et al. 2017. Spatio-temporal dynamics of a fish predator: Density-dependent and hydrographic effects on Baltic Sea cod population. PLOS ONE, 12: e0172004.

Berg, P.R., Jentoft, S., Star, B., et al. 2015. Adaptation to low salinity promotes genomic divergence in Atlantic Cod (*Gadus morhua* L.). Genome Biology and Evolution, 7: 1644–1663.

Berner, M. 1967. Results of cod taggings in the Western and Central Baltic in the period 1962–1965. ICES CM, 1967/F:05, 10 p.

Berner, M. 1971. Ergebnisse der Dorschmarkierungen des Jahres 1968 in der Bornholm- und Arkonasee. Fischerei-Forschung Wissenschaftliche Schriftenreihe, 9: 15–20.

Berner, M. 1974. Some results of the cod tagging experiments of the GDR in the Baltic 1968-1971. ICES CM, 1974/F:32.

Berner, M. 1985. The periodic changes in gonad weight and spawning cycle of the 'Baltic' and 'Belt cod' (*G. morhua callarias*/*G. morhua morhua*) in different regions of the Baltic. Fischerei-Forschung, 23: 49–57, 23 p.

Berner, M., and Vaske, B. 1985. Morphometric and meristic characters of cod stocks in the Baltic Sea. ICES CM, 1985/J:11.

Berner, M., and Müller, H. 1989. Discrimination between 'Baltic cod' (*Gadus morhua callarias L.*) and 'Belt Sea cod' (*Gadus morhua*) by means of morphometric and meristic characters. ICES CM, 1989/J:13: 19p.

Berner, M., and Müller, H. 1990. On the identification of cod as 'proper Baltic cod' (*Gadus morhua callarias* L.) & 'Belt Sea cod' (*Gadus morhua morhua* L.) by means of discriminant analysis. Fischerei-Forschung, 28: 46–49.

Birjukov, N.P. 1969. Spawning Communities of Baltic cod and the Extent of their Mixing. ICES CM, 1969/F:7, 17 p.

Björck, S. 1995, January 1. A review of the history of the Baltic Sea, 13.0-8.0 ka BP. Quaternary international, 27: 19–40.

Björck, S. 2008. The late Quaternary development of the Baltic Sea basin. pp. 398–407. *In:* Bolle, H.-J., Menenti, M., and Rasool, I. (eds). Assessment of climate change for the Baltic Sea basin. Springer.

Bleil, M. 1995. Rearing experiments with cod (*Gadus morhua morhua*) from the western Baltic. Part 2: Broodstock and hatching methods. Informationen aus der Fischwirtschaft und Fischerei-Forschung, Hamburg, 42: 133–146.

Bleil, M., and Oeberst, R. 1997. The timing of the reproduction of cod (*Gadus morhua morhua*) in the western Baltic and adjacent areas. ICES CM, 1997/CC:02, 27 p.

Bleil, M., and Oeberst, R. 1998. The spawning of cod (*Gadus morhua morhua*) under controlled conditions of captivity, quantity and quality of spawned eggs. ICES CM, 1998/DD:03, 27 p.

Bleil, M., and Oeberst, R. 2000. Reproduction areas of the cod stock in the western Baltic Sea. ICES CM, 2000/N:02, 20 p.

Bleil, M., and Oeberst, R. 2002. Spawning areas of the cod stock in the western Baltic Sea and minimum length at maturity. Archive of Fishery and Marine Research, 49: 243–258.

Bleil, M., and Oeberst, R. 2004. Comparison of spawning activities in the mixing area of both the Baltic cod stocks, Arkona Sea (ICES Sub-divisions 24), and the adjacent areas in the recent years. ICES CM, 2004/L:08, 22 p.

Bleil, M., Oeberst, R., and Urrutia, P. 2009. Seasonal maturity development of Baltic cod in different spawning areas: importance of the Arkona Sea for the summer spawning stock. Journal of Applied Ichthyology, 25: 10–17.

Börjesson, P., Jonsson, P., Pacariz, S., et al. 2013. Spawning of Kattegat cod (*Gadus morhua*) - Mapping spatial distribution by egg surveys. Fisheries Research, 147: 63–71.

Brander, K.M. 1994. The location and timing of cod spawning around the British Isles. ICES Journal of Marine Science, 51: 71–89.

Campana, S.E. 2001. Accuracy, precision and quality control in age determination, including a review of the use and abuse of age validation methods. Journal of Fish Biology, 59: 197–242.

Carstensen, J., Andersen, J.H., Gustafsson, B.G., et al. 2014. Deoxygenation of the Baltic Sea during the last century. Proceedings of the National Academy of Sciences of the United States of America, 111: 5628–5633.

Carstensen, J., and Conley, D.J. 2019. Baltic Sea hypoxia takes many shapes and sizes. Limnology and Oceanography Bulletin, 28: 125–129.

Casini, M., Cardinale, M., Hjelm, J., et al. 2005. Trends in cpue and related changes in spatial distribution of demersal fish species in the Kattegat and Skagerrak, eastern North Sea, between 1981 and 2003. ICES Journal of Marine Science, 62: 671–682.

Casini, M., Hjelm, J., Molinero, J.C., et al. 2009. Trophic cascades promote threshold-like shifts in pelagic marine ecosystems. Proceedings of the National Academy of Sciences of the United States of America, 106: 197–202.

Casini, M., Käll, F., Hansson, M., et al. 2016. Hypoxic areas, density-dependence and food limitation drive the body condition of a heavily exploited marine fish predator. Royal Society Open Science, 3: 160416.

Casini, M., Hansson, M., Orio, A., et al. 2021. Changes in population depth distribution and oxygen stratification are involved in the current low condition of the eastern Baltic Sea cod (*Gadus morhua*). Biogeosciences, 18: 1321–1331.

Claireaux, G., Webber, D.M., Lagardere, J.-P., et al. 2000. Influence of water temperature and oxygenation on the aerobic metabolic scope of Atlantic cod (*Gadus morhua*). Journal of Sea Research, 44: 257–265.

Conley, D.J., Bj̈orck, S., Bonsdorff, E., et al. 2009. Hypoxia-related processes in the Baltic Sea. Environmental Science and Technology, 43: 3412–3420.

Conley, D.J., Carstensen, J., Aigars, J., et al. 2011. Hypoxia is increasing in the coastal zone of the Baltic Sea. Environmental Science and Technology, 45: 6777–6783.

Dinesen, G.E., Neuenfeldt, S., Kokkalis, A., et al. 2019. Cod and climate: a systems approach for sustainable fisheries management of Atlantic cod (*Gadus morhua*) in coastal Danish waters. Journal of Coastal Conservation, 23: 943–958.

Eero, M., Köster, F.W., Plikshs, M., et al. 2007. Eastern Baltic cod (*Gadus morhua callarias*) stock dynamics: extending the analytical assessment back to the mid-1940s. ICES Journal of Marine Science, 64: 1257–1271.

Eero, M., Köster, F.W., and Vinther, M. 2012a. Why is the eastern Baltic cod recovering? Marine Policy, 36: 235–240.

Eero, M., Vinther, M., Haslob, H., et al. 2012b. Spatial management of marine resources can enhance the recovery of predators and avoid local depletion of forage fish. Conservation Letters, 5: 486–492.

Eero, M., Hemmer-Hansen, J., and Hüssy, K. 2014. Implications of stock recovery for a neighbouring management unit: experience from the Baltic cod. ICES Journal of Marine Science, 71: 1458–1466.

Eero, M., Hjelm, J., Behrens, J., et al. 2015. Eastern Baltic cod in distress: Biological changes and challenges for stock assessment. ICES Journal of Marine Science, 72: 2180–2186.

Eero, M., Hemmer-Hansen, J., Hüssy, K., et al. 2016. Optimal bæredygtig udnyttelse af tilgængelige torskebestande for dansk fiskeri. DTU Aqua Scientific Report (EMFF J.nr. 33010-13-k-0269): 54. Charlottenlund, Denmark.

Eero, M., Hinrichsen, H.-H., Hjelm, J., et al. 2019. Designing spawning closures can be complicated: Experience from cod in the Baltic Sea. Ocean and Coastal Management, 169: 129–136.

Fox, C.J., Taylor, M., Dickey-Collas, M., et al. 2008. Mapping the spawning grounds of North Sea cod (*Gadus morhua*) by direct and indirect means. Proceedings of the Royal Society B, 275: 1543–1548.

Friedland, R., Neumann, T., and Schernewski, G. 2012. Climate change and the Baltic Sea action plan: Model simulations on the future of the western Baltic Sea. Journal of Marine Systems, 105–108: 175–186.

Funk, S., Krumme, U., Temming, A., et al. 2020. Gillnet fishers' knowledge reveals seasonality in depth and habitat use of cod (*Gadus morhua*) in the western Baltic Sea. ICES Journal of Marine Science, 77: 1816–1829.

Gräwe, U., Friedland, R., and Burchard, H. 2013. The future of the western Baltic Sea: Two possible scenarios. Ocean Dynamics, 63: 901–921.

Grønkjær, P., and Schytte, M. 1999. Non-random mortality of Baltic cod larvae inferred from otolith hatch-check sizes. Marine Ecology Progress Series, 181: 53–59.

Haase, S., McQueen, K., Mion, M., et al. 2021. Short-term tagging mortality of Baltic cod (*Gadus morhua*). Fisheries Research, 234: 105804.

Hagström, O., Larsson, P.O., and Ulmestrand, M. 1990. Swedish cod data from the international young fish surveys 1981–1990. ICES C.M., 1990/G:65, 24 p.

Hansson, M., Viktorsson, L., and Andersson, L. 2019. Oxygen survey in the Baltic Sea 2019 - Extent of Anoxia and Hypoxia, 1960–2019. Report Oceanography 67, SMHI Scientific Publications, Göteborg, Sweden. 88 pp. URN: urn:nbn:se:smhi:diva-5643.

Heimbrand, Y., Limburg, K.E., Hüssy, K., et al. 2020. Seeking the true time: Exploring otolith chemistry as an age-determination tool. Journal of Fish Biology, 97: 552–565.

Helcom. 2009. Eutrophication in the Baltic Sea An integrated thematic assessment of the effects of nutrient enrichment in the Baltic Sea region. 152 pp.

Hemmer-Hansen, J., Hüssy, K., Baktoft, H., et al. 2019. Genetic analyses reveal complex dynamics within a marine fish management area. Evolutionary Applications, 12: 830–844.

Hemmer-Hansen, J., Hüssy, K., Albertsen, C.M., et al. 2020. Sustainable management of Kattegat cod; better knowledge of stock components and migration. DTU Aqua Report, no. 357-20: 42 pp. http://www.aqua.dtu.dk/publikationer.

Hinrichsen, H.-H., Böttcher, U., Oeberst, R., et al. 2001. The potential for advective exchange of the early life stages between the western and eastern Baltic cod (*Gadus morhua* L.) stocks. Fisheries Oceanography, 10: 249–258.

Hinrichsen, H.-H., Möllmann, C., Köster, F.W., et al. 2002. Biophysical modelling of larval Baltic cod (*Gadus morhua*) growth and survival. Canadian Journal of Fisheries and Aquaculture Sciences, 59: 1858–1873.

Hinrichsen, H.-H., Bottcher, U., Köster, F.W., et al. 2003. Modelling the influences of atmospheric forcing conditions on Baltic cod early life stages: distribution and drift. Journal of Sea Research, 49: 187–201.

Hinrichsen, H.-H., Kraus, G., Böttcher, U., et al. 2009. Identifying eastern Baltic cod nursery grounds using hydrodynamic modelling: knowledge for the design of Marine Protected Areas. ICES Journal of Marine Science, 66: 101–108.

Hinrichsen, H.-H., Huwer, B., Makarchouk, A., et al. 2011. Climate-driven long-term trends in Baltic Sea oxygen concentrations and the potential consequences for eastern Baltic cod (*Gadus morhua*). ICES Journal of Marine Science, 68: 2019–2028.

Hinrichsen, H.-H., Hüssy, K., and Huwer, B. 2012. Spatio-temporal variability in western Baltic cod early life stage survival mediated by egg buoyancy, hydrography and hydrodynamics. ICES Journal of Marine Science, 69: 1744–1769.

Hinrichsen, H.-H., Lehmann, A., Petereit, C., et al. 2016. Spawning areas of eastern Baltic cod revisited: Using hydrodynamic modelling to reveal spawning habitat suitability, egg survival probability, and connectivity patterns. Progress in Oceanography, 143: 13–25.

Hinrichsen, H.-H., von Dewitz, B., Lehmann, A., et al. 2017. Spatio-temporal dynamics of cod nursery areas in the Baltic Sea. Progress in Oceanography, 155: 28–40.

Hjelm, J., Simonsson, J., and Cardinale, M. 2004. Spatial distribution of cod in the Baltic Sea in relation to abiotic factors-a question of fish-age and area. Special Issue: Challenges and Opportunities for The EU Common Fisheries Policy application in The Mediterranean and Black Sea View project Waking. ICES C.M., 2004/L:16: 35.

Hüssy, K., Nielsen, B., Mosegaard, H., et al. 2009. Using data storage tags to link otolith macro- structure in Baltic cod *Gadus morhua* with environmental conditions. Marine Ecology Progress Series, 378: 161–170.

Hüssy, K. 2010. Why is age determination of Baltic cod (*Gadus morhua*) so difficult? ICES Journal of Marine Science, 67: 1198–1205.

Hüssy, K., Hinrichsen, H.-H.-H., Fey, D.P., et al. 2010. The use of otolith microstructure to estimate age in adult Atlantic cod *Gadus morhua*. Journal of Fish Biology, 76: 1640–1654.

Hüssy, K. 2011a. Review of western Baltic cod (*Gadus morhua*) recruitment dynamics. ICES Journal of Marine Science, 68: 1459–1471.

Hüssy, K. 2011b. Review of western Baltic cod (*Gadus morhua*) recruitment dynamics. ICES Journal of Marine Science, 68: 1459–1471.

Hüssy, K., Hinrichsen, H.-H., and Huwer, B. 2012. Hydrographic influence on the spawning habitat suitability of western Baltic cod (*Gadus morhua*). ICES Journal of Marine Science, 69: 1736–1743.

Hüssy, K., Mosegaard, H., Albertsen, C.M., et al. 2016a. Evaluation of otolith shape as a tool for stock discrimination in marine fishes using Baltic Sea cod as a case study. Fisheries Research, 174: 210–218.

Hüssy, K., Hinrichsen, H.-H., Eero, M., et al. 2016b. Spatio-temporal trends in stock mixing of eastern and western Baltic cod in the Arkona Basin and the implications for recruitment. ICES Journal of Marine Science, 73: 293–303.

Hüssy, K., Radtke, K., Plikshs, M., et al. 2016c. Challenging ICES age estimation protocols: lessons learned from the eastern Baltic cod stock. ICES Journal of Marine Science, 73: 2138–2149.

Hüssy, K., Eero, M., and Radtke, K. 2018. Faster or slower: has growth of eastern Baltic cod changed? Marine Biology Research, 14: 598–609.

Hüssy, K., Casini, M., Haase, S., et al. 2020. Tagging Baltic Cod – TABACOD. Eastern Baltic cod: Solving the ageing and stock assessment problems with combined state-of-the-art tagging methods. DTU Aqua Report no.: 368–2020: 94 pp., Technical University of Denmark, Lyngby, Denmark.

Hüssy, K., Krüger-Johnsen, M., Thomsen, T.B., et al. 2021a. It's elemental, my dear Watson: Validating seasonal patterns in otolith chemical chronologies. Canadian Journal of Fisheries and Aquatic Sciences, 78 : 551–566.

Hüssy, K., Albertsen, C.M., Hemmer-Hansen, J., et al. 2021b. Where do you come from, where do you go: Early life stage drift and migrations of cod inferred from otolith microchemistry and genetic population assignment. Canadian Journal of Fisheries and Aquatic Sciences, 79 : 300–313.

Huwer, B., Clemmesen, C., Grønkjær, P., et al. 2011. Vertical distribution and growth performance of Baltic cod larvae—Field evidence for starvation-induced recruitment regulation during the larval stage? Progress in Oceanography, 91: 382–396.

Huwer, B., Hinrichsen, H., Böttcher, U., et al. 2014. Characteristics of juvenile survivors reveal spatio-temporal differences in early life stage survival of Baltic cod. Marine Ecology Progress Series, 511: 165–180.

Huwer, B., Hinrichsen, H.-H., Hüssy, K., et al. 2016. Connectivity of larval cod in the transition area between North Sea and Baltic Sea and potential implications for fisheries management. ICES Journal of Marine Science, 73: 1815–1824.

ICES. 2014. Report of the Workshop on Scoping for Integrated Baltic Cod Assessment (WKSIBCA), 1–3 October 2014, Gdynia, Poland. ICES CM, 2014/ACOM, 51 pp.

ICES. 2015. Report of the Benchmark Workshop on Baltic Cod Stocks (WKBALTCOD), 2–6 March 2015, Rostock, Germany. ICES CM, 2015/ACOM, 172 pp.

ICES. 2017. Report of the Benchmark Workshop on Baltic Stocks (WKBALT), 7–10 February 2017, Copenhagen, Denmark. ICES C.M., 2017/ACOM: 108 pp.

ICES. 2019a. Benchmark Workshop on Baltic cod Stocks (WKBALTCOD2). ICES Scientific Reports, 1:9, 310 pp.

ICES. 2019b. Stock Annex: Cod (*Gadus morhua*) in Subdivisions 24–32, eastern Baltic stock. ICES Stock Annexes Reports. ICES Stock Annexes.

ICES. 2021a. Baltic Fisheries Assessment Working Group (WGBFAS). ICES Scientific Reports, 3:53, 717 pp.

ICES. 2021b. Inter-benchmark process on western Baltic cod (IBPWEB). ICES Scientific Reports, 3 :87, 76 pp.

ICES. 2021c. Stock Annex-Western Baltic cod in subdivisions 22-24. ICES Stock Annexes Reports. ICES Stock Annexes.

ICES. 2023. Baltic Fisheries Assessment Working Group (WGBFAS). ICES Scientific Reports, 5, 607 pp.

Jakobsson, M., Stranne, C., O'Regan, M., et al. 2019. Bathymetric properties of the Baltic Sea. Ocean Science, 15: 905–924.

Jonsson, P.R., Corell, H., André, C., et al. 2016. Recent decline in cod stocks in the North Sea-Skagerrak-Kattegat shifts the sources of larval supply. Fisheries Oceanography, 25: 210–228.

Kändler, R. 1949. Häufigkeit pelagischer Fischeier in der Ostsee als Massstab für die Zu- und Abnahme der Fischbestände. Kieler Meeresforschung, 6: 73–89.

Kändler, R. 1950. Jahreszeitliches Vorkommen und unperiodisches Auftreten von Fischbrut, Medusen und Dekapodenlarven im Fehmarnbelt in den Jahren 1934–1943. Berichte der Deutschen Wissenschaftlichen Kommission für Meeresforschung, 12: 49–85.

Knutsen, H., André, C., Jorde, P.E., et al. 2004. Transport of North Sea cod larvae into the Skagerrak coastal populations. Proceedings of the Royal Society of London. Series B: Biological Sciences, 271: 1337–1344.

Köster, F.W., Möllmann, C., Hinrichsen, H.-H., et al. 2005. Baltic cod recruitment—the impact of climate variability on key processes. ICES Journal of Marine Science, 62: 1408–1425.

Köster, F.W., Vinther, M., Mackenzie, B.R., et al. 2009. Environmental Effects on Recruitment and Implications for Biological Reference Points of Eastern Baltic Cod (*Gadus morhua*). J. Northw. Atl. Fish. Sci, 41: 205–220.

Köster, F.W., Huwer, B., Hinrichsen, H.-H., et al. 2017. Eastern Baltic cod recruitment revisited—dynamics and impacting factors. ICES Journal of Marine Science, 74: 3–19.

Köster, F.W., Huwer, B., Kraus, G., et al. 2020. Egg production methods applied to Eastern Baltic cod provide indices of spawning stock dynamics. Fisheries Research, 227: 105553.

Kraus, G., Müller, A., Trella, K., et al. 2000. Fecundity of Baltic cod: temporal and spatial variation. Journal of Fish Biology, 56: 1327–1341.

Kullenberg, G., and Jacobsen, T.S. 1981. The Baltic Sea: An outline of it's physical oceanography. Marine Pollution Bulletin, 12: 183–186.

Lehmann, A. 1995. A three-dimensional baroclinic eddy-resolving model of the Baltic Sea. Tellus, 47A: 1013–1031.

Lehmann, A., Hinrichsen, H.-H., Getzlaff, K., et al. 2014. Quantifying the heterogeneity of hypoxic and anoxic areas in the Baltic Sea by a simplified coupled hydrodynamic-oxygen consumption model approach. Journal of Marine Systems, 134: 20–28.

MacKenzie, B.R., Hinrichsen, H.-H., Plikshs, M., et al. 2000. Quantifying environmental heterogeneity: habitat size necessary for successful development of cod *Gadus morhua* eggs in the Baltic Sea. Marine Ecology Progress Series, 193: 143–156.

Mariani, P., Hemmer-Hansen, J., Le Moan, A., et al. 2020. KYSTFISK III. Population connectivity of cod and plaice in Danish coastal waters. DTU Aqua Report no. 356–2020, National Institute of Aquatic Resources, Technical University of Denmark, 58 pp.

Marteinsdottir, G., and Begg, G.A. 2002. Essential relationships incorporating the influence of age, size and condition on variables required for estimation of reproductive potential in Atlantic cod *Gadus morhua*. Marine Ecology Progress Series, 235: 235–256.

Matthäus, W., and Franck, H. 1992. Characteristics of major Baltic inflows - a statistical analysis. Continental Shelf Research, 12: 1375–1400.

Matthäus, W., and Schinke, H. 1999. The influence of river runoff on the deep water conditions of the Baltic Sea. Hydrobiologia, 393: 1–10.

McQueen, K., Eveson, J.P., Dolk, B., et al. 2019. Growth of cod (*Gadus morhua*) in the western Baltic Sea: estimating improved growth parameters from tag–recapture data. Canadian Journal of Fisheries and Aquatic Sciences, 76: 1326–1337.

McQueen, K., Casini, M., Dolk, B., et al. 2020. Regional and stock-specific differences in contemporary growth of Baltic cod revealed through tag-recapture data. ICES Journal of Marine Science, 77: 2078–2088.

Meier, H.E.M., Feistel, R., Piechura, J., et al. 2006. Ventilation of the Baltic Sea deep water: A brief review of present knowledge from observations and models. Oceanologia, 48: 133–164.

Mion, M., Thorsen, A., Vitale, F., et al. 2018. Effect of fish length and nutritional condition on the fecundity of distressed Atlantic cod *Gadus morhua* from the Baltic Sea. Journal of Fish Biology, 92: 1016–1034.

Mion, M., Hilvarsson, A., Hüssy, K., et al. 2020. Historical growth of Eastern Baltic cod (*Gadus morhua*): Setting a baseline with international tagging data. Fisheries Research, 223: 105442.

Mion, M., Haase, S., Hemmer-Hansen, J., et al. 2021. Multidecadal changes in fish growth rates estimated from tagging data: A case study from the Eastern Baltic cod (*Gadus morhua*, Gadidae). Fish and Fisheries, 22: 413–427.

Mion, M., Griffiths, C.A., Bartolino, V., et al. 2022. New perspectives on Eastern Baltic cod movement patterns from historical and contemporary tagging data. Marine Ecology Progress Series, 689: 109–126.

Mohrholz, V., Naumann, M., Nausch, G., et al. 2015. Fresh oxygen for the Baltic Sea - An exceptional saline inflow after a decade of stagnation. Journal of Marine Systems, 148: 152–166.

Møller, J.S., and Hansen, I.S. 1994. Hydrographic processes and changes in the Baltic Sea. Dana, 10: 87–104.

Möllmann, C., Kornilovs, G., Fetter, M., et al. 2005. Climate, zooplankton, and pelagic fish growth in the central Baltic Sea. ICES Journal of Marine Science, 62: 1270–1280.

Möllmann, C., Diekmann, R., Müller-Karulis, B., et al. 2009. Reorganization of a large marine ecosystem due to atmospheric and anthropogenic pressure: a discontinuous regime shift in the Central Baltic Sea. Global Change Biology, 15: 1377–1393.

Möllmann, C., Cormon, X., Funk, S., et al. 2021. Tipping point realized in cod fishery. Scientific Reports, 11: Article number: 14259 (2021).

Müller, H. 2002. The distribution of 'Belt Sea cod' and 'Baltic cod' in the Baltic Sea from 1995 to 2001 estimated by discriminant analysis of the number of dorsal fin rays. ICES CM, 200//L:16, 23 pp.

Naumann, M., Gräwe, U., Mohrholz, V., et al. 2020. Hydrographic-hydrochemical assessment of the Baltic Sea 2019. Meereswissenschaftliche Berichte, No 114.

Netzel, J. 1963. Polish cod tagging experiments in the Gdansk Area 1957-62. ICES CM, 1968/F:3, 15 pp.

Netzel, J. 1968. Polish cod tagging experiments in the region of Slupsk Furrow in the years 1957/1963. ICES CM, 1968/F:7, 13 pp.

Netzel, J. 1974a. Polish cod tagging experiments in the Baltic in 1969 and 1970. Rapports et Proces-Verbaux des Réunions du Conseil international pour l'Exploration de la Mer, 166: 40–46.

Netzel, J. 1974b. Polish investigations on juvenile cod in Gdansk Bay and in the southern part of the Bornholm Basin. Rapports et Proces-Verbaux des Réunions du Conseil international pour l'Exploration de la Mer, 166: 62–65.

Neuenfeldt, S., and Köster, F.W. 2000. Trophodynamic control on recruitment success in Baltic cod: the influence of cannibalism. ICES Journal of Marine Science, 57: 300–309.

Neuenfeldt, S., Bartolino, V., Orio, A., et al. 2020. Feeding and growth of Atlantic cod (*Gadus morhua* L.) in the eastern Baltic Sea under environmental change. ICES Journal of Marine Science, 77: 624–632.

Neumann, V., Schaber, M., Eero, M., et al. 2017. Quantifying predation on baltic cod early life stages. Canadian Journal of Fisheries and Aquatic Sciences, 74: 833–842.

Neumann, V., Köster, F.W., and Eero, M. 2018. Fish egg predation by Baltic sprat and herring: Do species characteristics and development stage matter? Canadian Journal of Fisheries and Aquatic Sciences, 75: 1626–1634.

Nielsen, B., Hüssy, K., Neuenfeldt, S., et al. 2013. Individual behaviour of Baltic cod *Gadus morhua* in relation to sex and reproductive state. Aquatic Biology, 18: 197–207.

Nielsen, E.E., Hansen, M.M., Ruzzante, D.E., et al. 2003. Evidence of a hybrid-zone in Atlantic cod (*Gadus morhua*) in the Baltic and the Danish Belt Sea revealed by individual admixture analysis. Molecular ecology, 12: 1497–1508.

Nielsen, E.E., Grønkjaer, P., Meldrup, D., et al. 2005. Retention of juveniles within a hybrid zone between North Sea and Baltic Sea Atlantic cod (*Gadus morhua*). Canadian Journal of Fisheries and Aquatic Sciences, 62: 2219–2225.

Nissling, A., Kryvi, H., and Vallin, L. 1994. Variation in egg buoayancy of Baltic cod *Gadus morhua* and its implications for egg survival in prevailing conditions in the Baltic Sea. Marine Ecology Progress Series, 110: 67–74.

Nissling, A., and Westin, L. 1997. Salinity requirements for successful spawning of Baltic and Belt Sea cod and the potential for cod stock interactions in the Baltic Sea. Marine Ecology Progress Series, 152: 261–271.

Oeberst, R. 2008. Distribution pattern of cod and flounder in the Baltic Sea based on international coordinated trawl surveys. ICES CM, 2008/J:09.

Otterlind, G. 1985. Cod migration and transplantation experiments in the Baltic. Zeitschrift für angewandte Ichthyologie, 1: 3–16.

Pacariz, S., Björk, G., Jonsson, P., et al. 2014a. A model study of the large-scale transport of fish eggs in the Kattegat in relation to egg density. ICES Journal of Marine Science, 71: 345–355.

Pacariz, S., Björk, G., and Svedäng, H. 2014b. Interannual variability in the transport of fish eggs in the Kattegat and Öresund. ICES Journal of Marine Science, 71: 1706–1716.

Petereit, C., Clemmesen, C., Kraus, G., et al. 2004. High resolution temperature influences on egg and early larval development of Baltic cod (*Gadus morhua*). ICES CM, 2004/K:03.

Petereit, C., Hinrichsen, H.H., Franke, A., et al. 2014. Floating along buoyancy levels: Dispersal and survival of western Baltic fish eggs. Progress in Oceanography, 122: 131–152.

Pihl, L., and Ulmestrand, M. 1993. Migration pattern of juvenile cod (*Gadus morhua*) on the Swedish west coast. ICES Journal of Marine Science, 50: 63–70.

Pihl, L., and Wennhage, H. 2002. Structure and diversity of fish assemblages on rocky and soft bottom shores on the Swedish west coast. Journal of Fish Biology, 61: 148–166.

Plambech, M., Van Deurs, M., Steffensen, J.F., et al. 2013. Excess post-hypoxic oxygen consumption in Atlantic cod *Gadus morhua*. Journal of Fish Biology, 83: 396–403.

Receveur, A., Bleil, M., Funk, S., et al. 2022. Western Baltic cod in distress: decline in energy reserves since 1977. ICES Journal of Marine Science, 79: 1187–1201.

Righton, D.A., Andersen, K.H., Neat, F., et al. 2010. Thermal niche of Atlantic cod *Gadus morhua*: Limits, tolerance and optima. Marine Ecology Progress Series, 420: 1–13.

Ryberg, M.P., Skov, P.V., Vendramin, N., et al. 2020. Physiological condition of Eastern Baltic cod, *Gadus morhua*, infected with the parasitic nematode *Contracaecum osculatum*. Conservation Physiology, 8: No. coaa093.

Schinke, H., and Matthäus, W. 1998. On the causes of major Baltic inflows - an analysis of long time series. Continental Shelf Research, 18: 67–97.

Schmidt, B., Wodzinowski, T., and Bulczak, A.I. 2021. Long-term variability of near-bottom oxygen, temperature, and salinity in the Southern Baltic. Journal of Marine Systems, 213: 103462.

Schurmann, H., and Steffensen, J.F. 1997. Effects of temperature, hypoxia and activity on the metabolism of juvenile Atlantic cod. Journal of Fish Biology, 50: 1166–1180.

Sokolova, M., Buchmann, K., Huwer, B., et al. 2018. Spatial patterns in infection of cod *Gadus morhua* with the seal-associated liver worm *Contracaecum osculatum* from the Skagerrak to the central Baltic Sea. Marine Ecology Progress Series, 606: 105–118.

Sparholt, H., Aro, E., and Modin, J. 1991. The spatial distribution of cod (*Gadus morhua* L.) in the Baltic Sea. Dana, 9: 45–56.

Stenseth, N.C., Jorde, P.E., Chan, K.-S., et al. 2006. Ecological and genetic impact of Atlantic cod larval drift in the Skagerrak. Proceedings of the Royal Society B: Biological Sciences, 273: 1085–1092. Støttrup, J G., Lund, H., Munk, P., et al. 2014. Coastfish I. Development in near-coastal fish stocks. DTU Aqua-report, No 281-201: 28. Technical University of Denmark, Lyngby, Denmark.

Svedäng, H., and Bardon, G. 2003. Spatial and temporal aspects of the decline in cod (*Gadus morhua* L.) abundance in the Kattegat and eastern Skagerrak. ICES Journal of Marine Science, 60: 32–37.

Svedäng, H., and Svenson, A. 2006. Cod *Gadus morhua* L. populations as behavioural units: inference from time series on juvenile abundance in the eastern Skagerrak. Journal of Fish Biology, 69: 151–164.

Svedäng, H., Righton, D., and Jonsson, P. 2007. Migratory behaviour of Atlantic cod *Gadus morhua*: natal homing is the prime stock-separating mechanism. Marine Ecology Progress Series, 345: 1–12.

Svedäng, H., André, C., Jonsson, P., et al. 2010. Migratory behaviour and otolith chemistry suggest fine-scale sub-population structure within a genetically homogenous Atlantic cod population. Environmental Biology of Fishes, 89: 383–397.

Svedäng, H., and Hornborg, S. 2017. Historic changes in length distributions of three Baltic cod (*Gadus morhua*) stocks: Evidence of growth retardation. Ecology and Evolution, 7: 6089–6102.

Thurow, F. 1970. Über die Fortpflanzung des Dorsches *Gadus morhua* (L.) in der Kieler Bucht. Berichte der Deutschen Wissenschaftlichen Kommission für Meeresforschung, 21: 170–192.

Tomkiewicz, J., Eriksson, M., Baranova, T., et al. 1997. Maturity ogives and sex rations for Baltic cod: establishment of a database and time series. ICES CM, 1993/C:52, 16 pp.

Tomkiewicz, J., and Köster, F.W. 1999. Maturation process and spawning time of cod in the Bornholm Basin of the Baltic Sea: Preliminary results. ICES CM, 1999/Y:25.

Tomkiewicz, J., Morgan, M.J., Burnett, J., et al. 2003. Available Information for Estimating Reproductive Potential of Northwest Atlantic Groundfish Stocks. Journal of Northwest Atlantic Fisheries Science, 33: 1–21.

Uzars, D., and Plikshs, M. 2000. Cod (*Gadus morhua* L.) cannibalism in the Central Baltic: interannual variability and influence of recruit abundance and distribution. ICES Journal of Marine Science, 57: 324–329.

Uzars, D., Baranova, T., and Yula, E. 2001. Inter annual variation in gonadal maturation of cod in the Gotland Basin of the Baltic Sea: Influence of environment and fish condition. ICES CM, 2001/V:29, 12 pp.

Vainikka, A., Gårdmark, A., Bland, B., et al. 2009. Two- and three-dimensional maturation reaction norms for the eastern Baltic cod, *Gadus morhua*. ICES Journal of Marine Science, 66: 248–257.

Vallin, L., and Nissling, A. 2000. Maternal effects on egg size and egg buoyancy of Baltic cod, *Gadus morhua*: Implications for stock structure effects on recruitment. Fisheries Research, 49: 21–37.

Viktorsson, L. 2018. Hydrography and oxygen in the deep basins. HELCOM Baltic Sea environmental fact sheets, Helsinki, Finland, 7 pp.

Vitale, F., Cardinale, M., and Svedäng, H. 2005. Evaluation of the temporal development of the ovaries in Gadus morhua from the Sound and Kattegat, North Sea. Journal of Fish Biology, 67: 669–683.

Vitale, F., Börjesson, P., Svedäng, H., et al. 2008. The spatial distribution of cod (*Gadus morhua* L.) spawning grounds in the Kattegat, eastern North Sea. Fisheries Research, 90: 36–44.

von Westernhagen, H. 1970. Erbrüten der Eier von Dorsch (*Gadus morhua*), Flunder (*Pleuronectes flesus*) and Scholle (*Pleuronectes platessa*) unter kombinierten Temperatur- und Salzgehaltsbedingungen. Helgoländer wissenschaftliche Meeresuntersuchungen, 21: 21–102.

von Westernhagen, H., Dethlefsen, V., Cameron, P., et al. 1988. Developmental defects in pelagic fish embryos from the western Baltic. Helgoland Marine Research, 42: 13–36.

Westerberg, H. 1994. The transport of cod eggs and larvae through Øresund. ICES CM, 1994/Q:4, 12 pp.

Wieland, K., Waller, U., and Schnack, D. 1994. Development of Baltic cod eggs at different levels of temperature and oxygen content. Dana, 10: 163–177.

Wieland, K., and Jarre-Teichmann, A. 1997. Prediction of vertical distribution and ambient development temperature of Baltic cod, *Gadus morhua* L., eggs. Fisheries Oceanography, 6: 172–187.

Wieland, K., Jarre-Teichmann, A., and Horbowa, K. 2000. Changes in the timing of spawning of Baltic cod: possible causes and implications for recruitment. ICES Journal of Marine Science, 57: 452–464.

Appendix 1

ICES Advice Sheets: Available at: https://www.ices.dk/advice/Pages/Latest-Advice.aspx

ICES survey data base DATRAS: Available at: https://www.ices.dk/data/data-portals/Pages/DATRAS.aspx

CHAPTER 9

Northeast Arctic Cod Stock

*Daniel Howell** and *Bjarte Bogstad*

Biology and Ecology

The Northeast Arctic cod stock (NEA cod, *Gadus morhua*) is currently the world's largest cod stock. The ICES term 'Northeast Arctic cod' is mostly used for this stock nowadays; previously, 'Arcto-Norwegian cod' was the most commonly used name. 'Barents Sea cod' is also used. We will use the ICES term 'Northeast Arctic cod' in this chapter. On the biological side, this chapter relies heavily on Yaragina et al. (2011), updated with more recent stock developments, while the management section draws on Gullestad et al. (2018).

Stock Distribution and Migration

The nursery and feeding areas of NEA cod are in the Barents Sea, with the main spawning grounds in the Lofoten Islands off the coast of Norway (Figure 1). The NEA cod is at the northern extreme of cod distributions globally thanks to a significant influx of warm water from the North Atlantic Current. Physically, the Barents Sea is characterised by a series of shallow (100–200 m) banks separated by channels up to 400–500 m deep. The oceanographical conditions are driven by the polar front, which separates warmer Atlantic water in the south and west from colder Arctic water in the north and east. During the summer, the Barents Sea is ice-free; in the winter, the northern and eastern regions are ice-covered, while the south and west remain largely ice-free.

Northeast Arctic cod is widely distributed along the Norwegian coast and throughout the Barents Sea as far north as 81° N (north of Svalbard) and occasionally into the northwestern part of the Kara Sea (Figure 1). The maximum distribution to the north and east in the Barents Sea is reached in August-September when the Barents Sea is ice-free, and the cod can spread across the whole Barents Sea. In

Institute of Marine Research, P.O. box 1870 Nordnes, NO-5817 Bergen, Norway.
* Corresponding author: daniel.howell@hi.no

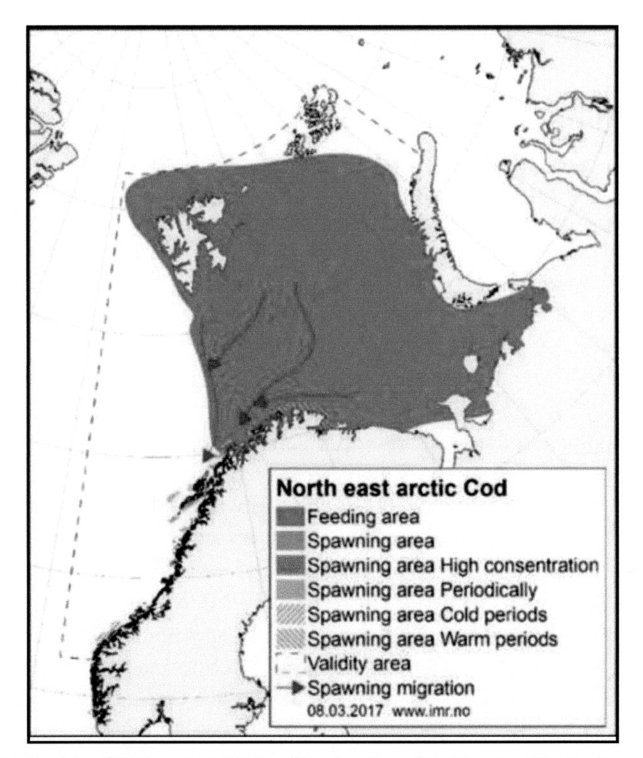

Figure 1. Distribution of NEA cod, Institute of Marine Research, Norway. The feeding area (blue) is in the Barents Sea and around Svalbard. Spawning areas (orange) along the Norwegian coast, predominantly near the Lofotens.

winter, the distribution area is considerably smaller and concentrated in the ice-free areas in the south and west, as cod is normally not distributed in ice-covered areas. Off the west and northwest coast of Svalbard, cod has in recent years been found in January-February with a similar distribution as in August-September, in 2021 cod was found as far east along the north coast of Svalbard as 81°23' N 25°17' E (Fall et al. 2022).

The southernmost distribution is found during spawning, which in recent years mainly has taken place in Norwegian coastal areas between 67°30' N and 70° N (Lofoten-Sørøya), but spawning has been observed as far south as 60° N (off the island of Sotra close to Bergen) and as far northeast as Motovsky Bay (west of the Kola Bay). Spawning has extended south during colder periods during the 1900s (Sundby and Nakken 2008).

NEA cod is mainly found in temperatures ranging from 0–7°C, but during feeding migrations in summer and autumn, it can be found at temperatures below 0°C (Righton et al. 2010). Cods occur at the bottom, mainly at depths of 100–300 m, but they can be found down to 600 m. They can be found over even deeper waters close to the shore, as described by Ingvaldsen et al. (2018).

Eggs and larvae drift passively with the currents from the spawning grounds, predominantly in northern and eastern directions. In August-September, they are distributed as a 0-group throughout most of the area on the warm side of the Polar Front, typically with the largest concentrations in the central Barents Sea (Eriksen et al. 2012). In September-October, the 0-group settles to the bottom.

Stock Identity—Cod Stocks in Nearby Areas

The coastline of Norway consists of numerous fjords that favour the formation of a complex system of cod subpopulations. In the Barents Sea region and along the coast of Norway south to 62° N, there are two main types of cod: Norwegian Coastal cod (NCC) and NEA cod. These species have spatial overlap in coastal areas. The degree of intermixing of these groups is little known.

Rollefsen (1933, 1935) observed that the otoliths of coastal cod differed from those of NEA cod. Using otolith shape for population structure analysis (Stransky et al. 2008) and other studies, including SNP analysis, suggest that NCC consists of several separate populations (e.g., Salvanes et al. 2004; Breistein et al. 2022).

It is believed that the exchange of individuals with other stocks than the coastal cod stocks is minimal, although tagging experiments and genetic studies have shown this to be possible. Tåning (1934) reported that of 4939 cod tagged in 1924–1925 off the SW coast of Iceland, two were recaptured off Northern Norway. Andrews et al. (2019) report that in total, ten young cod (all specimens < 1 kg) were caught on the shelf off Northeast Greenland and in the Fram Strait in the period 2007–2017. Genetic analyses showed these to be NEA cod. These observations are much further north than previous reports of cod off Northeast Greenland. Most likely, they have drifted northwards along the west coast of Svalbard as 0-group and then entered a branch of the current, which goes westwards to Northeast Greenland. Also, some cod specimens fished around Jan Mayen in 2019–2020 were similar to NEA cod when using otoliths and genetics to identify cod type (Bogstad 2023).

There is a separate sub-species of cod (*Gadus morhua marisalbi*) in the nearby White Sea, which likely is not connected with NEA cod and has a shorter life span, smaller body size, earlier maturation and is adapted to lower temperatures than NEA cod (see, e.g., Stroganov 2015). It is currently unknown whether cod found in the southwestern part of the Kara Sea and fjords on the eastern coast of the Southern Island of Novaya Zemlya (e.g., Heldal et al. 2018) belong to separate stocks.

Growth

NEA cod grow rapidly in their first year of life: in August-September, the average length is about 7 cm (Ottersen et al. 2014). The length of growth of immature cod is approximately linear, and the average annual length increments of age 1–7 cod are 8–10 cm. Thus, length at age in cm at the beginning of the year (survey data from January–March) is approximately 10* age up to age 7. Length growth slows after maturation but continues throughout life. Kvamme and Bogstad (2007) used the following length/weight relationship in their studies based on available data: $W = 7.904 * 10^{-6} * L^{3.03}$

Here, weight W is in kg and length L is in cm.

Weight is, however, also impacted by environmental conditions. During the collapse of capelin (the main forage fish in the Barents Sea) during the 1980s, cod weight and condition declined significantly. More recently, high-stock biomasses led to reduced weight at the age of the larger cod due to density-dependent competition for food (Bogstad et al. 2015).

The longest and oldest specimen recorded was 169 cm and 24 years, caught in June 1940 (Boitsov et al. 1996). This fish had a fairly low weight (40 kg) in relation to length, as it was caught a couple of months after spawning. The largest fish recorded in recent years was a specimen weighing 55.5 kg (of which 14 kg was roe) and aged 20 years caught off the coast of Finnmark in 2018 (IMR 2018). The length of this specimen was not recorded.

Maturation and Reproduction

The sex ratio of immature cod is approximately 1:1. However, males mature approximately one year earlier than females (Ajiad et al. 1999). Males thus dominate amongst mature cod up to age 8, while females prevail among larger and older cod; the oldest age groups and the largest fish are mainly females (Ajiad et al. 1999; Marshall et al. 2006).

From the 1940s to the 1980s, the age at 50% maturation shifted more than two years downwards from 9–10 to 7 years (Rørvik et al. 2022). The mean length at first maturation was reduced by 12 cm in the same period (Yaragina 2006) and is now around 65 cm for males and 75 cm for females. Since the 1980s, however, maturation parameters have remained relatively stable.

Cod fecundity ranges from 117,000 to 19.2 million eggs per female. The mean relative fecundity is around 600 eggs g^{-1}. The ratio between total egg biomass and body weight increases as body weight increases. The peak spawning time in the Lofotens is around 1 April. The egg development takes around 25 days, and the duration of the larval period from hatching to metamorphosis is about two months (Otterlei et al. 1999), by which time the fry has all the morphological characteristics of adult cod.

NEA cod recruitment is highly variable and positively related to temperature, although the correlation with temperature varies over time (Bogstad et al. 2013). Recruitment was found to be low at low temperatures and variable at medium/high temperatures during the first year of life. The temperature during the first winter of life correlates positively with cod recruitment residuals. This correlation is weakened towards the end of the period 1913–2007. The relationship between spawning stock and recruitment is fairly weak (Figure 2), but there is a concentration of points in the lower left corner of the stock-recruitment plot, indicating that the recruitment has been negatively affected by a low spawning stock in some years. The reference point B_{lim}, below which recruitment is considered to be impaired, has been calculated to be 220,000 tonnes (ICES 2003).

Year-class strength is for this stock, as for many other stocks, thought to be established during the first half-year of life. An analysis by Bogstad et al. (2016), however, gives a more nuanced view of this issue. They analysed data from the

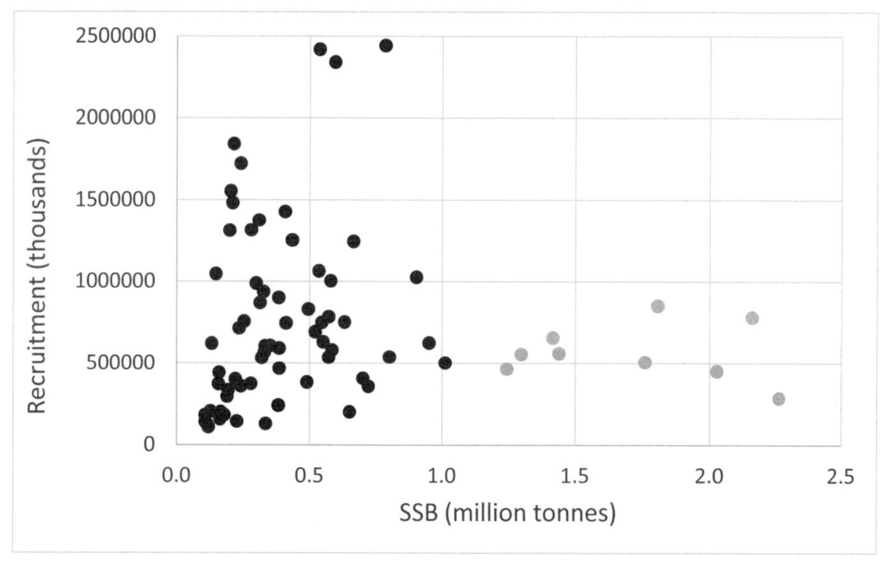

Figure 2. NEA cod stock-recruitment relationship at age 3 (ICES 2021a). Grey dots indicate cohorts 2010–2018.

1983–2009 year-classes, where comprehensive data from total egg production (TEP), surveys, and stock assessments were available to examine these cohorts thoroughly. On average, only six out of 1 million new generations at the TEP stage reach the age of recruitment to the fishery. The between-cohort variability in abundance is greatest at the ages 0–1 stage. Although the mortality is highest during the first months of life, the year-class strength can also be affected considerably by processes taking place between the 0-group stage (6 months) and age 3. The mortality in this period of life seems to be strongly density-dependent, and cannibalism is an important source of mortality.

Natural Mortality

The instantaneous natural mortality rate M for NEA cod is assumed to be a base 0.2 year^{-1} with cannibalism mortality in addition for ages 3–6 in the stock assessment (based on diet data, Yaragina et al. 2009, 2018; ICES 2021a). Also, cannibalism is assumed to be the main source of mortality for age 1–3 cod. However, investigations by Tretyak (1984) indicate that natural mortality is age-dependent and has a minimum at age 8–9 years, which on average of ages 3–15 is lower than 0.2.

Feeding

Small cod feed mainly on crustaceans, while between 20 and 30 cm NEA cod becomes piscivorous, with capelin as the main prey. The diet of NEA cod has been extensively studied since the 1930s, and there is a large body of literature on this topic. The most recent review of the cod diet was made by Holt et al. (2019). The role of cod in the Barents Sea ecosystem, including the cod diet, is discussed in the chapter on trophic interactions.

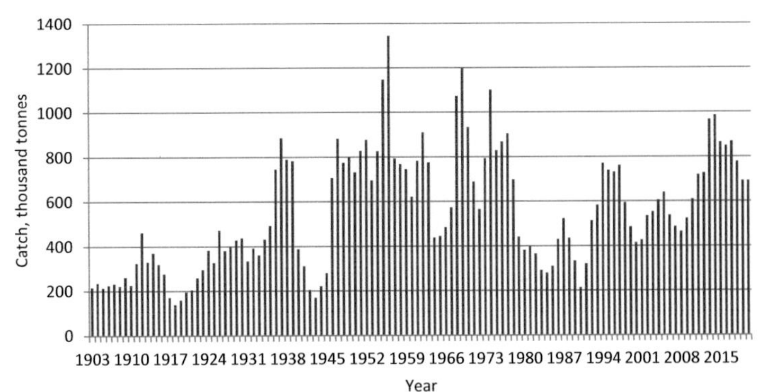

Figure 3. Northeast Arctic cod catches 1903–2020 (ICES 2021a).

Stock and Catch History

The NEA cod fishery has existed since at least Viking times, with an export fishery from Norway documented over 1000 years ago (Anon c. 1240). By the Middle Ages, the export of dried cod and herring was thriving through the Hansa trading network. Cod fishing in the coastal zone of Murman was first mentioned in the 15th century when the first settlements of Novgorodians, who used the cod longline for fishing, appeared here. In the 18th century, Russian fishers worked not only off the coast of Murman but also off Bear Island and on Grumant (Svalbard), where there were up to 20 camps (Boitsov et al. 1996). The Russian fisheries statistics off the coast of Murman go back to 1880, and the Murmansk fishery averaged over 1880–1899, giving annually 9–10 thousand tonnes of cod (Maslov 1944).

As a result of this history, active research has been ongoing for over 100 years, and the assessment of this stock extends back to 1946 (ICES 2021a), and a tentative assessment has been made back to 1913 (Hylen 2002). Catches increased in the 1930s (Figure 3) due to favourable climatic conditions, which caused the stock to increase, as well as the introduction of large-scale trawl fisheries. Following severe limitations of fishing activities during World War II, stock size peaked by the end of the war. Increased fishing activity and catches (peaking at 1.3 million tonnes) following the Second World War led to a decline in the stock, which remained at a somewhat low (but not collapsed) state until the 1990s (Figure 3–5). The decline was slowed by occasional strong year classes (Figure 6). The stock exhibited other signs of high fishing pressures during this period, with earlier maturity and a truncated age structure (Figure 7). The stock size increased again in the early 1990s following a period with good recruitment but was then reduced again due to renewed high fishing pressure. From 2007 onwards, the stock size increased considerably due to strong year classes in 2004 and 2005 (Figure 6) and low fishing pressure following the introduction of a harvest control rule (Kjesbu et al. 2014). In 2013, the total stock size was around the level at the end of World War II, while the spawning stock size was at an all-time high due to the earlier maturation. The age composition of the stock has also been rebuilt and is now similar to the 1940s (Figure 7), but age and size at maturation have not changed. It has, however, been suggested that the genetic

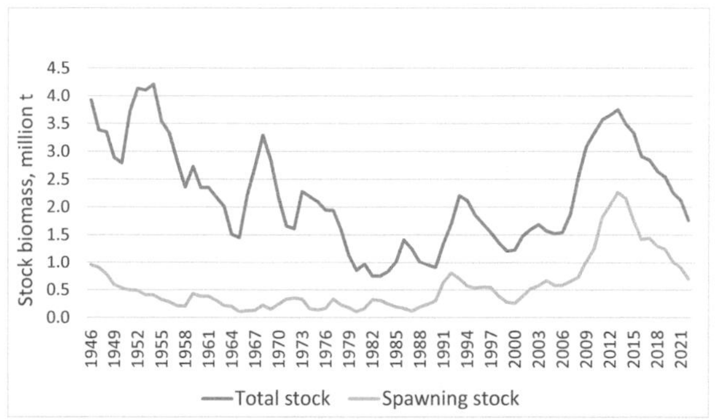

Figure 4. Northeast Arctic cod stock development 1946–2021. Total stock (age 3+) and spawning stock 1946–2022 (ICES 2021a).

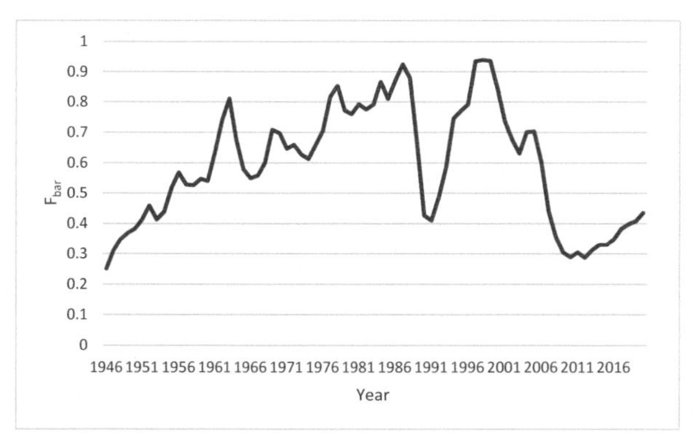

Figure 5. Northeast Arctic cod fishing mortality (ages 5–10) 1946-–2020 (ICES, 2021a).

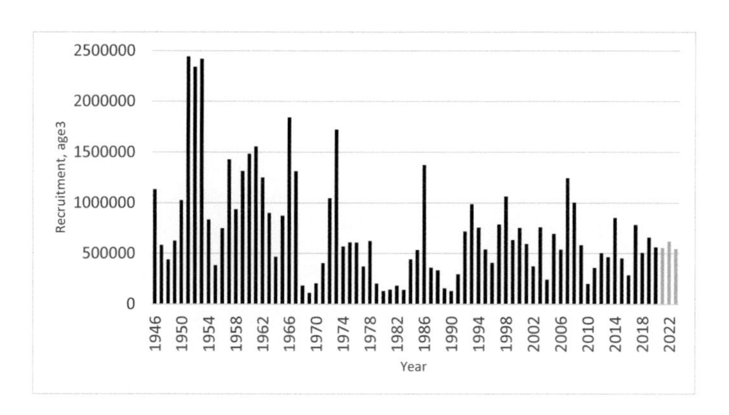

Figure 6. NEA cod recruitment (age 3). Data from ICES 2021a. Grey indicates forecasts of recruits not yet at age 3.

composition of the stock has been changed due to a long period of high fishing pressure (Rørvik et al. 2022). After 2013, the stock size decreased due to recruitment being below average. The fishing mortality has increased slightly but did not exceed

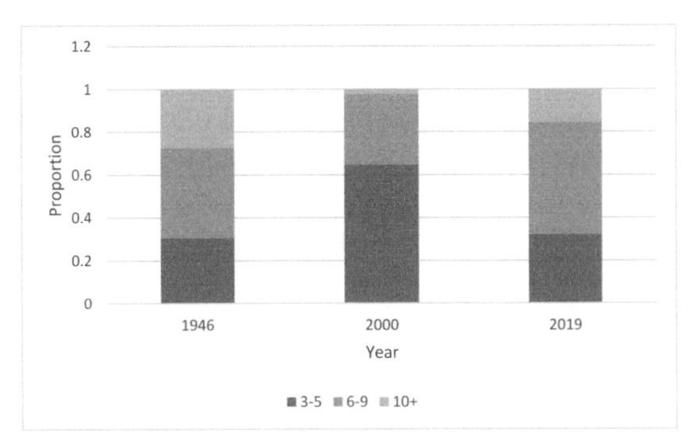

Figure 7. Age composition of the cod stock (biomass) in 1946, 2000 and 2019 (Howell et al. 2023).

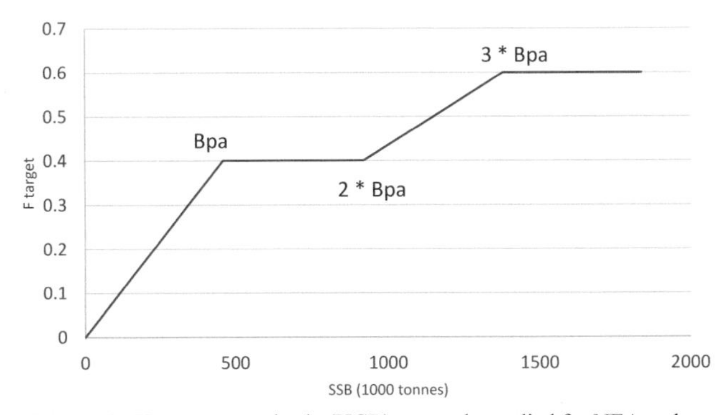

Figure 8: Harvest control rule (HCR) currently applied for NEA cod.

the 0.4 level at the first plateau in the harvest control rule (HCR; Figure 8) until 2020. The stock distribution area has also decreased due to either lower temperatures or reduced adult biomass (ICES 2021c).

Fisheries Management

Until the 1970s, the fishery was operated as open access, with limited government ability to regulate the fishery. The only restrictions were on trawl gears to protect local fisheries from competition from larger vessels (Gullestad et al. 2018). The collapse of the herring fishery in the 1960s prompted the Norwegian government to take action, and in 1972, the 'Act on Participation in Fisheries' gave the Norwegian government the ability to limit access to fisheries when this was required to protect the fish resources. With subsequent amendments and revisions, this law remains the basis of Norwegian fisheries management. Initial implementation of these powers was focussed on the herring stock, with reductions in fishing pressure for NEA cod not being implemented until the 1990s. In 1977, Norway (along with other coastal states) extended their exclusive economic zone, resulting in a decline in catches from distant waters fleets (especially from the UK and Germany, Figure 3) and transforming the fishery into a largely Norwegian and Russian one. The Joint Norwegian-Russian Fisheries Commission (JNRFC) was established in 1975 between Norway and the USSR (Russia since 1991) to manage the fisheries. The share of catches between the two countries (50–50) was established at this time and has remained stable ever since, thus avoiding the often contentious issue of zonal attachments and quota sharing. Other ('third') countries have, through various agreements, obtained a share of the TAC; this share is currently around 14%, meaning that the TAC is split approximately as 43–43–14% between Norway, Russia and third countries. It is also important that the fishermen of the two countries have the right to fish for cod (up to 60–70% of the quota) in each other's zones.

The central feature of NEA cod management has been maintaining moderate fishing pressure since the early 2000s, with fishing pressure based on the HCR since 2002. The rule was revised in 2009 and again in 2017. The final revision accounts for the fact that the management has been successful in protecting the large-year classes discussed above and that the stock was, therefore, entering a region of reduced productivity due to density-dependent effects. The current rule employs a 'double hockey stick' methodology. Allowable fishing pressure rises from F= 0 at 0 SSB to F = 0.4 at Bpa. F remains at 0.4 between Bpa and 2 * Bpa but then rises linearly to F = 0.6 at and above 3 * Bpa. This allows for higher fishing pressure at very high stock sizes where density-dependent effects are likely to reduce productivity, combined with a more moderate fishing pressure when the stock is closer to the presumed Bmsy levels (ICES 2021a; Gullestad et al. 2018).

In addition to efforts to control overall fishing pressure, a long series of measures have been aimed to reduce the catch of small fish and thus maximise the overall yield. As well as technical regulations on fishing gear and grounds, there are agreements on international access to facilitate this. Cods found in Russian waters are typically younger and smaller than those found further west; as early as 1977, Russian vessels were therefore allowed to fish their quotas in Norwegian waters outside 12 nautical miles. As a result of the different sizes of fish, there has been tension between pressure for different gear regulations in the two countries and a desire for uniformity across the fishery. A series of changes within Norway and agreed bilaterally has led to a compromise accepted across the two countries (full details in Gullestad et al. 2018). The stock is currently fished with 130 mm mesh, a 55 mm sorting grid, and a minimum landing size of 44 cm, with these regulations being stable since 2011.

However, gear regulations alone do not prevent unwanted bycatch of undersized fish or high grading resulting in discards. The fishery is therefore regulated with a regime of area closures, both permanent and Real-Time Area closures on regions reporting high catches of undersized fish. Individual vessels must move to fish elsewhere if they get more than 15% of undersized fish by number in a haul, and managers monitor the fishery through inspections and are free to close an area as required. This dynamic management has successfully diverted fishing efforts away from concentrations of young fish (Gullestad et al. 2015).

Since the early 1990s, ICES assessed the Norwegian coastal cod (NCC) separately from the NEA cod stock, but separate advice was given for the first time in 2001. In 2021, the NCC stock was split into two parts for assessment purposes (ICES 2021b), one for the area between 62° and 67° N and one for the area north of 67° N.

Cod in the Barents Sea and coastal cod along the Norwegian coast south to 62° N are managed together, although they are assessed as different units. From the mid-1970s to 2003, an expected catch of 40,000 tonnes of NCC was added annually to the quota for NEA. In 2004, this quantity was set to 20,000 tonnes; from 2005 onwards, it has been 21,000 tonnes. However, actual catches of NCC have, from 2004 onwards, been higher than the quantity added to the NEA cod quota.

References

Anon. c. 1240. Egils saga Skallagrímssonar. Stofnun Árna Magnússonar, University of Reykjavik, Iceland.

Ajiad, A.M., Jakobsen, T., and Nakken, O. 1999. Sexual differences in maturation of Northeast Arctic cod. Journal of Northwest Atlantic Fishery Science, 25: 1–15.

Andrews, A.J., Christiansen, J.S., Bhat, S., et al. 2019. Boreal marine fauna from the Barents Sea disperse to Arctic Northeast Greenland. Sci. Rep., 9, 5799. https://doi.org/10.1038/s41598-019-42097-xDOI: 10.1101/394346.

Boitsov, V.D., Lebed, N.I., Ponomarenko, V.P., et al. 1996. The Barents Sea Cod: Biological and Fisheries Outline. PINRO press, Murmansk, 285 pp. (In Russian).

Bogstad, B., Dingsør, G.E., Gjøsæter, H., et al. 2013. Changes in the relations between oceanographic conditions and recruitment of cod, haddock and herring in the Barents Sea. Marine Biology Research 9(9): 895–907.

Bogstad, B., Yaragina, N.A., and Nash, R.D.M. 2016. The early life-history dynamics of Northeast Arctic cod: levels of natural mortality and abundance during the first three years of life. Canadian Journal of Fisheries and Aquatic Science, 73(2): 246–256. Doi:10.1139/cjfas-2015-0093.

Bogstad, B., Gjøsæter, H., Haug, T., et al. 2015. A review of the battle for food in the Barents Sea: Cod vs. marine mammals. Frontiers in Ecology and Evolution, 3. doi: 10.3389/fevo.2015.00029 10.3389/fevo.2015.00029.

Bogstad, B. 2023. Jan Mayen—a new spawning and fishing area for Atlantic cod *Gadus morhua*. Polar Biology, 46: 103–109. https://doi.org/10.1007/s00300-022-03102-8.

Breistein, B., Dahle, G., Johansen, T., et al. 2022. Geographic variation in gene flow from a genetically distinct migratory ecotype drives population genetic structure of coastal Atlantic cod (*Gadus morhua* L.). Evolutionary Applications, 15: 1162–1176. https://doi.org/10.1111/eva.13422.

Eriksen, E., Ingvaldsen, R., Stiansen, J.E., et al. 2012. Thermal habitat for 0-group fish in the Barents Sea; how climate variability impacts their density, length, and geographic distribution. ICES Journal of Marine Science, 69: 870–879.

Fall, J., Wenneck, T., de Lange, et al. 2022. Fish investigations in the Barents Sea winter 2021. IMR-PINRO Joint Report Series 1-2022, 100 pp.

Gullestad, P., Blom, G., Bakke, G., et al. 2015. The "Discard Ban Package": experiences in efforts to improve the exploitation pattern in Norwegian fisheries. Marine Policy, 54(5): 1–9. https://doi.org/10.1016/j.marpol.2014.09.025.

Gullestad, P., Howell, D., Stenevik, E.K., et al. 2018. Management and rebuilding of herring and cod in the Northeast Atlantic. pp. 12–37. *In:* Garcia, S., and Ye, Y. (eds.). Rebuilding of Marine Fisheries. FAO Fisheries and Aquaculture Technical Paper, 630, 2. Rome, Italy: FAO.

Heldal, H.E., Bogstad, B., Dolgov, A.V., et al. 2018. Observations of biota in Stepovogo Fjord, Novaya Zemlya, a former dumping site for radioactive waste. Polar Biology, 41(1): 115–124, Doi: 10.1007/s00300-017-2175-3.

Holt, R.E., Bogstad, B., Durant, J.M., et al. 2019. Barents Sea cod (*Gadus morhua*) diet composition: long-term interannual, seasonal, and ontogenetic patterns. ICES Journal of Marine Science 76(6): 1641–1652, doi:10.1093/icesjms/fsz082.

Howell, D., Bogstad, B., Chetyrkin, A. et al. 2023. Report of the Joint Russian-Norwegian Working Group on Arctic Fisheries (JRN-AFWG) 2023. IMR-PINRO Report Series 7-2023, 213 pp.

Hylen, A. 2002. Fluctuations in the abundance of Northeast Arctic cod during the 20th century. ICES Mar. Sci. Symp. 215: 543–550.

ICES 2003. ICES C. M. 2003/ACFM:11.

ICES 2021a. Arctic Fisheries Working Group (AFWG). ICES Scientific Reports 3: 58. 817 pp. https://doi.org/10.17895/ices.pub.8196.

ICES. 2021b. Benchmark Workshop for Barents Sea and Faroese Stocks (WKBARFAR 2021). ICES Scientific Reports. 3: 21. 205 pp. https://doi.org/10.17895/ices.pub.7920.

ICES 2021c. Working Group on the Integrated Assessments of the Barents Sea (WGIBAR). ICES Scientific Reports. 3: 77. 236 pp. https://doi.org/10.17895/ices.pub.8241.

Ingvaldsen, R.B., Gjøsæter, H., Ona, E., et al. 2018. Atlantic cod (*Gadus morhua*) feeding over deep water in the high Arctic. Polar Biology, 40: 2105–2111.

IMR. 2018. https://www.hi.no/hi/nyheter/2018/juni/slik-bestemte-de-alderen-til-kjempetorsken.

Kjesbu, O.S., Bogstad, B., Devine, J.A., et al. 2014. Synergies between climate and management for Atlantic cod fisheries at high latitudes. Proceedings National Academy of Science, 111(9): 3478–3483.

Kvamme, C., and Bogstad, B. 2007. The effect of including length structure in yield-per-recruit estimates for northeast Arctic cod. ICES Journal of Marine Science, 64: 357–368.

Marshall, C.T., Needle, C.T., Thorsen, A., et al. 2006. Systematic bias in estimates of reproductive potential of cod stocks: implication for stock/recruit theory and management. Canadian Journal of Fisheries and Aquatic Sciences, 63: 980–994.

Maslov, N.A. 1944. The demersal fishes of the Barents Sea and their fisheries. Trudy PINRO, 8: 3–186. (In Russian).

Otterlei, E., Nyhammer, G., Folkvord, A., et al. 1999. Temperature- and size-dependent growth of larval and early juvenile Atlantic cod (*Gadus morhua*): a comparative study of Norwegian coastal cod and northeast Arctic cod. Canadian Journal of Fisheries and Aquatic Sciences, 56(11): 2099–2111. https://doi.org/10.1139/f99-168.

Ottersen, G., Bogstad, B., Yaragina, N.A., et al. 2014. A review of early life history dynamics of Barents Sea cod (*Gadus morhua*). ICES Journal of Marine Science, 71(8): 2064–2087.

Righton, D.A., Andersen, K.H., Neat, F., et al. 2010. Thermal niche of Atlantic cod *Gadus morhua*: limits, tolerance and optima. Marine Ecology Progress Series, 420: 1–13.

Rollefsen, G. 1933. The otoliths of cod. Fiskeridirektoratets skrifter, serie Havundersøkelser, 4(3): 1–14.

Rollefsen, G. 1935. The spawning zone in cod otoliths and prognosis of the stock. Fiskeridirektoratets skrifter, serie Havundersøkelser, 4(11): 3–10.

Rørvik, C.J., Bogstad, B., Ottersen, G. et al. 2022. Long-term interplay between harvest regimes and biophysical conditions may lead to persistent changes in age-at-sexual maturity of Northeast Arctic cod (*Gadus morhua*). Canadian Journal Fisheries and Aquatic Sciences, 79(4): 576–586. https://cdnsciencepub.com/doi/full/10.1139/cjfas-2021-0068.

Salvanes, A.G.V., Skjæraasen, J.E., and Nilsen, T. 2004. Sub-populations of coastal cod with different behavior and life-history strategies. Marine Ecology Progress Series, 267: 241–251.

Stransky, C., Baumann, H., Fevolden, S-E., et al. 2008. Separation of Norwegian coastal cod and Northeast Arctic cod by outer otolith shape analysis. Fisheries Research, 90: 26–35.

Stroganov, A.N. 2015. Genus Gadus (Gadidae): Composition, Distribution and Evolution of Forms. Journal of Ichthyology, 55(3): 319–336.

Sundby, S., and Nakken, O. 2008. Spatial shifts in spawning habitats of Arcto-Norwegian cod related to multidecadal climate oscillations and climate change. ICES Journal of Marine Science, 65: 953–962.

Tretyak, V.L. 1984. A method of estimating the natural mortality of cod at different ages (exemplified by the Arcto-Norwegian cod stock). pp. 241–274. *In:* Godø, O.R. and Tilseth, S. (eds.). Reproduction and Recruitment of Arctic cod. The Proceedings of the Soviet-Norwegian symposium, Leningrad, 26–30 September 1983. Institute of Marine Research, Bergen, Norway.

Tåning, A.V. 1934. Survey of long distance migrations of cod in the North Western Atlantic according to marking experiments. Rapp. P-v Reun int Cons Explor. Mer., 9: 5–11.

Yaragina, N.A. 2006. Biology of reproduction of Atlantic cod (on example of Barents Sea cod population). Synopsis of thesis. Doctor of biological sciences. Petroskoi. 47 pp. (In Russian).

Yaragina, N.A., Bogstad, B., and Kovalev, Yu. A. 2009. Variability in cannibalism in Northeast Arctic cod (*Gadus morhua*) during the period 1947–2006. *In:* Haug, T., Røttingen, I., Gjøsæter, H., et al. (Guest eds.). Fifty Years of Norwegian-Russian Collaboration in Marine Research. Thematic issue No. 2, Marine Biology Research, 5(1): 75–85. Doi: 10.1080/17451000802512739.

Yaragina, N.A., Aglen, A., and Sokolov, K.M. 2011. Cod. Chapter 5.4. pp. 225–270 *In:* Jakobsen, T., and Ozhigin, V.K. (eds.). The Barents Sea. Ecosystem, Resources, Management. Half a century of Russian-Norwegian cooperation. Tapir Academic Press, Trondheim, Norway.

Yaragina, N.A., Kovalev, Y., and Chetyrkin, A. 2018. Extrapolating predation mortalities back in time: an example from North-east Arctic cod cannibalism, Marine Biology Research, 14(2): 203–216, DOI: 10.1080/17451000.2017.1396342.

Chapter 10

Comparison of the Atlantic Cod Stocks

Biology, Fisheries, and Management

Arni Magnusson,[1,*] *Ingibjörg G. Jónsdóttir,*[2] *Jacob M. Kasper*[2,3,4] and *Peter J. Wright*[5]

Introduction

Atlantic cod (*Gadus morhua*) are found across the shelf sea areas of the North Atlantic Ocean from North Carolina and the English Channel in the south to Greenland and Barents Sea in the north. Throughout much of this range, cod are the most abundant mid-sized piscivores and are one of the keystone species in the ecosystem (Bogstad 2024). Cods are found from nearshore to off the edge of continental shelf down to 600 m (Righton et al. 2010). They occupy temperatures from –1.5 to 20°C (Righton et al. 2010; Le Bris et al. 2013), although, during the spawning phase, cod experience a narrower temperature range (Righton et al. 2010; Morgan et al. 2013). Variation in environmental conditions has a large impact on cod development and survival.

Across their physical range, two ecotypes are commonly found, migratory and resident, which were first defined based on tag-recapture data (Robichaud and Rose 2004) but have since been found to reflect a genetic basis (Berg et al. 2017; Matschiner et al. 2022). While some cod stocks are composed of a single ecotype, such as the migratory Northeast Arctic cod, many stocks are comprised of both ecotypes, such as the Icelandic cod stock (Pálsson and Thorsteinsson 2003; Matschiner et al. 2022;

[1] Oceanic Fisheries Programme, The Pacific Community, Nouméa, New Caledonia.
[2] Marine and Freshwater Research Institute, Fornubúðum 5, 220 Hafnarfjörður, Iceland.
[3] Faculty of Agricultural Sciences, Agricultural University of Iceland, Keldnaholt, Árleynir 22, 112, Reykjavík, Iceland.
[4] Department of Ecology and Evolutionary Biology, University of Connecticut, 75 North Eagleville Road, Unit 3043, Storrs, Connecticut 06269-3043, USA.
[5] Marine Ecology and Conservation Consultancy. Ellon, AB41 8YH, UK.
* Corresponding author: arnim@spc.int

Pampoulie et al. 2023). The migratory ecotype can move thousands of kilometres between spawning and feeding grounds (Rose 1993; Neuenfeld et al. 2013), while some resident ecotypes have a home range of 10–100 km (Neat et al. 2014). While cod ecotypes overlap seasonally, they can be quite dispersed in other seasons and thus experience significant temperature differences. For example, Le Bris et al. (2013) found that in the Gulf of St. Lawrence, residents displayed a prolonged period of immobility in near-freezing (–1.5°C) shallow (< 100 m) coastal waters of western Newfoundland while migratory cod overwintered in relatively deep (300–500 m) and warm (5°C) waters. Conversely, migratory northern stock ecotypes (e.g., Icelandic) seek polar fronts during feeding season and experience far colder temperatures than the resident ecotype (Pálsson and Thorsteinsson 2003).

Due to the commercial importance across its range, fishing nations regularly monitor cod with annual demersal surveys that provide fisheries-independent information on abundance, age, size, growth, maturity, and recruitment. Furthermore, fisheries-dependent data is collected throughout the year from commercial vessels, typically including landings and effort (measured as, e.g., days at sea or towing time), a measurement which can be used to calculate catch-per-unit-effort (CPUE) and follow changes in stock status. These data are used to provide advice for the cod stocks in the International Council for the Exploration of the Sea (ICES) region (ICES 2023e) as well as within the Northwest Atlantic Fisheries Organisation (NAFO) region (NAFO 2023b) in the Northwest Atlantic.

In this chapter, fisheries-independent and fisheries-dependent parameters of different cod stocks in the North Atlantic are compared. We include cod stocks ranging from the northernmost stock in the Barents Sea to the southernmost stock on Georges Bank. For each of the parameters, we describe similarities and differences and possible causes for the observed differences. The quantitative data come primarily from two sources: stock assessment tables compiled by the ICES Methods Working Group (ICES 2023f; Magnusson et al. 2019) (see citations in Table 2) and ICES/NAFO advice sheets (see citations in Table 3). All cod stocks on both sides of the Atlantic, where age-specific data and stock assessment results were available, are included in this comparison chapter. The stock codes and names of the 15 stocks are listed in Table 1.

Biological Factors

Growth

The biological state of cod stocks is assessed based on numbers and weight at age. There are substantial differences in weight at age among stocks that are generally greater than intra-stock differences. Interannual variation in growth can have a severe impact on the estimate of stock size. Cod weight at age varies considerably among the stocks (Figures 1 and 3b; Table 2). Cods from southern stocks, such as the North Sea and adjacent areas, along with New England, experience higher rates of growth and mortality than northern stocks, rarely exceeding the age of 6–7 years compared to > 14 in Icelandic and Northeast Arctic stocks. However, the magnitude of present-day stock differences may reflect long-term fishing pressure as the historical record of North Sea cod includes cod aged > 20 years (Graham 1934).

Table 1. The 15 cod stocks compared in this chapter.

Stock Code	Short Name	Full Name
ICES 21.1	Greenland West	West Greenland inshore cod
ICES 27.1-2	Northeast Arctic	Northeast Arctic cod
ICES 27.5a	Iceland	Icelandic cod
ICES 27.5b1	Faroe Plateau	Faroe Plateau cod
ICES 27.1-2coast	Norway	Norwegian coastal waters cod
ICES 27.47d20	North Sea	North Sea, eastern English Channel, and Skagerrak cod
ICES 27.7a	Irish Sea	Irish Sea cod
ICES 27.7e-k	Southern Celtic	Western English Channel and sSouthern Celtic Seas cod
ICES 27.22-24	Western Baltic	Western Baltic Sea cod
ICES 27.24-32	Eastern Baltic	Eastern Baltic Sea cod
NAFO 2J3KL	Northern	Northern cod
NAFO 3M	Flemish Cap	Flemish Cap cod
NAFO 3NO	Grand Bank	Grand Bank cod
NOAA GM	Gulf of Maine	Gulf of Maine cod
NOAA GB	Georges Bank	Georges Bank cod

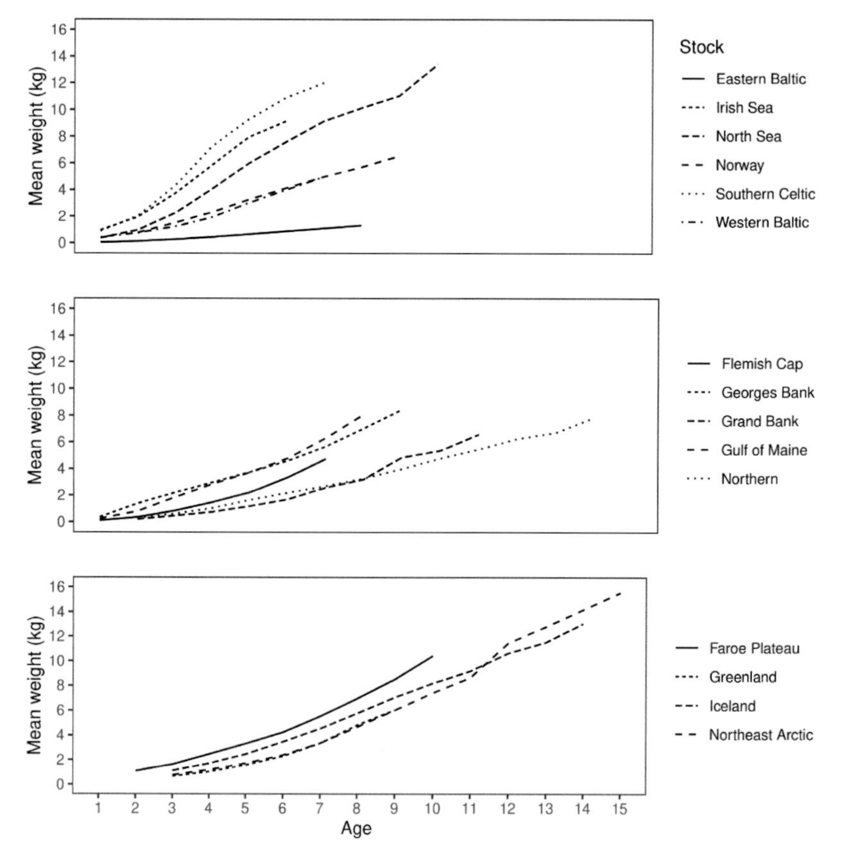

Figure 1. Mean weight at age in catches from the 15 cod stocks. The weights are averaged over the last decade, indicated in Table 2.

Table 2. Quantitative comparison of the cod stocks, averaged over the last decade in recently published stock assessments. The average annual catch is measured in tonnes and the weight a age 5 is measured in kg. The data come from stock assessment reports indicated in the last column.

Stock	Decade	Annual catch	Average Age in the Catch	Age at 50% Maturity	Weight at Age 5	Source
Greenland	2008–2017	17100	5.1	4.3	1.6	ICES (2018c) NWWG
Northeast Arctic	2008–2017	758000	6.7	7.0	1.2	ICES (2018b) AFWG
Iceland	2008–2017	204000	5.9	6.4	2.7	ICES (2018c) NWWG
Faroe Plateau	2008–2017	7600	3.9	2.7	3.3	ICES (2018c) NWWG
Norway	2008–2017	44400	5.4	5.2	2.9	ICES (2018b) AFWG
North Sea	2008–2017	33900	2.3	2.5	6.0	ICES (2018e) WGNSSK
Irish Sea	2008–2017	500	2.0	1.7	7.9	ICES (2018d) WGCSE
Southern Celtic	2008–2017	4200	2.3	2.2	9.2	ICES (2018d) WGCSE
Western Baltic	2008–2017	15100	2.9	1.7	3.0	ICES (2018a) WGBFAS
Eastern Baltic	2008–2017	51200	3.9	2.0	0.6	ICES (2019) WGBFAS
Northern	2006–2015	3700	6.7	5.3	1.0	Brattey et al. (2018) NAFO 2J3KL
Flemish Cap	2008–2017	9300	4.6	4.0	2.2	González-Troncoso et al. (2018) NAFO 3M
Grand Bank	2008–2017	800	4.6	5.5	0.9	Rideout et al. (2018) NAFO 3NO
Gulf of Maine	2002–2011	6400	4.2	2.5	2.8	NOAA (2013a) GM
Georges Bank	2002–2011	5900	4.0	2.3	3.3	NOAA (2013b) GB

Mean weight at age is higher for stocks at lower latitudes than those at higher latitudes. The mean weight at the age of 2-year-old cod stock ranges between 0.5 and 2 kg, and with increasing age, the weight difference among stocks increases. The mean weight of 3-year-old cod in Greenland, Iceland and the Barents Sea is around 1 kg compared with around 6 kg in the Celtic Sea, where it is highest. The difference in the mean weight of 3-year-old cod is, therefore, about sixfold at this age.

Temperature is an important factor influencing cod growth, with average annual differences explaining a substantial proportion of the inter-stock variation (Brander 1995). The temperature may limit growth by affecting metabolism, hence the difference between the northern and southern cod stocks. Based on differences in the growth rate of archive-tagged cod in the Northeast Atlantic, the optimum

Table 3. Stock status relative to reference points. The current spawning stock biomass (Bcurrent) is shown in thousands of tonnes. Blim is the limit reference point that the spawning stock biomass should not go below. Bpa is a precautionary biomass level that takes into account the uncertainty about the estimated stock size. The data come from published advice sheets and stock assessments indicated in the last column.

Stock	Bcurrent	Blim	Bpa	Status	Source
Greenland	22	4	6	Above Bpa	ICES (2022a) 21.1
Northeast Arctic	902	220	460	Above Bpa	ICES (2021a) 27.1–2
Iceland	368	125	160	Above Bpa	ICES (2023a) 27.5a
Faroe Plateau	10	18	25	Below Blim †	ICES (2022c) 27.5b1
Norway	34	-	60	Below Btarget	ICES (2020) 27.1–2coast
North Sea	54	70	98	Below Blim	ICES (2022b) 27.47d20
Irish Sea	12	12	17	Below Blim †	ICES (2023b) 27.7a
Southern Celtic	1	4	6	Below Blim †	ICES (2023c) 27.7e–k
Western Baltic	6	15	23	Below Blim	ICES (2022d) 27.22–24
Eastern Baltic	77	109	122	Below Blim †	ICES (2023d) 27.24–32
Northern	411	790	-	Below Blim	DFO (2022) 2J3KL
Flemish Cap	28	15	-	Above Blim	NAFO (2023a) 3M
Grand Bank	7	60	-	Below Blim †	NAFO (2021) 3NO
Gulf of Maine	2	20	40	Below Blim	NOAA (2021b) GM
Georges Bank	2.7*	-	-	Overfished*	NOAA (2021a) GB

-: Reference point not defined.
*: Spawning stock biomass proxy for Georges Bank is based on a three3-year smoothed survey index. NOAA considers the stock status of Georges Bank cod to be overfished.
†: Scientific advice is zero catch

temperature for growth was 10°C, although there were population differences, and this study did not account for prey availability (Righton et al. 2010). Experimental studies on Icelandic cod fed ad libitum suggest an optimum temperature of 7°C for a 2000 g fish (around age 4) (Björnsson et al. 2001).

However, other factors affect growth, such as prey availability, density-dependent competition, size, genetic predisposition, life stage and reproductive investment (Björnsson et al. 2001; Yoneda and Wright 2005) as well as population differences (Dutil et al. 2008; Harrald et al. 2010). These other influences are evident from the negative temperature growth relationships in the northern latitudinal ranges. Conversely, high summer to autumn temperatures appear super-optimal for cod growth in the southern North Sea (Righton et al. 2010). A longer annual growth period may further explain the large size in the south.

Food availability and density-dependent competition linked to stock size will likely impact annual variability within rather than among stocks. In the northern areas, where capelin (*Mallotus villosus*) is present, growth is influenced by the availability of this prey (Mehl and Sunnanå 1991; Krohn et al. 1997; Pálsson and Björnsson 2011), which subsequently impacts cod condition (Rose and O'Driscoll 2002). Accelerated growth under a lower cod biomass would be expected due to lower competition for food.

Maturity

Both age and size at maturity vary among cod stocks (Lambert et al. 2003). Cods mature from ages 1–8 years old with a total length of 50% maturity ranging from 35–78 cm (Olsen et al. 2004; Thorsen et al. 2010). Males can mature younger and smaller than females, and maturation and spawning of males may begin as early as their first year (Armstrong et al. 2004; Yoneda and Wright 2004; Nash et al. 2010). In the southern stocks, most individuals have reached maturity by four years old (Figures 2 and 3, Table 2), whereas in the Icelandic and Northeast Arctic cod stocks, < 15% of cod reach maturity by that age, and most are mature by age 9. The Greenland and coastal Norway stocks fall in between, with a high proportion of mature at ages 5–6 years old.

Maturation is a key energetic constraint to somatic growth, with a high energy commitment in the later stages of secondary gametogenesis and spawning (Wright and Rowe 2019). Consequently, size and pre-maturation growth play a large role in the onset of maturation. Differences in age at maturity among stocks may reflect phenotypic plasticity, as faster-growing cod can mature at smaller sizes (Godø and Haug 1999), and this can explain differences in maturity at age between the Barents Sea and Norwegian coastal cod (Svåsand et al. 1996). Age at maturity tends to decrease with increasing temperature (Drinkwater 2005). However, selection on life history should favour a delay in maturation if it results in a gain in fecundity or reduced juvenile mortality (Stearns and Crandall 1984; Roff 1992). This life history trade-off is evident in southern cod stocks that would not persist if they delayed maturity to ages older than that observed (Wright 2014).

Maintaining up-to-date information on maturity at age and weight at age is important because these parameters change over time (Trippel 1999). Failures to account for this variation have led to major revisions in spawning stock biomass (SSB) estimates for some stocks (ICES 2015). For many cod stocks, a general trend towards smaller size and younger age at maturity with time has been widely reported (Trippel 1999; Olsen et al. 2004), although interannual variation is often high (Wright and Rowe 2019). These changes are largely associated with increased fishing mortality (Devine et al. 2012), which is expected to be selected for maturation at a younger age (Law 2000).

Intra-stock population differences in maturity complicate stock maturity at age estimation. For example, in the North Sea, the offshore migratory population centred around Viking Bank matures at an older age than other shallow resident groups (Wright et al. 2011). As these shallow water components of the North Sea stock declined, the relative contribution of this Viking population increased (Holmes et al. 2014), which increased the mean age at maturity (ICES 2015). Recent variation in stock maturity at the age of cod reflects the average age taken by the fishery, which probably reflects landing minimum size limits intended to reduce the catch of juveniles and, indirectly, a long-term selection against maturing much above these sizes (Figure 3).

The size-fecundity relationship in cod is an allometric rather than isometric relationship with the exponent ranging between 1.15 and 5.46 (Lambert et al. 2005; Thorsen et al. 2010). Thus, cod produce more eggs per weight as they get larger and

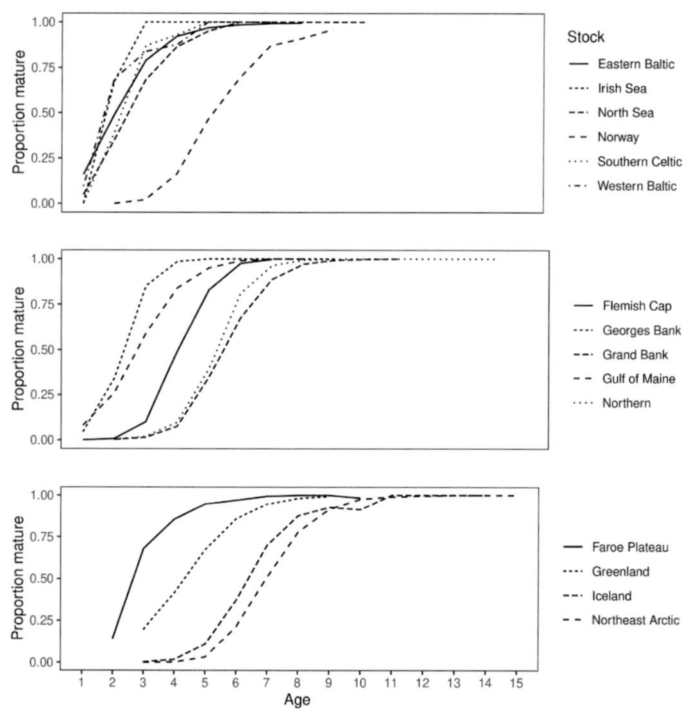

Figure 2. Maturity at the age of the 15 cod stocks. The proportion of mature is averaged over the last decade, as indicated in Table 2.

older, and it may be beneficial for total production to have larger and older females in the stock. In addition to this positive effect on fecundity, the eggs and larvae produced by older and larger females tend to be more viable (Marteinsdottir and Thorarinsson 1998), and male fertilisation success increases (Politis et al. 2014). This demographic effect of older spawners may explain why indices of age diversity are positively correlated with recruitment in a few cod stocks (Marteinsdottir and Thorarinsson 1998; Wright 2014; Ohlberger et al. 2022). However, for other stocks, no effect of age distribution on recruitment was identified (Morgan et al. 2007; Ottersen 2008). Compared with other cod stocks, those in the southern areas have a high potential fecundity relative to their size, with the highest reported in the Irish Sea (Thorsen et al. 2010). Some cod do not spawn yearly, presumably due to the high energy cost of spawning (Rideout et al. 2005). As noted from depth and temperature profiles from electronic data tags, individuals who skip spawning do not migrate to the spawning area in the year of skipped spawning (Jónsdóttir et al. 2014). Hence, these individuals may save energy otherwise used for spawning migrations. Such skip spawning is far more prevalent in northern cod stocks. Life-time fecundity varies considerably among cod stocks and is related to individual and population growth rates. By ages 4 or 5, cod in the Irish and North Sea have already attained the lifetime fecundity equivalent to that of many later maturing and long-lived cod stocks (Thorsen et al. 2010; Wright and Rowe 2019).

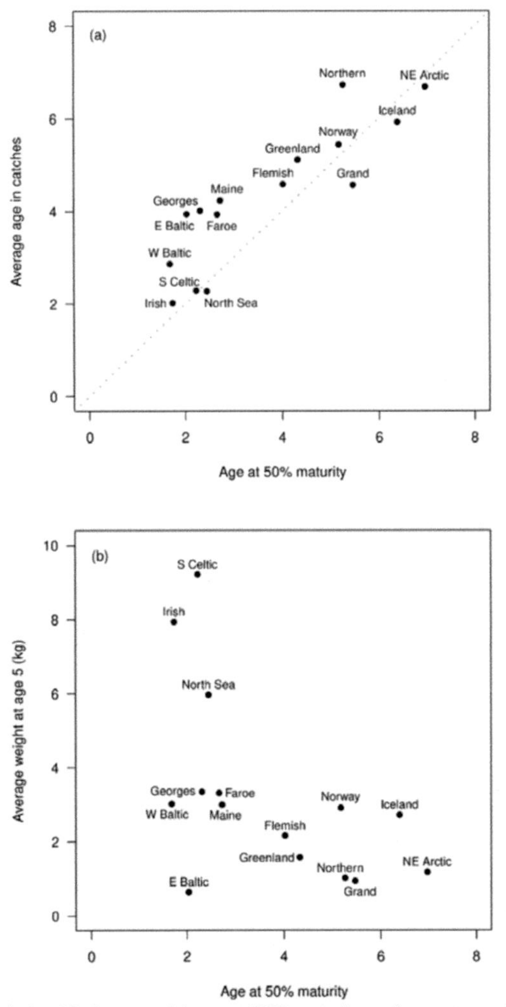

Figure 3. Bivariate relationship between (a) age at 50% maturity and average age in the catches, and (b) age at 50% maturity and average weight at age 5. The measurements are averaged over the last decade, indicated in Table 2.

The average age in the catches, age at 50% maturity and the weight at age 5 is provided in Table 2. The North Sea, southern Celtic, Irish Sea, and Western Baltic stocks are characterised by a young age at maturity, around two years old, and the fisheries also catch them around that age (Figure 3a). Conversely, the Northeast Arctic, Iceland, Norway, northern, and Grand Bank stocks are characterised by old age at maturity, around six years old, and the fisheries also catch them around that age (Figure 3a). The average weight at age five varies greatly between the stocks, with the North Sea, Irish Sea, and southern Celtic stocks being the heaviest, around 6–9 kg, while the Northeast Arctic, northern, Grand Bank, and Eastern Baltic stocks are lighter, around 1 kg (Figure 3b). Generally, the stocks that reach a large body weight by age five are also the stocks that mature early, often in southern waters. The exception is Eastern Baltic cod, a southern stock with a young age at maturity but has the lowest body weight of all the stocks (Figure 3b).

Spawning Stock Biomass

Spawning stock biomass (SSB) estimates are available from analytical age-based stock assessments since at least the 1960s for most cod stocks. Population size in cod is proportional to the area of available habitat, given the extent of migration, where migratory stocks occupy a larger area (Robichaud and Rose 2004). In contrast, sedentary populations within stocks tend to occupy a much smaller area than migratory populations. The SSB of larger stocks such as northern, Iceland, Northeast Arctic, North Sea and Eastern Baltic has reached over half a million tonnes (Figure 4). Faroe Plateau, Flemish Cap, Georges Bank, Grand Banks and Norway are of intermediate size (approximately 100,000 t), while the other stocks are considerably smaller (less than 50,000 t).

The state of the stocks has varied considerably, but, in general, SSB has decreased in the past ten years (Figure 4). The Flemish Cap stock exhibited the greatest fluctuations, with the lowest SSB being only 0.06% of the highest value. For the northern, Greenland, South Celtic, Grand Bank, Northeast Arctic, Faroe Plateau, and Irish Sea stocks, the SSB is 1–10% of the maximum, while the other stocks are 10–17% of the maximum. Fishing pressure is one of the most influential factors that affects SSB and is the focus of most management regulations. In Iceland, for example, SSB increased in response to a decreased harvest rate after 2000 (MFRI 2018). Furthermore, recruitment fluctuations affect year-to-year variability in SSB, with larger year-classes contributing to increased SSB (recruitment variability is further discussed in the next section).

Recruitment

Recruitment is a major driver of population growth, especially in heavily fished and early maturing cod stocks (Wright 2014) and is of great importance for stock assessment as it provides the means for predicting the number of individuals available for fishing. Recruitment can be defined as the number of fish which survive to a certain age and is estimated at different ages depending on stock, with three years old being common in the northern stocks and ages 0 (six months) or one in the southern stocks. Differences in recruitment age reflect age distribution and age at maturity among the stocks. As recruitment is the major cause of variability in biomass, the younger year-classes are closely monitored from as early an age since it is possible to estimate the size of the coming year-classes.

Recruitment is influenced by a variety of density-dependent and density-independent effects. Models of the stock-recruitment relationships (SRR) using Ricker, Beverton-Holt, and smooth hockey stick have all suggested a significant effect of SSB in many cod stocks, although large variability around the SRR fits suggest a strong influence of other factors in determining recruitment success. Spawners influence recruitment not only through the number of eggs released but also their location and timing relative to the onset of larval prey production, favourable sea circulation,

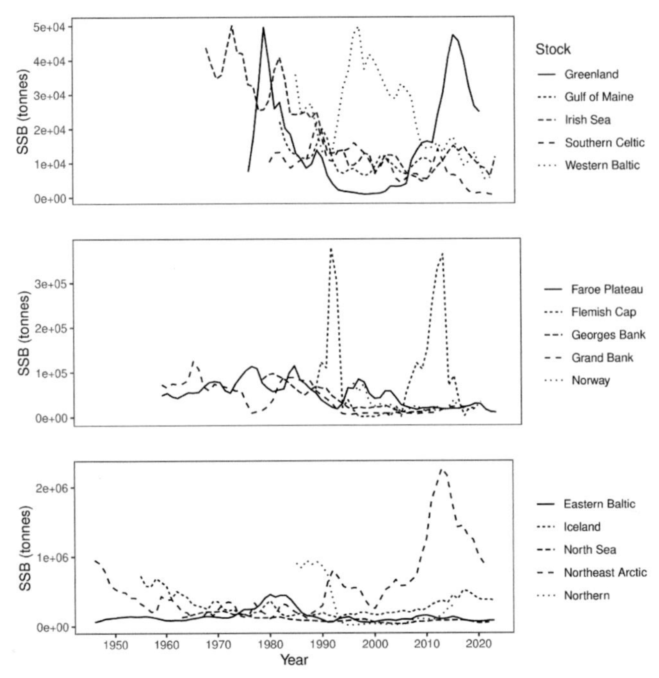

Figure 4. Spawning stock biomass (SSB) for the 15 cod stocks.

and offspring provisioning (maternal effects; Marteinsdóttir and Steinarsson 1998; Trippel 1999; Wright and Trippel 2009). Estimates of the productivity in terms of survivors per spawner biomass (Minto et al. 2014) or survivors per egg (Wright 2014) show little long-term change in the West of Scotland, Irish and Celtic Seas stocks, but in the North Sea productivity has generally declined since the mid-1980s.

Recruitment is mainly determined during the first year (Campana et al. 1989); during this time, eggs, larvae and juveniles are exposed to high predation pressure, and growth and survival are highly dependent on food availability and suitable habitat (Fjøsne and Gjøsæter 1996). As mortality in the first year can be very high, small changes in survival can result in highly variable year-class sizes (Hjort 1914; Beyer 1989; Houde 1989). Environmental factors may have a major impact on recruitment and large-scale physical forcing, including the North Atlantic Oscillation and Atlantic Multidecadal Oscillation, has been implicated in year-class variation (Stige et al. 2006; Bentley et al. 2020) with correlated recruitment among the Celtic, Irish Sea and to some extent West of Scotland stocks (Myers et al. 1995; Minto et al. 2014). Even though recruitment variability has received the attention of fisheries scientists for over a century (Hjort 1914), we may never fully understand what controls it due to the interaction between different mechanisms in the life history of cod.

Cod recruitment can vary greatly from year to year, and stock-specific patterns are present (Figure 5). For most stocks, recruitment has generally declined over time, and many stocks have had below-average recruitment since 2000 or even earlier. The difference between the lowest and highest values for each stock is high; the lowest value ranges between 0.05 and 5% for all stocks except Eastern Baltic, Iceland,

and Norwegian, where the values are 11%, 21%, and 26%, respectively. Despite increased SSB in the Icelandic and Northeast Arctic stocks in the past two decades, recruitment has not increased to levels seen between 1950 and 1980. However, the risk of very low recruitment seems to have decreased.

Fisheries

Landings

Cod was the most important demersal commercial species in the North Atlantic for centuries (Kurlansky 1997). Cods are caught yearly using multiple fishing gears; demersal trawl and longline are the most common, but Danish Seine, gillnets, and jiggers are also routinely used (ICES 2022b). Landings have varied among stocks, with the highest landings obtained from the largest stocks, with maximum landings of 1.3 million tonnes for the Barents Sea stock in 1956. For the smaller stocks, the maximum annual landings have been around 20 thousand tonnes.

Landings have fluctuated in all stocks (Figure 6). For the southern stocks and the Baltic stocks, landings started to decrease soon after 1980 and reached very low values. Landings from the Northern Cod, Grand Bank, Georges Bank, and the Gulf of Maine decreased around 1990 and have been very low since then, except

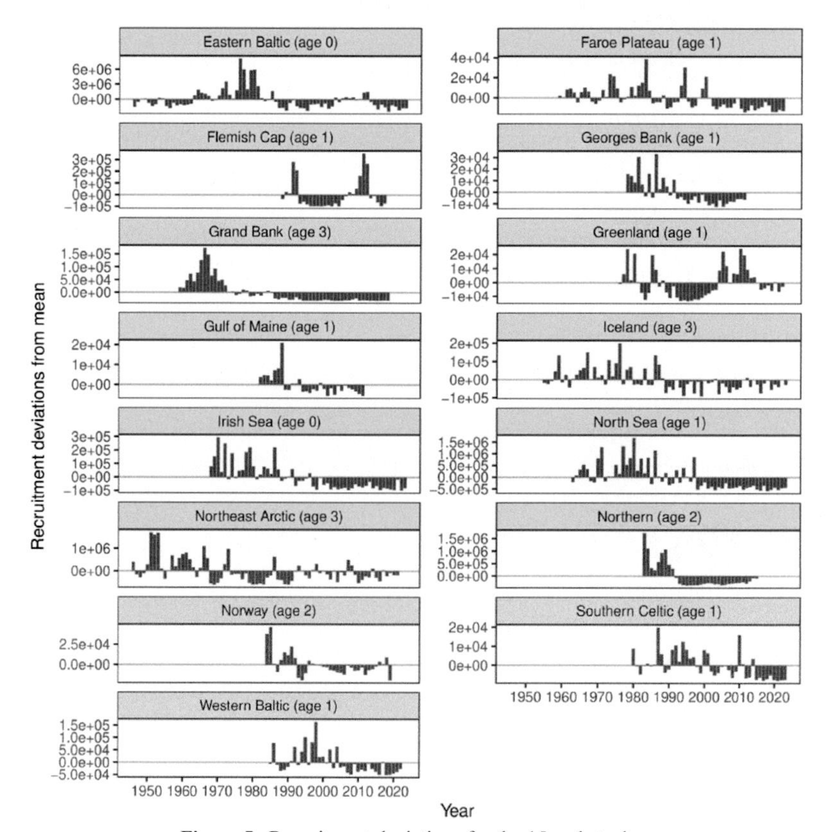

Figure 5. Recruitment deviations for the 15 cod stocks.

at Flemish Cap, where landings have increased again since 2010. For the northern stocks, landings reached a minimum at some point between 1990 and 2010 but have increased again. Landings follow stock size trends in each stock, leading to lower catches in the past years due to the poor state of the stocks.

Fishing Mortality

Many cod stocks have been subjected to excessive fishing pressure for decades, leading to declines in stock abundance. Fishing mortality (F) is a major driver of stock changes across the North Atlantic, with some stocks still subject to unsustainable levels.

Temporal trends in fishing mortality for different stocks are shown in Figure 7. The age range for fishing mortality considered in managing these stocks varies between stocks and is based on either the main age classes in the catches or the ages that are fully selected by the fishing fleet. The general trend is that fishing mortality is estimated for older ages in the northern stocks compared to younger ages in the southern stocks. As such, the average fishing mortality is calculated for

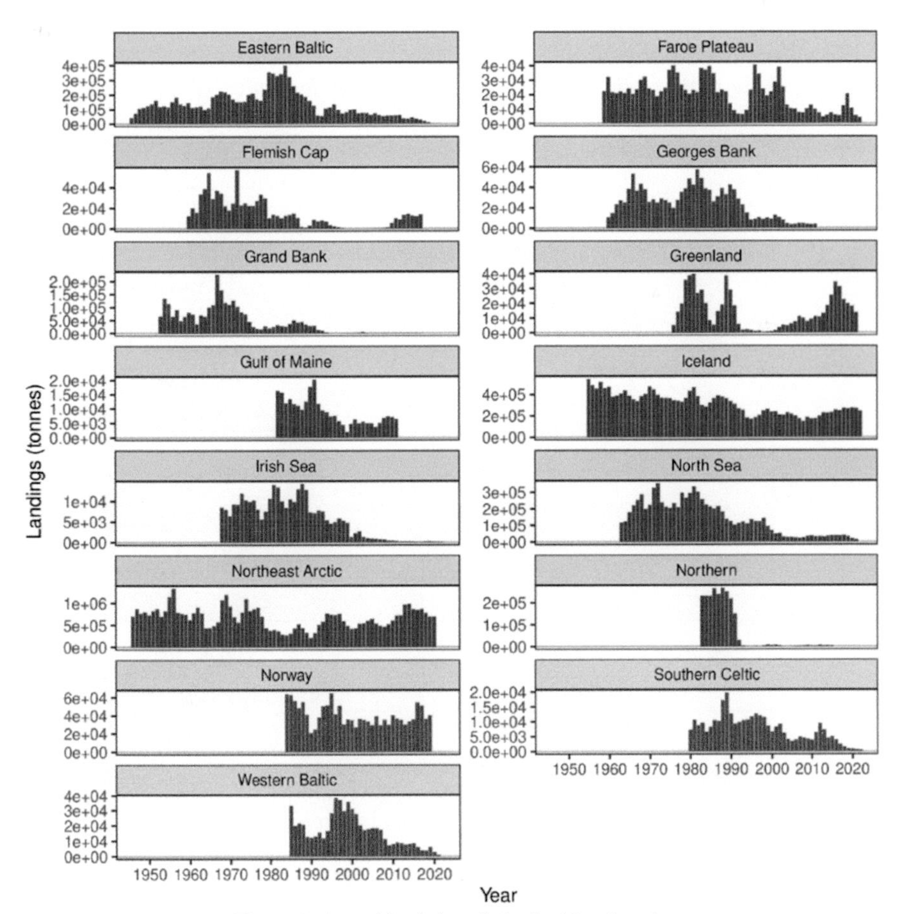

Figure 6. Annual landed catch for the 15 cod stocks.

ages 5–10, 3–7 and 2–4 for Iceland, the Faroe Plateau and the Irish Sea, respectively. Variability in fishing mortality is seen both within and between the stocks, where some of the stocks have endured dramatic changes in fishing mortality. Constant high fishing mortality was observed for the West Baltic and some southern stocks (Irish Sea and North Sea), whereas fishing mortality was constantly at lower levels for the Norway and Iceland stocks. There is a general trend towards increasing F from the 1960s up to the 1980s, but in more recent years, fishing mortality has decreased. However, stock recovery has been slow despite lower fishing mortality in the past decades. Fishing mortality has led to substantial age truncation in many stocks, and this has sometimes been linked to low recruitment in cod stocks (Marteinsdottir and Thorarinsson 1998; Wright 2014), which might be linked to greater sensitivity to environmental conditions, although this has been debated (Stige et al. 2017; Ottersen and Holt 2023). Historically, high fishing mortality has also been associated with selection for smaller sizes, ages at maturation, and generation time (Devine et al. 2012). As has occurred in many cod stocks, prolonged intense overexploitation has been found to delay rebuilding and substantially increase the uncertainty in recovery times (Neubauer et al. 2013).

For some cod stocks, uncertainty in fishing mortality estimation has arisen because it is derived from direct estimates of total mortality from stock assessments and natural mortality that is either estimated from measures of predation rate,

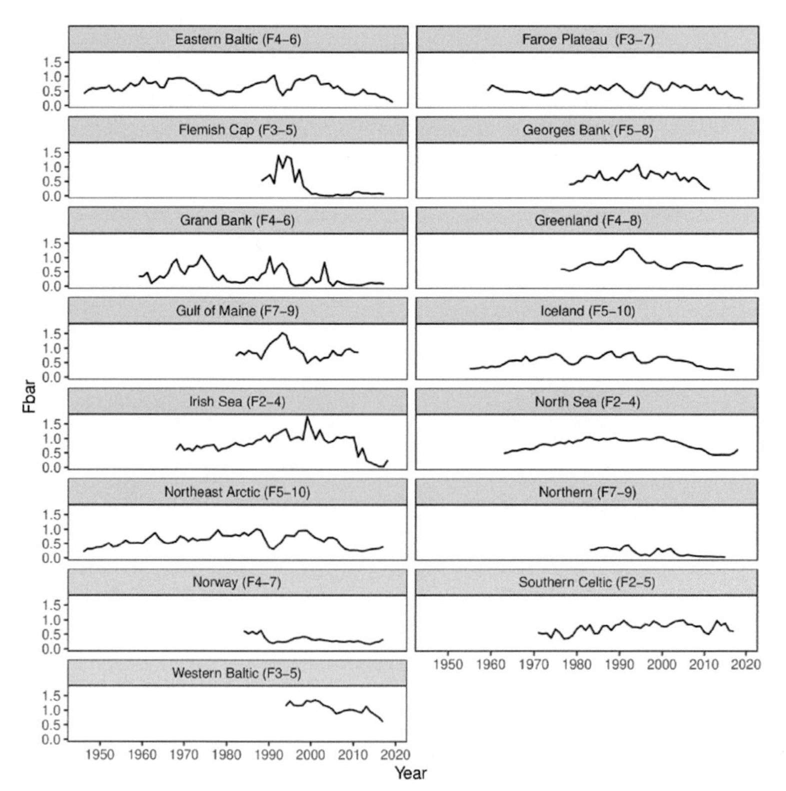

Figure 7. Fishing mortality averaged over the indicated age groups (Fbar) for the 15 cod stocks.

size-predation relationships, or simply assumed values. Hence, higher uncertainty in fishing mortality is likely seen in cod stocks with higher predation pressure. For example, increased predation from seals may have affected fishing mortality estimates for the Gulf of St. Lawrence (Benoît et al. 2011) and West of Scotland stocks (Cook and Trijoulet 2016). However, as fishing mortality in many cod stocks is often higher than natural mortality for all but the younger ages, overfishing has often been the major contributor to stock declines.

Management

Atlantic cod fisheries are managed by ministries, departments, commissions, and councils representing coastal states involved in the fisheries. Scientific advice and reviews are provided by national agencies and international bodies such as ICES and NAFO. General management objectives include the conservation of fish stocks and protecting the marine environment while optimising the long-term economic and social benefits of the fisheries (European Commission 2023; NOAA 2007). A key instrument in fisheries management is the restriction of the amount of fishing that takes place within each year, usually by limiting the total allowable catch in tonnes, or in the case of the Faroe Plateau cod, by limiting the number of fishing days in the year. Other management instruments include permanent and temporary area and seasonal closures, as well as mesh size restrictions.

Limit reference points indicate the minimum spawning stock size to ensure sustainable levels of recruitment for a given stock. Precautionary reference points are sometimes also defined slightly above the limit reference points to take into account the imperfect information and statistical uncertainty about the estimated current stock size (ICES 2021b). Target reference points may also be defined, although they often refer to an optimal constant fishing pressure rather than the stock size subject to natural fluctuations.

In general, the scientific advice is to apply optimal fishing pressure but to reduce the fishing pressure when the stock is below a precautionary reference point and to reduce fishing to a minimum if the stock falls below the limit reference point. A common definition of an overfished stock is when the spawning stock biomass is below the limit or precautionary reference point.

Four out of the 15 cod stocks compared in this chapter are in a healthy state. The Northeast Arctic, Iceland and Greenland inshore cod stocks are above the precautionary reference points, and the Flemish Cap stock is well above the limit reference point (Table 3). The other 11 are all below the reference point, defining them as overfished, and the scientific advice is either zero catch or rebuilding the stock with reduced fishing pressure.

Each stock is managed separately, even though some overlap between them may occur. Examples of overlap are cod eggs and larvae drifting from Iceland to East Greenland, where individuals stay until they reach maturation, or the Northeast Arctic and Norwegian coastal cod stocks that utilise the same spawning grounds. While taking these migrations into account in management would be challenging, evidence of larval and adult exchange has recently been used to change the stock units in the West of Scotland and the North Sea (Wright et al. 2024).

In many stocks, cod is taken as part of a mixed demersal fishery involving bottom trawling and seining. Restrictions to promote cod stock recovery have led to it becoming a choke species in New England and northwest European shelf cod stocks, limiting the catches of more abundant species taken by the same gear. This has sometimes led to high levels of cod discards (Kerr and Cadrin 2024) and area misreporting of cod landings (Wright et al. 2024).

The Eastern Baltic cod is a special case where the scientific advice has a special emphasis on the degradation of ecosystem status resulting from cumulative anthropogenic pressures and climate change (ICES 2023d). The recommendation is habitat restoration efforts, focusing on reducing eutrophication to improve bottom oxygen content. Low growth, poor condition, and high natural mortality of Eastern Baltic cod are linked to poor oxygen status, reduced availability of fish prey, and high levels of parasite infections (ICES 2023d).

A severely overfished cod stock can take decades to rebuild and can also be affected by a major restructuring of marine food webs and fisheries (Frank et al. 2011). A positive sign of a stock recovery comes from the northern Cod stock, where after a moratorium of three decades, the spawning stock size is currently estimated at around 400 thousand tonnes (DFO 2022), approaching levels that could allow sustainable commercial fishing to resume.

Acknowledgements

The authors collected the original tables from published stock assessments as a part of an ICES Methods Working Group project in 2018, expanded and updated for this study. We thank ICES for providing the platform to conduct a comparison of cod stocks across the Atlantic and the working group members for stimulating discussions.

References

Armstrong, M.J., Gerritsen, H.D., Allen, M., et al. 2004. Variability in maturity and growth in a heavily exploited stock: cod (*Gadus morhua* L.) in the Irish Sea. ICES Journal of Marine Science, 61(1): 98–112.

Benoît, H.P., Swain, D.P., Bowen, W.D., et al. 2011. Evaluating the potential for grey seal predation to explain elevated natural mortality in three fish species in the southern Gulf of St. Lawrence. Marine Ecology Progress Series, 442: 149–167.

Bentley, J.W., Serpetti, N., Fox, C.J., et al. 2020. Retrospective analysis of the influence of environmental drivers on commercial stocks and fishing opportunities in the Irish Sea. Fisheries Oceanography, 29: 415–435.

Berg, P.R., Star, B., Pampoulie, C., et al. 2017. Trans-oceanic genomic divergence of Atlantic cod ecotypes is associated with large inversions. Heredity, 119: 418–428.

Beyer, J.E. 1989. Recruitment stability and survival-simple size-specific theory with examples from the early life dynamics of marine fish. Dana, 7: 45–147.

Björnsson, B., Steinarsson, A., and Oddgeirsson, M. 2001. Optimal temperature for growth and feed conversion of immature cod (*Gadus morhua* L.). ICES Journal of Marine Science, 58(1): 29–38.

Bogstad, B. 2024. Trophic interactions. This book.

Brander, K.M. 1995. The effects of temperature on growth of Atlantic cod (*Gadus morhua* L.). ICES Journal of Marine Science, 52: 1–10.

Brattey, J., Cadigan, N., Dwyer, K.S., et al. 2018. Assessment of the Northern Cod (*Gadus morhua*) stock in NAFO Divisions 2J3KL in 2016. DFO Canadian Science Advisory Secretariat Research Document 2018/018. 107 pp.

Campana, S.E., Frank, K.T., Hurley, P.C.F., et al. 1989. Survival and abundance of young Atlantic cod (*Gadus morhua*) and haddock (*Melanogrammus aeglefinus*) as indicators of year-class strength. Canadian Journal of Fisheries and Aquatic Sciences, 46: 171–182.

Cook, R.M., and Trijoulet, V. 2016. The effects of grey seal predation and commercial fishing on the recovery of a depleted cod stock. Canadian Journal of Fisheries and Aquatic Sciences, 73: 1319–1329.

DFO. 2022. Stock assessment of Northern cod (NAFO Divisions 2J3KL) in 2021. DFO Canadian Science Advisory Secretariat Research Document 2022/041. 22 pp.

Devine, J.A., Wright, P.J., Pardoe, H.E., et al. 2012. Comparing rates of contemporary evolution in life-history traits for exploited fish stocks. Canadian Journal of Fisheries and Aquatic Sciences, 69: 1105–1120.

Drinkwater, K.F. 2005. The response of Atlantic cod (*Gadus morhua*) to future climate change. ICES Journal of Marine Science, 62: 1327–1337.

Dutil, J.D., Jabouin, C., Larocque, R., et al. 2008. Atlantic cod (*Gadus morhua*) from cold and warm environments differ in their maximum growth capacity at low temperatures. Canadian Journal of Fisheries and Aquatic Sciences, 65: 2579–2591.

European Commission. 2023. The common fisheries policy today and tomorrow: a Fisheries and Oceans Pact towards sustainable, science-based, innovative and inclusive fisheries management. Communication from the Commission to the European Parliament and the Council COM (2023) 103. 20 pp.

Fjøsne, K., and Gjøsæter, J. 1996. Dietary composition and the potential of food competition between 0-group cod (*Gadus morhua* L.) and some other fish species in the littoral zone. ICES Journal of Marine Science, 53: 757–770.

Frank, K., Petrie, B., Fisher, J.A.D., et al. 2011. Transient dynamics of an altered large marine ecosystem. Nature 477: 86–89.

Godø, O.R., and Haug, T. 1999. Growth rate and sexual maturity in cod (*Gadus morhua*) and Atlantic halibut (*Hippoglossus hippoglossus*). Journal of Northwest Atlantic Fisheries Sciences, 25: 117–125.

González-Troncoso, D., Fernández, C., and González-Costas, F. 2018. Assessment of the cod stock in NAFO Division 3M. NAFO Scientific Council Research Document 18/042. 45 pp.

Graham, M. 1934. Report on the North Sea cod. Ministry of Agriculture and Fisheries, Fisheries Investigations Series II, 13. 160 pp.

Harrald, M., Neat, F.C., Wright, P.J., et al. 2010. Population variation in thermal growth responses of juvenile Atlantic cod (*Gadus morhua* L.). Environmental Biology of Fishes, 87: 187–194.

Hjort, J. 1914. Fluctuations in the great fisheries of northern Europe, viewed in the light of biological research. Rapports et Procès-Verbaux des Réunions du Conseil Permanent International pour l'Exploration de La Mer, 20: 1–228.

Holmes, S.J., Millar, C.P., Fryer, R.J., et al. 2014. Gadoid dynamics: differing perceptions when contrasting stock vs. population trends and its implications to management. ICES Journal of Marine Science, 71: 1433–1442.

Houde, E.D. 1989. Subtleties and episodes in the early life of fishes. Journal of Fish Biology, 35: 29–38.

ICES 2015. Report of the benchmark workshop on North Sea stocks (WKNSEA). ICES CM 2015/ACOM:32. 253 pp.

ICES 2018a. Baltic Fisheries Assessment Working Group (WGBFAS). ICES CM 2018/ACOM:11. 727 pp.

ICES 2018b. Report of the Arctic Fisheries Working Group (AFWG). ICES CM 18/ACOM:06. 863 pp.

ICES 2018c. Report of the North Western Working Group (NWWG). ICES CM 2018/ACOM:09. 662 pp.

ICES 2018d. Report of the Working Group on Celtic Seas Ecoregion (WGCSE). ICES CM 2018/ACOM:13. 1873 pp.

ICES 2018e. Report of the Working Group on the Assessment of Demersal Stocks in the North Sea and Skagerrak (WGNSSK). ICES CM 2018/ACOM:22. 1250 pp.

ICES 2019. Baltic Fisheries Assessment Working Group (WGBFAS). ICES Scientific Reports 1:20. 653 pp.

ICES 2020. Cod (*Gadus morhua*) in subareas 1 and 2 (Norwegian coastal waters cod). ICES advice on fishing opportunities, catch, and effort. https://doi.org/10.17895/ices.advice.5893.

ICES 2021a. Cod (*Gadus morhua*) in subareas 1 and 2 (Northeast Arctic). ICES advice on fishing opportunities, catch, and effort. https://doi.org/10.17895/ices.advice.7741.

ICES 2021b. ICES fisheries management reference points for category 1 and 2 stocks. Technical Guidelines. https://doi.org/10.17895/ices.advice.7891.

ICES 2022a. Cod (*Gadus morhua*) in NAFO Subarea 1, inshore (West Greenland cod). ICES advice on fishing opportunities, catch, and effort. https://doi.org/10.17895/ices.advice.19447835.

ICES 2022b. Cod (*Gadus morhua*) in Subarea 4, Division 7.d, and Subdivision 20 (North Sea, eastern English Channel, Skagerrak). ICES advice on fishing opportunities, catch, and effort. https://doi.org/10.17895/ices.advice.21406881.

ICES 2022c. Cod (*Gadus morhua*) in Subdivision 5.b.1 (Faroe Plateau). ICES advice on fishing opportunities, catch, and effort. https://doi.org/10.17895/ices.advice.19772368.

ICES 2022d. Cod (*Gadus morhua*) in subdivisions 22–24, western Baltic stock (western Baltic Sea). ICES advice on fishing opportunities, catch, and effort. https://doi.org/10.17895/ices.advice.19447868.

ICES 2023a. Cod (*Gadus morhua*) in Division 5.a (Iceland grounds). ICES advice on fishing opportunities, catch, and effort. https://doi.org/10.17895/ices.advice.21828315.

ICES 2023b. Cod (*Gadus morhua*) in Division 7.a (Irish Sea). ICES advice on fishing opportunities, catch, and effort. https://doi.org/10.17895/ices.advice.21840786.

ICES 2023c. Cod (*Gadus morhua*) in divisions 7.e–k (western English Channel and southern Celtic Seas). ICES advice on fishing opportunities, catch, and effort. https://doi.org/10.17895/ices.advice.21840789.

ICES 2023d. Cod (*Gadus morhua*) in subdivisions 24–32, eastern Baltic stock (eastern Baltic Sea). ICES advice on fishing opportunities, catch, and effort. https://doi.org/10.17895/ices.advice.21820497.

ICES 2023e. ICES advice 2023. A collection of all ICES advice released in 2023. https://doi.org/10.17895/ices.pub.c.6398177.

ICES 2023f. ICES Methods Working Group (MGWG). https://www.ices.dk/community/groups/Pages/mgwg.aspx (last accessed 11 December 2023).

Jónsdóttir, I.G., Thorsteinsson, V., Pálsson, Ó.K., et al. 2014. Evidence of spawning skippers in Atlantic cod from data storage tags. Fisheries Research, 156: 23–25.

Kerr, L.A. and Cadrin, S.X. 2024. New England cod stocks. This book.

Krohn, M., Reidy, S., and Kerr, S. 1997. Bioenergetic analysis of the effects of temperature and prey availability on growth and condition of northern cod (*Gadus morhua*). Canadian Journal of Fisheries and Aquatic Sciences, 54: 113–121.

Kurlansky, M. 1997. Cod: The biography of the fish that changed the world. Penguin, New York.

Lambert, Y., Yaragina, N.A., Kraus, G., et al. 2003. Correlation between reproductive characteristics and environmental and biological indices as alternative methods of estimating egg and larval production. Journal of the Northwest Atlantic Fisheries Sciences, 33: 115–159.

Lambert, Y., Kjesbu, O.S., Kraus, G., et al. 2005. How variable is the fecundity within and between cod stocks? ICES CM 2005/Q:13. 19 pp.

Law, R. 2000. Fishing, selection, and phenotypic evolution. ICES Journal of Marine Science, 57: 659–668.

Le Bris, A., Fréchet, A., Galbraith, P.S., et al. 2013. Evidence for alternative migratory behaviours in the northern Gulf of St Lawrence population of Atlantic cod (Gadus morhua L.). ICES Journal of Marine Science, 70: 793–804.

Magnusson, A., Kasper, J., Pinto, C., et al. 2019. Should we fish younger or older cod? ICES CM/P: 590. 1 p.

Marteinsdottir, G., and Steinarsson, A. 1998. Maternal influence on the size and viability of Iceland cod Gadus morhua eggs and larvae. Journal of Fish Biology, 52: 1241–1258.

Marteinsdottir, G., and Thorarinsson, K. 1998. Improving the stock-recruitment relationship in Icelandic cod (*Gadus morhua*) by including age diversity of spawners. Canadian Journal of Fisheries and Aquatic Sciences, 55: 1372–1377.

Matschiner, M., Barth, J.M.I., Torresen, O.K., et al. 2022. Supergene origin and maintenance in Atlantic cod. Nature Ecology and Evolution, 6: 469–481.

Mehl, S., and Sunnanå, K. 1991. Changes in growth of Northeast Arctic cod in relation to food consumption in 1984–1988. ICES Marine Science Symposium, 109: 93–112.

Minto, C., Flemming, J.M., Britten, G.L., et al. 2014. Productivity dynamics of Atlantic cod. Canadian Journal of Fisheries and Aquatic Sciences, 71: 203–216.

MFRI (2018). Cod *Gadus morhua*. State of Marine Stocks and Advice 2018. Marine and Freshwater Institute Iceland.

Morgan, M.J., Shelton, P.A., and Brattey, J. 2007. Age composition of the spawning stock does not always influence recruitment. Journal of Northwest Atlantic Fishery Science, 38: 1–12.

Morgan, M.J., Wright, P.J., and Rideout, R.M. 2013. Effect of age and temperature on spawning time in two gadoid species. Fisheries Research, 138: 42–51.

Myers, R., Mertz, G., and Barrowman, N. 1995. Spatial scales of variability in cod recruitment in the North Atlantic. Canadian Journal of Fisheries and Aquatic Sciences, 52: 1849–1862.

NAFO 2021. Cod in Divisions 3NO. Advice June 2021 for 2022–2024. https://www.nafo.int/Portals/0/PDFs/Advice/2021/cod3no.pdf (last accessed 11 December 2023).

NAFO 2023a. Cod in Divisions 3M. Advice June 2023 for 2024. https://www.nafo.int/Portals/0/PDFs/Advice/2023/cod3m.pdf (last accessed 11 December 2023).

NAFO 2023b. Stock advice: summary sheets and assessment tables. https://www.nafo.int/Science/Stocks-Advice (last accessed 13 November 2023).

Nash, R.D.M., Pilling, G.M., Kell, L.T., et al. 2010. Investment in maturity-at-age and -length in northeast Atlantic cod stocks. Fisheries Research, 104: 89–99.

Neat, F.C., Bendall, V., Berx, B., et al. 2014. Movement of Atlantic cod around the British Isles: implications for finer scale stock management. Journal of Applied Ecology, 51: 1564–1574.

Neubauer, P., Jensen, O.P., Hutchings, J.A., et al. 2013. Resilience and recovery of overexploited marine populations. Science, 340: 347–349.

Neuenfeldt, S., Righton, D., Neat, F., et al. 2013. Analysing migrations of Atlantic cod Gadus morhua in the north-east Atlantic Ocean: then, now and the future. Journal of Fish Biology, 82: 741–763.

NOAA 2007. Magnuson-Stevens Fishery Conservation and Management Act, as amended through January 12, 2007. https://media.fisheries.noaa.gov/dam-migration/msa-amended-2007.pdf (last accessed 11 December 2023).

NOAA 2013a. Gulf of Maine Atlantic cod. pp. 16–639. *In:* 55th Northeast Regional Stock Assessment Workshop (55th SAW) Assessment Report,

NOAA 2013b. Stock assessment of Georges Bank Atlantic cod (*Gadus morhua*) for 2012. pp. 640–845. *In:* 55th Northeast Regional Stock Assessment Workshop (55th SAW) Assessment report

NOAA 2021a. Georges Bank Atlantic cod 2021 management track assessment report. https://apps-st.fisheries.noaa.gov/sis/docServlet?fileAction=download&fileId=7470 (last accessed 11 December 2023).

NOAA 2021b. Gulf of Maine Atlantic cod 2021 update assessment report. https://apps-st.fisheries.noaa.gov/sis/docServlet?fileAction=download&fileId=7550 (last accessed 11 December 2023).

Ohlberger, J., Langangen, Ø., and Stige, L. Chr. 2022. Age structure affects population productivity in an exploited fish species. Ecological Applications, 32: e2614.

Olsen, E.M., Knutsen, H., Gjøsæter, J., et al. 2004. Life-history variation among local populations of Atlantic cod from the Norwegian Skagerrak coast. Journal of Fish Biology, 64: 1725–1730.

Ottersen, G., 2008. Pronounced long-term juvenation in the spawning stock of Arcto-Norwegian cod (*Gadus morhua*) and possible consequences for recruitment. Canadian Journal of Fisheries and Aquatic Sciences, 65, 523–534.

Ottersen, G., and Holt, R.E. 2023. Long-term variability in spawning stock age structure influences climate–recruitment link for Barents Sea cod. Fisheries Oceanography, 32: 91–105. https://doi.org/10.1111/fog.12605.

Pálsson, Ó.K., and Björnsson, H. 2011. Long-term changes in trophic patterns of Iceland cod and linkages to main prey stock sizes. ICES Journal of Marine Sciences, 68: 1488–1499.

Pálsson, Ó.K., and Thorsteinsson, V. 2003. Migration patterns, ambient temperature, and growth of Icelandic cod (*Gadus morhua*): evidence from storage tag data. Canadian Journal of Fisheries and Aquatic Sciences, 60(11): 1409–1423.

Pampoulie, C., Berg, P.R., and Jentoft, S. 2023. Hidden but revealed: After years of genetic studies behavioural monitoring combined with genomics uncover new insight into the population dynamics of Atlantic cod in Icelandic waters. Evolutionary Applications, 16: 223–233.

Politis, S.N., Dahlke, F.T., Butts, I.A., et al. 2014. Temperature, paternity and asynchronous hatching influence early developmental characteristics of larval Atlantic cod, *Gadus morhua*. Journal of Experimental Marine Biology and Ecology, 459: 70–79.

Rideout, R.M., Rogers, B., and Ings, D.W. 2018. An Assessment of the Cod Stock in NAFO Divisions 3NO. NAFO Scientific Council Research Document 18/028. 52 pp.

Rideout, R.M., Rose, G.A., and Burton, M.P.M. 2005. Skipped spawning in female iteroparous fishes. Fish and Fisheries, 6: 50–72.

Righton, D.A., Andersen, K.H., Neat, F., et al. 2010. Thermal niche of Atlantic cod *Gadus morhua*: Limits, tolerance and optima. Marine Ecology Progress Series, 420, 1–13. https://doi.org/10.3354/meps08889.

Roff, D.A. 1992. The Evolution of Life Histories, Theory and Analysis. Chapman and Hall, New York.

Robichaud, D., and Rose, G.A. 2004. Migratory behaviour and range in Atlantic cod: inference from a century of tagging. Fish and Fisheries, 5: 185–214.

Rose, G.A. 1993. Cod spawning on a migration highway in the North-West Atlantic. Nature, 366: 458–461.

Rose, G.A., and O'Driscoll, R.L. 2002. Capelin are good for cod: can the northern stock rebuild without them? ICES Journal of Marine Science, 59: 1018–1026.

Stearns, S., and Crandall, R. 1984. Plasticity for age and size at sexual maturity: a life-history response to unavoidable stress. pp. 13–34. *In:* Potts, G. and Wootton, R. (eds.). Fish reproduction: Strategies and tactics. Academic Press, London.

Stige, L.C., Ottersen, G., Brander, K., et al. 2006. Cod and climate: effect of the North Atlantic Oscillation on recruitment in the North Atlantic. Marine Ecology Progress Series, 325: 227–241.

Stige, L.C., Yaragina, N.A., Langangen, Ø., et al. 2017. Effect of a fish stock's demographic structure on offspring survival and sensitivity to climate. Proceedings of the National Academy of Sciences, 114: 1347–1352.

Svasand, T., Jorstad, K.E., Ottera, H., et al. 1996. Differences in growth performance between Arcto-Norwegian and Norwegian coastal cod reared under identical conditions. Journal of Fish Biology, 49: 108–119.

Thorsen, A., Witthames, P.R., Marteinsdóttir, G., et al. 2010. Fecundity and growth of Atlantic cod (Gadus morhua L.). along a latitudinal gradient. Fisheries Research, 104: 45–55.

Trippel, E.A. 1999. Estimation of stock reproductive potential: history and challenges for Canadian Atlantic gadoid stock assessments. Journal of Northwest Atlantic Fishery Science, 25: 61–81.

Wright, P.J. 2014. Are there useful life history indicators of stock recovery rate in gadoids? ICES Journal of Marine Science, 71: 1393–1406.

Wright, P.J., and Rowe, S. 2019. Reproduction and Spawning. pp. 87–132. *In:* Rose, G. (ed.). Atlantic Cod. Wiley Blackwell.

Wright, P.J., Millar, C.P., and Gibb, F.M. 2011. Intrastock differences in maturation schedules of Atlantic cod, Gadus morhua. ICES Journal of Marine Science, 68: 1918–1927.

Wright, P.J., Dobby, H., and Fox, C. 2024. Northwest European shelf cod stocks; North Sea, West of Scotland, Irish Sea and Celtic Sea. This book.

Yoneda, M., and Wright, P.J. 2004. Temporal and spatial variation in reproductive investment of Atlantic cod Gadus morhua in the northern North Sea and Scottish west coast. Marine Ecology Progress Series, 276: 237–248.

Yoneda, M., and Wright, P.J. 2005. Effect of temperature and food availability on reproductive investment of first-time spawning male Atlantic cod, *Gadus morhua*. ICES Journal of Marine Science, 62: 1387–1393.

Index

M

N

O

P

Q

R

S

T